Effects of Atomic Radiation

Effects of Atomic Radiation

A Half-Century of Studies from Hiroshima and Nagasaki

William J. Schull
Human Genetics Center
University of Texas Health Science Center
Houston, Texas 77225

A John Wiley & Sons, Inc., Publication
New York   Chichester   Brisbane   Toronto   Singapore

Address All Inquiries to the Publisher
Wiley-Liss, Inc., 605 Third Avenue, New York, NY 10158-0012

Copyright © 1995 Wiley-Liss, Inc.

Printed in the United States of America.

Under the conditions stated below the owner of copyright for this book hereby grants permission to users to make photocopy reproductions of any part or all of its contents for personal or internal organizational use, or for personal or internal use of specific clients. This consent is given on the condition that the copier pay the stated per-copy fee through the Copyright Clearance Center, Incorporated, 27 Congress Street, Salem, MA 01970, as listed in the most current issue of "Permissions to Photocopy" (Publisher's Fee List, distributed by CCC, Inc.), for copying beyond that permitted by sections 107 or 108 of the US Copyright Law. This consent does not extend to other kinds of copying, such as copying for general distribution, for advertising or promotional purposes, for creating new collective works, or for resale.

Library of Congress Cataloging-in-Publication Data

Schull, William J.
    Effects of atomic radiation : a half-century of studies from
  Hiroshima and Nagasaki / William J. Schull.
      p. cm.
    Includes index.
    ISBN 0-471-12524-5 (cloth : alk.paper)
    1. Atomic bomb victims—Health and hygiene—Japan. 2. Atomic Bomb
  Casualty Commission. 3. Ionizing radiation—Health aspects.
  4. Atomic bomb victims—Health and hygiene—Japan—Research.
  I. Title.
  RA648.3.S38   1995
  616.9'897—dc20   95-14938
                                                                                        CIP

The text of this book is printed on acid-free paper.

This book is dedicated to the tens of thousands of survivors whose lives have been irreparably affected as a result of their exposure to atomic radiation, but through whose travail and selfless cooperation in these studies a better, peaceful, and humane world may yet be constructed.

Contents

Preface / xi

1. **Japan: Summer and Autumn 1945** / 1
   Initial attempts to Assess Biological and Physical Damage / 10
2. **Establishment of the Atomic Bomb Casualty Commission** / 18
   Creation of the Radiation Effects Research Foundation / 32
3. **The First Decade** / 38
   Hematological Investigations / 38
   The Surveillance of the Outcome of Pregnancy / 39
   Completeness and Accuracy of Information / 45
   Abortions and Miscarriages / 49
   The Search for Chromosomal Abnormalities / 51
   Growth and Development of Exposed Children / 52
   Radiation-Related Changes in the Lens / 54
   The Appearance of Leukemia Among the Atomic Bomb Survivors / 58
   The Effects of Prenatal Irradiation / 65
   The Master File / 68
4. **A New Research Strategy** / 72
   The Life Span Study / 79
   Establishment of Tumor and Tissue Registries / 83
   The Adult Health Study / 87
   Other Studies of General Health Import / 93
   Consanguineous Marriages and the Child Health Survey / 94
   Cardiovascular Disease and the Ni-Hon-San Study / 98

5. **Exposure and Dose** / 102
   Radiation from the Bombs / 103
   Prompt Radiation / 107
   Delayed Radiation / 118
   Fallout / 118
   Residual Radiation / 120
   Dose Estimation / 121
   Diagnostic and Therapeutic Doses / 127

6. **The Postnatally Exposed Survivors** / 130
   The Estimation of Risk / 132
   Premature Death / 136
   Cancer / 142
   Leukemia / 148
   Death from Cancers Other Than Leukemia / 153
   Factors Affecting Cancer Risk / 159
   The Impact of Cancer on the Population of Survivors / 160
   The Incidence of Solid Malignant Tumors / 161
   Uncertainties in the Estimates of Cancer Risk / 162
   Benign Tumors and Related Disease / 168
   Psychosocial Concerns / 169
   Fertility / 170
   Radiation-Related Cellular and Molecular Events / 172
   Chromosomal Abnormalities / 172
   Somatic Mutations / 179
   The Outcome of Pregnancy among Exposed Pregnant
     Women / 184
   The Future / 185

7. **The Prenatally Exposed Survivors** / 188
   Mental Retardation / 189
   Small Head Size / 192
   Other Evidence of Brain Damage / 195
   Uncertainties in the Estimates of Risk / 209
   The Biological Nature of the Damage to the Brain / 214
   Development of the Brain / 215
   Summary of the Developmental Features of the Brain / 223
   Developmental Abnormalities of the Brain / 225
   Physical Growth and Development / 232
   Ocular Damage / 233
   Chromosomal Abnormalities / 233
   Somatic Mutations / 234
   Cancer among the Prenatally Exposed / 235
   Fertility of the Prenatally Exposed / 238
   The Future / 239

Contents

8. **The Survivors' Children** / 243
   Pregnancy Outcome / 244
   Mortality after Birth / 248
   Sentinel Phenotypes / 252
   Chromosomal Abnormalities / 253
   Biochemical Studies / 256
   The Growth and Development of These Children / 259
   The Sex Ratio / 260
   An Overall Estimate of Genetic Risk / 262
   The Future / 267

9. **Summary** / 272

10. **Epilogue** / 274
    Projections of Risk at Low Doses and Lose Dose Rates / 278
    The Projection of Risk of Cancer to a Lifetime / 279
    Choice of the Projection Model / 280
    The Use of Risk Estimates Derived from One Population to
       Project the Lifetime Experience of Another Population / 285
    Low Doses and Dose Rates / 288
    Current Estimates of Lifetime Risk / 289
    Perceptions of Risk / 292
    The Challenge / 294

Chronology of Major Events / 297

Notes / 303

Glossary / 333

References / 347

Index / 375

Preface

On Monday, the 16th of July 1945, at 0529:45 AM, the firing circuit closed on the first test of a nuclear weapon in an arid region of New Mexico, prophetically known to early colonists as the Jornado del Muerto—the Journey of Death. Trinity, as this test was codenamed, was a formidable achievement. It was, in a sense, the ultimate proof of the truths of particle physics. But it was also the starting point of a trail of widespread apprehension and public concern about the health hazards of exposure to ionizing radiation, which was further exacerbated as the news of the atomic bombing of Hiroshima and Nagasaki emerged some three to four weeks later.

Much has been written about the survivors of the bombings of Hiroshima and Nagasaki. These accounts have poignantly revealed their individual tragedies—the dislocation of lives and the loss of loved ones.[1] An equally large, if not larger literature describes the biological consequences of their exposure to atomic radiation, but these findings are poorly known to all save specialists. Unfortunately, they have generally appeared in scientific publications inaccessible or unintelligible to most persons. Nevertheless, these findings impinge on all our lives, dictating acceptable radiation exposures in medical practice, the workplace, national defense, and even manned space flight. As a consequence, they warrant wider dissemination and understanding. It is this that I seek to achieve by detailing pertinent events and findings as they have unfolded over the half-century that studies have been under way in Hiroshima and Nagasaki.

A study of the size, intricacy, and duration of the one to be described did not spring forth fully formed; it grew and evolved, adjusting to new knowledge and investigative techniques. The studies of the atomic bomb survivors have been dynamic from their outset, and to present them otherwise is to suggest a clarity or rigidity of thought and study design that diminishes the inherently questioning nature of scientific research. The untidiness of science is implicit in its purpose—to bring order to otherwise chaotic observations. Science is, however, inseparable from scientists, who are not infallible. They are limited by the knowledge and fashions of their times as well as their own preconceptions. These can cloud or color their interpretations of their observations. It has become fashionable, therefore, to impute to them self-serving motives, or worse, a conscious malevolence where none surely existed. As Robert Crease and Nicholas Samios have written, "the beneficial effects of scientific discoveries often become so thoroughly integrated into the world that they are taken for granted, as part of its warp and woof, whereas the pernicious applications of scientific discoveries are often portrayed as representative of scientific activity itself."[2]

Perceptions of the acceptability of the risks associated with ionizing radiation, particularly as they concern the use of nuclear energy, are often dominated by a fear of nuclear war or accidents at nuclear power plants. But the use of nuclear power must be examined in a context different from atomic or nuclear weapons. Personal and national needs must be reconciled with the risks. This requires the factual bases on which an intelligent judgment can be made. Unfortunately, individual decisions are compromised by erroneous and often inflammatory articles in the press, and a growing distrust of the evaluations of the hazard made by federal agencies—occasionally construed as serving their own interests—where political machinations loom more important than the public weal. This estrangement serves neither the public nor our nation well.

There are data pertinent to evaluating the effects of ionizing radiation other than those to be described; these include the experiences of individuals with the deforming rheumatic disease known as ankylosing spondylitis who were exposed to X-irradiation therapeutically, patients given thorium dioxide (thorotrast), children exposed because of enlargement of the thymus or because of ringworm of the scalp, women treated for cervical cancers, children exposed prenatally to diagnostic irradiation or during radio-therapy of their mothers, uranium miners, radium dial painters, individuals living in areas with higher than "normal" levels of naturally occurring radioactivity, servicemen exposed in the course of military exercises involving nuclear weapons, industrial radiographers, and employees in the nuclear power industry or the nuclear weapons complex. However, the purpose of this book is not to synthesize all of the available information on the risks of ionizing radiation—a function better served by various national and international bodies charged with this responsibility—but to present the atomic bomb survivor story in all of its complexity.

To aid the reader I have eschewed the use of technical terms wherever possible, but where this could not be done I have routinely defined or described such terms on their first use, and as a further aid a glossary and a chronology of the major events will be found near the end of the book.

Some errors of fact or emphasis are inevitable in what is presented. I alone am responsible for these. To avoid egregious mistakes, many colleagues and friends, who are more competent in certain subject areas than I, have read and commented on this text. In particular, I am indebted to Seymour Abrahamson, Gilbert W. Beebe, Randolph L. Carter, John B. Cologne, Charles W. Edington, William H. Ellett, Stuart C. Finch, Howard B. Hamilton, Seymour Jablon, George D. Kerr, Charles E. Land, Mortimer L. Mendelsohn, Robert W. Miller, William C. Moloney, James V. Neel, Dale L. Preston, Frank Putnam, Michael E. Rappaport, Joop W. Thiessen, Robert E. Thomas, and James E. Trosko, all of whom have played prominent roles in the work I recount. I am also indebted to Mitoshi Akiyama, Akio A. Awa, Eric Boerwinkle, Cherie Bourgin, Stephen P. Daiger, Tommy C. Douglas, Kevin and Kim Dunn, Shoichiro Fujita, Kevin Grayson, Craig L. Hanis, Kenji Joji, Hiroo Kato, Nori Nakamura, Masanori Otake, Lief Peterson, Chiyoko Satoh, Yukiko Shimizu, Emoke Szathmary, Louis K. Wagner, and Kenji Yorichika, who graciously commented on parts or all of the text, and to Sara Barton, Beth Magura, Michael Edington, and my wife, Vicki, who did much to improve the manuscript.[3]

Finally, the interpretation of the data is a personal one, and is not an official statement of the Atomic Bomb Casualty Commission or its successor, the Radiation Effects Research Foundation, or the opinions of my collegues who have been so helpful.

Effects of Atomic Radiation

1

Japan: Summer and Autumn 1945

> Tis not onely the mischiefe of disease, and the villanie of poysons that make an end of us, we vainly accuse the fury of gunnes, and the new inventions of death; 'tis in the power of every hand to destroy us, and wee are beholding unto every one wee meete hee doth not kill us. There is therefore but one comfort left, that though it be in the power of the weakest arme to take away life, it is not in the strongest to deprive us of death.
> —Sir Thomas Browne, *Religio Medici*, 1643

Monday, August 6, 1945, began like any other hot, sultry, summer morning in western Japan. But this day would be etched deeply into the world's collective consciousness. It started innocuously with the appearance over Hiroshima of a single American Tinian-based weather observation plane shortly after 7 AM. A gentle wind was blowing out of the north-northeast at about 2–3 miles per hour and visibility was fair. After radioing this information the observation plane turned and sped southward. With its departure, the air raid alert that sounded earlier was lifted and everyone resumed their activities, many preparing for work. However, within half an hour another plane, a Marianas-based B-29, the Enola Gay, accompanied by an escort of two observation planes, appeared out of the northeast high above the city. Unknown to the inhabitants, the Enola Gay was carrying a new, fearsome bomb, unlike any previously known, one capable of destroying humankind.

Three days later, on Thursday August 9th, another B-29, whimsically named Bock's Car after its captain, Frederick Bock, lifted off a runway on Tinian and turned north-northwest toward Kyushu, over a thousand miles away, and Kokura, its target of preference. When Bock's Car arrived over this industrial city, with its arsenal and steel fabricating complex at Yahata, the sky was heavy with clouds and visibility was poor. After some fruitless minutes spent circling waiting for the clouds to part, the aircraft headed toward Nagasaki, its second target, a hundred miles to the southwest. Here, too, the cloud cover was heavy, but a brief gap appeared and below could be seen the stadium to the north of the Mitsubishi Heavy Industries Nagasaki Arsenal. The bomb was hurriedly released, albeit some distance from the intended target near the Arsenal, and as it tumbled earthward the bomber, short of fuel, headed quickly for Okinawa.

The foul and awesome emanations of the weapons these aircraft carried posed a threat to human integrity beyond the knowledge then—and still—available to the world. Lay and scientific communities alike were poorly informed about the short- as well as the long-term consequences of exposure to atomic radiation; and during the days and weeks following the bombings, the print media worsened matters by publishing many ill-founded speculations. Allegations were trumpeted in national and international newspapers: Hiroshima and Nagasaki would be uninhabitable for decades, if not centuries; an epidemic of misshapen monsters—animal and plant—would occur; a lingering death faced those individuals who had fortuitously survived the immediate aftermath of the detonation of these weapons.

Although none of these allegations came to pass, against this intimidating background, it is understandable that anxiety, if not fear, haunts those persons who experienced the bombings of Hiroshima and Nagasaki, distorting their lives and perceptions of the future. Every death of a survivor, regardless of cause, heightens apprehension in those survivors who continue to live. Some of this concern is exaggerated, though uncompromising answers to all of their festering uncertainties are not at hand. The reasons are many. Survivors of the bombings of these cities were exposed not only to ionizing radiation but to the blast (or shock wave) and the thermal pulse (the heat wave) that emanated from the detonation of the bomb. This multiplicity of exposures has complicated the assessment of their individual effects. Injuries from fire, collapsing buildings, and flying debris were threats to life too, diminishing the ability of a survivor to cope with the hazard that radiation exposure itself constituted.

At the time of the bombings, both of these cities were contributing significantly to the Japanese war effort. Hiroshima was a major industrial and rail center, a staging area for many of the troops sent overseas, and the fulcrum of a military complex extending some fifty-odd miles from the naval aircraft manufacturing facilities at Hiro to the east and south of the city, to Iwakuni around the bay to the west. Its citizens, spared the systematic fire-bombing of Japan in the spring and early summer of 1945, found comfort in the erroneous thought that they were unscathed because the city was the ancestral home of so many individuals who lived in the United States or were the relatives of American citizens. Nagasaki was not only the site of a major shipyard, but its large Mitsubishi complex provided the war machine with various munitions, notably torpedoes. But the inhabitants of both cities were unaware of the search to find suitable military targets for the newly developed atomic weapons. They also did not know how imminent the Allied invasion of Japan was and that the possibly hundreds of thousands of Japanese and American lives that would be lost were thought to warrant steps to reduce Japan's will and ability to further the war effort. The newly developed plutonium and uranium weapons provided the means to this end.

It is impossible to portray adequately the subsequent destruction of life and property; words fail to convey the dimensions of the catastrophe. One hundred thousand human beings, civilian or military, perished instantly or died within a few weeks in Hiroshima and tens of thousands were injured.[1] Correspondingly large numbers of individuals were killed or severely injured in Nagasaki. The physical devastation was staggering; within moments, thousands of buildings were leveled or obliterated, vehicles destroyed, streetcars incinerated, and bridges collapsed or moved sideways on their pilings, rupturing the mains carrying the water essential to the fighting of the hundreds of fires ignited simultaneously. These raged relentlessly across the city; little withstood their insatiable demands. Churches, shrines, temples, and schools vanished along with thousands of homes and places of business (see Figs. 1–5). When the fires were finally curbed, few structures remained standing within two kilometers ($1\frac{1}{4}$ miles) of the point (the hypocenter) beneath the exploding bomb in either city—only the reinforced concrete ones, and these stood stark, gutted, and scarred amidst ash and charred timbers that had once been business or dwelling places. Those few areas of the city that remained relatively unscathed, such as the region known as Ujina to the southeast of the hypocenter, did so either because of their distance from the burst point or because of a capricious shift in the wind as the fires approached.

Cities once vital with life were now seas of desolation across which the maimed and dying drifted like human flotsam, aimlessly seeking a solace not to be found. Some were incoherent, befuddled; some were like automata, merely responding to a primeval drive to survive; and still others, too weakened by their injuries, collapsed by the roadside as they tried to flee the fires. Some more fortunate than others were wheeled to an aid station or to safety on simple carts. Especially pitiful were the

Fig. 1. Partially collapsed wooden structures near the Yokogawa railway station in Hiroshima, some 2 km to the west of the hypocenter. (Courtesy of the Atomic Bomb Casualty Commission Archives of the Texas Medical Center Library)

Fig. 2. A street scene in Hiroshima showing incinerated vehicles and streetcars and the absence of structures to the sides of the street. An ambulance can be seen in the distance collecting casualties. (Courtesy of the Atomic Bomb Casualty Commission Archives of the Texas Medical Center Library)

Fig. 3. A scene near one of the branches of the Ota River in Hiroshima, showing the extent of the physical devastation produced by the blast and subsequent fires. (Courtesy of the Atomic Bomb Casualty Commission Archives of the Texas Medical Center Library)

young children, many instantly orphaned. They were ill-prepared to cope for themselves in a culture where their every need and whim had been traditionally met by an extended family, one that now no longer existed. Over the young and the old alike hung the acrid smell of erupting fires and burnt wood, and worse, the stench of death and the cries of the maimed and dying.

Other post-bombing events contributed to the misery and ill-health of the survivors. Hiroshima, for example, was struck by a severe typhoon on September 17th. Already limited water supplies were contaminated, the remaining structures were further damaged, and more people died. Eleven members of the University of Kyoto medical team who were sent to Ohno, a town near Hiroshima, to aid in treating the injured, plus almost 100 hospitalized atomic bomb survivors lost their lives in a landslide that swept them across the Osaka-to-Shimonoseki rail line into the sea. The line itself, the thin artery along which aid came, was severed at several places between Mihara to the east and Iwakuni to the south and west of Hiroshima. Heroic efforts notwithstanding, days elapsed before rail travel could be restored.

Fig. 4. The gutted Chugoku Newspaper building in Kami-Nagarekawa cho in Hiroshima, some 871 m to the east of the hypocenter. Note the burned-out streetcar and bent power pole on this the main street of the city. (Courtesy of the Atomic Bomb Casualty Commission Archives of the Texas Medical Center Library)

Fig. 5. A Japanese military ambulance on one of the devastated streets of Hiroshima collecting casualties. Note the denuded trees and the absence of buildings. (Courtesy of the Atomic Bomb Casualty Commission Archives of the Texas Medical Center Library)

These events also impaired health. They could produce an association between presence in either city at the time of the bombing and the later occurrence of some biological phenomenon, an association not causally related to radiation. Contaminated water, for instance, increased the likelihood of outbreaks of typhoid fever, a potentially fatal bacterial disease. Moreover, its accompanying high fever usually leads to the loss of hair, a symptom associated with acute radiation sickness. Similarly, the malaise and the watery or bloody diarrhea seen with radiation illness are also symptoms of typhoid fever, complicating efforts to distinguish one cause from the other.

Medical help—physicians and nurses—began to arrive in Hiroshima from neighboring communities within hours of the bombing to succor the injured and dying, and within a day after the explosion some 33 relief stations were in operation and over 150 physicians were attending the injured. Their principal activity was the extending of first aid, but they labored amidst the most chaotic conditions—a demoralized, stunned population, rubble everywhere, and still-smoldering fires. Planning had been limited and inadequate for a disaster of this monumental size. Essential services were virtually nonexistent, particularly in Hiroshima. Electrical power was interrupted. The fire-fighting equipment that had been spared was insufficient, many of the firemen were dead, and the debris-strewn streets made it virtually impossible to move the still-functioning equipment. Destruction or damage of the city's bridges further complicated matters, making a journey from one part of the city to another a time-consuming, frustrating task. Fires could only be fought on their margins. The planned fire breaks were useless, since fires had often broken out simultaneously on both sides. Pressure in the water mains had fallen to zero as a result of the innumerable water pipes that had been damaged or broken as buildings collapsed. Under the best of circumstances, no municipal fire department could have coped with the myriad erupting fires or the ensuing firestorm. Containing the conflagration was the only reasonable aspiration, and achieving this was daunting.

Most hospitals, save those in outlying areas, had disappeared. Only three of some 45 civilian hospitals in Hiroshima were usable, and neither of the two most modern medical installations, the Red Cross Hospital (*Niiseki Byoin*, located 1,470 m south of the point immediately beneath the exploding bomb) and the Communications Hospital (*Teishin Byoin*, 1,400 m northeast of the point), could do more than function as shelters for the severely injured. Their beds had been destroyed, their bedding burned, their food and medical supplies were soon exhausted, and many of their professional and nonprofessional staff had been injured, some severely. Dr. Michihiko Hachiya, the Director of the Communications Hospital, for example, had been so severely injured that he had to be hospitalized and was unable to supervise patient care. Much of the burden of caring for the casualties fell to the two hospitals maintained by Mitsubishi Heavy Industries, which were located some 4 km away and were largely undamaged, and to the Hiroshima Army Welfare Hospital which, despite its name, functioned as a public facility. The military hospitals in the city had fared no better than most of the civilian ones. Two of the largest (known as *Daiichi* and *Daini*, that is as the First and Second Hiroshima Army Hospitals) both simple, extended wooden structures, were within 1,000 m from the point beneath the exploding bomb; they collapsed and burned and all of the approximately 1,150 patients died. However, some of the more distantly placed military medical installations, such as the hospital in Mitaki, could and did offer succor not only to the military casualties but to civilians as well.

Adequate medical care was complicated by the sheer number of injured individuals, the shortage of supplies and equipment, and the condition in which they arrived at the care facilities. Most were covered with grime and dust, and their wounds were contaminated by filth. Wounds of many survivors went untended for days and became infested with unsightly translucent maggots industriously removing rotten and rotting flesh. Swarms of large houseflies filled the air, and the fear of insect-spread or water-borne epidemics prompted the immediate disposal of the dead—many burned beyond identification—either in mass graves on some of the islands immediately offshore, such as Ninoshima, or through cremation in large pyres scattered about the city using what little burnable timber remained. Although necessary, this complicated later efforts to assess the loss of life accurately.

Essential public health measures were further worsened by the absence of means for relief agencies to communicate effectively with the survivors. The local radio station and the printing facilities of Hiroshima's major newspaper, the *Chugoku Shinbun*, had been destroyed, and many of the professional staff killed (see Fig. 4). Although printing was resumed two or three days later at a subsidiary plant in Nukushina to the northeast of the city, this facility was destroyed by the September 17th typhoon. It was not until the beginning of November 1945 that the paper could start anew at its original plant in Nagarekawa, about 870 m east of the hypocenter. Normally, the authorities would have employed the system of neighborhood groups, the *chonaikai* or *tonari-gumi*, to distribute instructions to the city's households, but these were in disarray. Many of the leaders as well as members of the various groups were dead, and the surviving ones dispersed. Conditions were obviously such as to further rumors, some of which undoubtedly heightened apprehension and concern and impeded access to care.

The situation in Nagasaki was marginally less desperate. Although this bomb released more energy than the one detonated over Hiroshima, the peculiarities of the local terrain and the population distribution limited the fire damage and the number of casualties. The mountains intervening between the Urakami Valley, where the bomb fell, and the densely settled valley through which the Nakagawa River flows protected more than half of the city's population and structures. Some of its smaller hospitals remained functional, although the city's largest medical facility, the University Hospital, with more than 75% of the city's patient beds, was badly damaged and largely inoperable, and the basic science departments of the Medical School, which were located in less sturdy buildings closer to the hypocenter, were devastated. Some 892 individuals at the Medical School lost their lives, and of the 20 senior faculty members, including the heads of the clinical departments, 12 had been killed and four were seriously injured. As a result, many of the survivors had to be sent to hospitals elsewhere, either to the Naval Hospital in Omura to the north or to Isahaya to the east. These installations had adequate supplies and equipment to meet immediate needs, but the first aid stations within the city lacked nearly everything required to manage the casualties and were soon overwhelmed. Nagasaki's geographical isolation also worked to its disadvantage; it was further from the central government and its assistance, and the only rail line to the city, which passed through the Urakami Valley had been damaged—the tracks spread or twisted through the shifting of bridges displaced by the blast.

One of the survivors, Raisuke Shirabe, Professor of Surgery at the Medical School, has written of his own experiences in some detail.[2] He had been working in his office on the second floor of the surgery building when the bombing took place. He recalled the faint sounds of the bomber as it approached, the subsequent flash and following darkness, and the turmoil as the blast wave coursed through his office setting the papers on which he had been working askew. Fearful of another detonation, he hastily sought his glasses, which had been lying on his desk but couldn't find them. When he emerged from the building to seek a safer location, his eyes swept the devastation and the bodies scattered about the entrance. Among these was his dean, Susumu Tsunoo, who ironically had survived the bombing of Hiroshima three days earlier.[2] Shirabe found him alive though injured; he helped him to his feet and carried him to the top of the mountain behind the school where other survivors had gathered. While they watched, fires continued to erupt in the valley beneath them, coalesced and marched relentlessly toward the bay a kilometer or more to the south. Dr. Tsunoo's condition steadily worsened and he would eventually die.

Once the immediate needs of the survivors about him were met, Shirabe made his way to the school again to aid the injured and to search for other survivors of his staff. Soon he was so deeply immersed in the care of the injured that the fatigue plaguing him seemed unimportant and not especially unusual. Days later, when minute hemorrhages, one of the symptoms of acute radiation sickness, began to appear beneath his skin, he watched apprehensively as his white blood cell count fell to a third of its normal value. Fortunately, he recovered. When able, he and his surviving colleagues began a series of seminal studies of mortality among the survivors and of the onset and distribution of the symptoms associated with acute radiation sickness. Working with limited means and faced with the continuing need to minister to the sick and dying, they accumulated an important body of data on the catastrophe that had overwhelmed the city on August 9th.

Although Japanese physicists were not aware of the Manhattan Project (the secret U.S. atomic weapons project) at the time of the bombing of Hiroshima, enough of the research on nuclear physics had reached them before the war that they too had contemplated the construction of a nuclear weapon and quickly surmised the probable nature of the bomb that had destroyed Hiroshima. Indeed, Yoshio Nishina, an internationally respected scientist, who had studied with the Danish physicist and Nobel laureate Niels Bohr, made a preliminary survey of the city on August 8th and concluded that the bomb was a nuclear one. Teams of investigators were sent to corroborate his belief, the first arriving on August 10th. Although initially skeptical of his conclusion, in a series of visits extending over several weeks, Nishina's judgment was confirmed, and these investigators estimated the nuclear weapon's yield to be 20 kilotons (kT). They also located the area of fallout in the western limits of the city, and collected materials that would be crucial to all subsequent efforts to estimate the doses received by the survivors.[3]

Initial Attempts to Assess Biological and Physical Damage

Japanese scientists had begun to study the biological and physical damage associated with the bombing almost immediately after the bombs fell, but several weeks elapsed before the first of the United States Army, Navy, and Manhattan District Project investigating groups arrived in Japan. A team of individuals had been sent from the Manhattan District the day after General MacArthur entered Tokyo but did not reach Hiroshima until September 8 and Nagasaki on the following day, about a week after the signing of the articles of capitulation on board the USS Missouri on September 2. The members of the study group included William G. Penney, a British mathematician and theoretical physicist, Robert Serber, and George T. Reynolds, all of whom had figured in the development of the weapons that had been detonated. Penney had been an observer on one of the planes that accompanied Bock's Car on its mission. Their charge was to report on the physical damage and to present estimates of the probable yield of each of the two bombs. These were impossible tasks for only three investigators, since they were allotted but a few days in which to achieve these ends. Moreover, when they arrived in the devastated cities it was not clear how the yield might best be estimated. As Lord Penney has written, "After a day or two of almost pointless work, a new idea emerged. This was to look systematically for any simple damaged object which would enable calculations to be made about the parameters of the blast wave."[4] Accordingly, they sought the locations of snapped telegraph poles, bent flagpoles, overturned memorial stones, and crushed drums. Their aim was to draw contours of the blast damage in the two cities that could then be compared. Lord Penney deplored the fact that the limited time and the inaccuracies in the maps used to obtain the distances from the burst point of the weapons made impossible more than a superficial analysis. However, they concluded that the yield of the Hiroshima weapon was approximately 10 kT and the Nagasaki one was greater, three to four times so, but they noted that these estimates could be in error substantially.

Later, in 1964, at the request of the U.S. Atomic Energy Commission (AEC), Lord Penney (who was then the Chairman of the U.K. Atomic Energy Agency), working with Dennis Samuels and Guy Scorgie, re-estimated the yield of the Hiroshima weapon. They were persuaded that the best estimate of the yield was afforded by the data collected earlier on the bending of I-beam utility poles by the blast wind. The estimate derived for this bomb was 12 kT (with an error of 1 kT); the Nagasaki weapon was estimated to be 22 kT (with an error of 2 kT). They noted, however, that the interpretation of the observations (and hence the yield) is critically dependent on the time variation in the dynamic pressure, and direct measurements of this variation were not available.[4]

Another month passed before the Joint Commission for the Investigation of the Effects of the Atomic Bomb in Japan, under the direction of Colonel Ashley Oughterson, was formed on October 12th by order of General Douglas MacArthur Supreme Commander of the Allied Powers (SCAP).[5] The immediate purpose of this Commission was to provide a unified control and focus to the efforts of various groups of investigators interested in the medical and physical effects. But Oughterson and his staff quickly recognized that the American group sent to Japan, although including more than 60 individuals in total, could not mount a study of the scope that was obviously needed and that the active participation of Japanese professionals was essential if the study was to succeed. Fortunately, through Dr. Masao Tsuzuki, who represented the Japanese government on the Commission, and with the unstinting support of the Japan Science Council and the Tokyo Imperial University, it was possible to enlist the help of some 90 Japanese professionals whose efforts contributed enormously to the success of the Joint Commission's activities.

These investigative groups had little notion of what to expect. In a memorandum to Brigadier General Guy Denit, the Chief Surgeon at the General Headquarters of the U.S. Army Forces–Pacific, Oughterson, aboard the S.S. General Sturgis en route to Japan on 28 August 1945, wrote an almost apologetic defense of the need for careful study of the casualties. He also noted "since the effects of atomic bombs are unknown, the data should be collected by investigators who are alert to the possibility of death and injury due to as yet unknown causes." While it is reasonable to believe that Colonel Oughterson assumed he was writing to a privileged audience, this was a candid admission of how little was known about the health effects of exposure to atomic radiation at the time.

The coordinated American and Japanese studies focused on the physical damage wrought by the bombing and the immediate biological consequences of exposure to radiation. Included were detailed measurements and descriptions of the blast damage, attempts to determine the probable amount of shielding from ionizing radiation afforded by the various buildings that had withstood the blast, and estimates of the number of individuals killed or injured in each of these cities. Contrary to what is commonly supposed, the bulk of the fatalities in Hiroshima and Nagasaki were due to burns caused either by the flash at the instant of the explosion or from the numerous fires that were kindled, and were not a direct consequence of the amount of atomic radiation received. Indeed, the Joint Commission estimated that over half the total deaths were due to burns and another 18% due to blast injury. Nonetheless, ionizing radiation accounted for a substantial number of deaths, possibly 30%. Their medical findings have been exhaustively described in their report.[6]

Briefly, much of what was learned rested on a survey of some 13,503 individuals who were known to have been alive 20 days after the bombing and on the information Japanese investigators had independently assembled before the initiation of the Joint Commission's study. The value of the latter information was compromised to some degree since it had not been systematically collected, was often understandably incomplete, and was dispersed through many academic institutions and governmental agencies around the country. Collection, collation, and interpretation of these data proved formidable, and would not have been possible without Dr. Tsuzuki's influence and assistance.

Many of the individuals examined by the study group had developed what has been termed acute radiation sickness. In the days and weeks following the bombing they experienced fever, nausea, vomiting, lack of appetite, bloody diarrhea, loss of hair (epilation), bleeding under the skin (purpura or petechiae), sores in their throat and mouth (nasopharyngeal ulcers), and decay and ulceration of the gums about the teeth (necrotic gingivitis). The onset of these symptoms varied, but it generally occurred sooner among the heavily exposed. More subtle events, less recognizable to the individual survivor, occurred too. Many lapsed into a period of pancytopenia—a pronounced reduction in the number of red and white cells and platelets in the blood—and their bodily defense mechanisms to infections were lowered by other blood changes. Although we do not know precisely the time course of these events in the survivors, we can presume it followed a pattern similar to

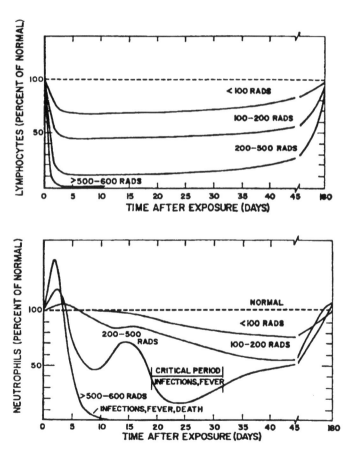

Fig. 6. The changes with time following exposure to various doses of ionizing radiation in two hematologic parameters—lymphocytes and neutrophils. (After Langham, 1967)

that seen in Fig. 6, which is based on other sources of information.[7] It will be noted that initially there is a rise in the nonlymphocytic white cells (neutrophils), much as would occur in any infection, but this initial response exhausts the hematopoietic reservoirs and then these cells fall precipitously in number. At a dose of 100 rads (1 gray), the ability to produce these cells and platelets is only half the normal value.

Efforts to ameliorate these hematologic effects, which were poorly understood at the time, were limited. None of the hospitals had blood banks that could have been used to mitigate these effects, and when transfusions were used they invariably involved very small amounts of blood, generally drawn from the survivor, and reintroduced into the circulation by syringe. These efforts were ineffective in reducing mortality. While in this pancytopenic state, the survivors were prone to infectious disease, often a secondary result of the injuries sustained from flying glass and other debris. Many were badly burned from the fires ignited by the bomb, the hot ash that fell from the sky, or the thermal radiation that was intense enough to etch into the skin the pattern of the garments they wore. Often, as these injuries healed, extensive scars (keloids) formed, limiting the movement of the survivor's arms, legs, and hands, and hindering their full recovery.

Mortality in this early period, the first nine weeks or so following the bombings, was correlated with the severity of the pancytopenia. But it is not known whether the responses to radiation seen in later years are related to the severity of the bone marrow damage and this may never be known with certainty.[8] The available records do not contain enough detail to determine which individuals examined in the Joint Commission studies have been participants in the subsequent health studies in Hiroshima and Nagasaki. Ideally, each record would contain the individual's full name represented in ideograms (*kanji*) with accompanying romanization (since transliteration of the ideograms can often occur in more than one way), birth date, and address at the time of the bombing. Unfortunately, many of the records do not include ideograms; the location of the individual at the time of the bombing is given, but often not the residential address, and instead of birth date, age in years was recorded.[9] Whether the incompleteness of the identifying information reflected the fact that the histories were hastily collected, or an unfamiliarity with Japanese conventions on the part of the American investigators, or a failure to realize that these records could have uses other than the immediate ones of the Joint Commission is moot. But whatever the explanation, the long-term usefulness of these early clinical examinations was compromised.

Recovery from acute radiation sickness, if it occurred, was often slow. The limited supplies of antibiotics in Hiroshima were of low potency and high toxicity, and in Nagasaki they were not available at all.

Other potentially useful medications, such as the sulfonamides, were scarce. Treatment was largely palliative (see Figs. 7–9). Recovery was further impeded by the poor nutritional status of the population. The food situation in Japan had steadily worsened from 1943 onward, and by August of 1945 was tenuous, particularly in the larger cities. Crops were affected by the shortage of farm labor and fertilizers, and even the supply of fish was limited by the commandeering of all sizable fishing vessels to haul military cargo and the drafting of their crews. Moreover, as food supplies grew smaller, farmers and fishermen kept more of their produce or catch for themselves and their families. Food was rationed, but the daily allowance was small. The caloric intake from cereals and soybeans at the time of the bombing was only 1,200 calories per day in Nagasaki and 1,600 in Hiroshima, well below the daily requirement of 2,160 calories projected by the Japanese Institute of Nutrition. Nagasaki residents could supplement this meager ration with produce grown in household gardens or with fish, but there was less opportunity for this in more densely settled Hiroshima, despite its proximity to the sea.

Much of the effort of the joint study group went into the assessment of the frequency and nature of the early effects, and out of their studies has grown the most comprehensive, albeit still incomplete, picture of the dreadfulness of a nuclear fire. It is not known, for example, precisely how much radiation was required to produce death within 60 days in

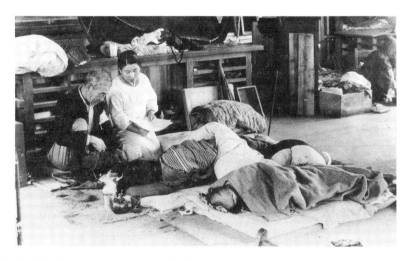

Fig. 7. A husband and nurse comforting several survivors lying on futons on the floor in one of the first aid stations in Hiroshima. Note the primitiveness of the facilities available to care for the injured. The survivor on the right has suffered epilation. (Courtesy of the Atomic Bomb Casualty Commission Archives of the Texas Medical Center Library)

Fig. 8. An injured, treated survivor being transported on a simple two wheeled cart from a treatment center. (Courtesy of the Atomic Bomb Casualty Commission Archives of the Texas Medical Center Library)

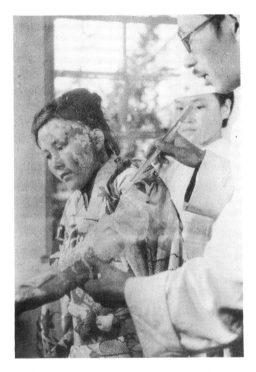

Fig. 9. Treatment of a burned survivor. Note the extensive burns to the left arm and left side of the face. The physician appears to be placing sterile gauze dressings on the arm. This is the same survivor seen in Fig. 8. (Courtesy of the Atomic Bomb Casualty Commission Archives of the Texas Medical Center Library)

50% of the individuals exposed to the bombing of Hiroshima or Nagasaki—a dose radiation biologists call the $LD_{50/60}$.[10] This dose is more than just a scientific curiosity—an arcane number—since it and the survival curve of which it is a simple descriptor can be of immense importance in the planning of responses to nuclear accidents such as that at Chernobyl, where difficult decisions must be made with respect to the allocation of medical resources. When such accidents occur, these decisions should not be capricious, but should be based upon the intelligent use of all available means to mitigate the immediate health consequences of exposure. To achieve this end requires some way to identify those individuals most in need and knowledge of the $LD_{50/60}$ can contribute to a systematic allocation of treatment that will maximize the number of survivors.

2

Establishment of the Atomic Bomb Casualty Commission

As a result of their findings, the Joint Commission urged that a continuing study of the possible late or delayed effects of exposure be initiated under civilian, nongovernmental auspices. In May 1946, Major General Norman Kirk, who was then the Surgeon General of the Army, wrote to the Chairman of the Division of Medical Sciences of the National Research Council (NRC), Lewis H. Weed, pointing out the government's desire to have the survivors studied further, and requesting that the Council examine the possible methods of achieving this end.[1] On May 28, 1946, the Council convened a conference of civilians and governmental representatives to consider this request. Out of their deliberations grew a series of recommendations that were sent to General Kirk. Among these was the suggestion that several civilians be appointed as consultants to the Secretary of War and sent to Hiroshima and Nagasaki as soon as possible to assess the problems attending the establishment of a permanent organization to conduct the proposed long-term medical and biological studies. Accordingly, two civilian consultants, Drs. Austin M. Brues of the University of Chicago and Paul S. Henshaw of the Clinton Laboratory of the Manhattan Project, accompanied by three military officers, Lt. Melvin A. Bloch, Lt. James V. Neel, and Lt. (j.g.) Frederick M. Ullrich, were dispatched to Japan in the autumn of 1946. This group chose to call themselves the Atomic Bomb Casualty Commission (ABCC), and since their activities related to matters of health, they operated under the Public Health and Welfare Section of SCAP.

While the fact-finding group was still in Japan, other events were already shaping the future. James Forrestal, the Secretary of the Navy, in response to the recommendations forwarded to General Kirk earlier and at the urging of his advisors—the Surgeon General of the Navy, the

Vice Chief of Naval Operations, and the Chief of Naval Research—suggested to President Harry Truman that he issue a Directive to initiate a long-term study of the medical and biological consequences of exposure to atomic radiation. Specifically, Forrestal and his advisors recommended:

> That the Presidental Directive instruct the National Academy of Sciences–National Research Council to undertake a long range, continuing study of the biological and medical effects of the atomic bomb on man. That in this directive the Council be authorized to enlist the aid of governmental agencies and personnel, and such civilian agencies and personnel as may be needed. Further, that those governmental agencies whose aid is requested by the Council be authorized and requested to provide the needed cooperation.

In support of this delegation of scientific responsibility, Forrestal argued that the study needed was "beyond the scope of military and naval affairs, involving as it does humanity in general, not only in war but in anticipated problems of peaceful industry and agriculture."[2] But he also noted, presciently, as history attests, that "Such a study should continue for a span of time as yet undeterminable." On November 26, 1946, President Truman endorsed Forrestal's recommendations and once he had done so planning of this larger study of still uncertain duration began. Specifically identified as matters of concern in the planning documents to carry out this directive were "cancer, leukemia, shortened life span, reduced vigor, altered development, sterility, modified genetic pattern, changes in vision, 'shifted epidemiology,' abnormal pigmentation, and epilation."[3]

Traditionally, the Academy and its National Research Council have been independent advisory bodies to the government, not agencies involved in the conduct of field studies, and this new charge lay outside their previous experience. Moreover, studies could not proceed in Japan without the support of Douglas MacArthur, the Commander of the occupying forces, who was noted for his hubris and independent bent of mind. This support was soon forthcoming, since he saw the political benefits, if not the scientific importance of the studies. But the distance intervening between the Academy and its field offices would remain a formidable obstacle to effective administration and implementation of the proposed investigations.

In early 1947, after the Academy had designated a standing committee to supervise the long-term studies under the chairmanship of Thomas M. Rivers, an eminent virologist, the Atomic Bomb Casualty Commission (ABCC) was formally created.[4] As planning for the studies began, three questions presented themselves. First, how were the doses received by individual survivors to be determined? Without such information it would be difficult, if not impossible, to quantify the effects in a manner permitting their use under other circumstances of exposure. Second, how was a representative group of survivors to be identified for study? There were no rosters of individuals present in these cities at the time of the bombing that might serve to define a suitable group for investigation, nor was it known how many survivors had moved elsewhere. And third, what measures, clinical or otherwise, were to be used to assess the extent and nature of the radiation damage? The potentially adverse effects embraced almost all aspects of health, but it was obvious that there had to be some focus to the investigations. Without this, the study could drift aimlessly and, in an effort to answer all conceivable questions, would answer none.

At the time of the establishment of the Commission, although it was recognized that exposure to ionizing radiation was hazardous, there was little evidence that would permit estimating the risk of any of the longer-term effects in a quantitative manner. It was not known whether the occurrence of a specific effect was proportional to the amount of radiation received, that is, whether the effect increased linearly (as a straight line) as dose increased, or whether the effect increased in frequency in some other fashion. Nor was it certain whether a threshold of exposure existed below which radiation was not harmful. Similarly, it was unclear how pervasive the harmful effects might be, and among those effects that had been reported, none was known that could be unambiguously attributed to exposure. They also occurred in individuals and populations that were not exposed to amounts of ionizing radiation above those naturally occurring. The Manhattan Project, which had given rise to the development of the atomic bomb, had fostered a series of experimental investigations, including genetic ones, that sought to provide some insight into the possible biological hazards. But the results of these studies were not widely known, and it was uncertain whether the experimental results could be extrapolated to the human situation. It was centrally important, therefore, to establish appropriate comparison groups so it would be clear whether a particular event was or was not occurring more frequently among the survivors.

Initially, it was thought that Hiroshima and Nagasaki would rebuild very slowly, and there might be too few nonexposed immigrants to provide the needed comparison population. There was concern too whether radioactive contamination of the cities might make these immigrants inappropriate subjects even if they were numerous enough. These considerations led to the supposition that comparison populations would have to be sought elsewhere. Kure, near Hiroshima, and Sasebo, near Nagasaki, were selected as the sites where such groups might be found. No systematic studies were ever initiated in Sasebo, although some data on growth and development were collected there, but modest genetic and pediatric programs did begin in Kure. These were terminated in 1950 when it was apparent that many individuals who were either not exposed or exposed to very small amounts of ionizing radiation could be found within the study cities themselves, and ABCC had been assured by the Academy's Advisory Committee that fallout and residual radiation would not prevent their use as comparison subjects.

Before these matters could be addressed, suitable research facilities had to be established, which was no easy undertaking. This meant finding working space, personnel, and the ancillary facilities needed. In the spring of 1947, space was limited and ABCC began in Hiroshima with a dozen or so individuals in rented rooms in the Red Cross Hospital, immediately to the south of the city's center. Although damaged, this reinforced-concrete building had withstood the blast and subsequent fires. ABCC's origins in Nagasaki were equally unpretentious: in borrowed classrooms in a school. However, as quickly as practical, a search was initiated for more suitable space in both cities. In Nagasaki, arrangements were made to occupy a building, the Kaikan, owned by the local teachers association; whereas in Hiroshima, additional space was obtained in a former military building, the Gaisenkan (ironically translated as the Hall of Triumphant Return), adjacent to which were built temporary clinical and laboratory facilities, and a motor pool. However, once architectural plans could be completed, construction was begun in the summer of 1949 on new clinical and research facilities in a park on Hijiyama, a hill some 2 km to the east of the hypocenter. When construction was completed late in 1950, the need for better working facilities was eased, but the choice of the site brought an unanticipated problem—one of public relations. Most of the land the facilities occupied had been a playground, but the area did impinge on a military cemetery, which had to be moved to permit construction of the new buildings, and one corner was once the location of a shrine to the Emperor Meiji. The latter had been destroyed in the bombing, but the land was considered sacred, although this had not prohibited its use as the site of numerous anti-aircraft guns during the war. Nonetheless, the laboratories do look down on the city, and this fact has been repeatedly used to infer a "looking down" on the population of survivors as well. In retrospect, the choice was not wise, but any location would have encountered some opposition, since the Commission was long seen as the last vestige of the Occupation.

Most importantly, professional and nonprofessional personnel, Japanese and American, had to be identified, recruited, housed, and, where necessary, trained. At that time, for example, Japanese nurses were educated differently from nurses in the United States or in Japan now, and most were only accustomed to menial work. These differences necessitated training "on-the-job," largely under the supervision of American nurses, initially Louise Cavagnaro. Similarly, locally recruited technicians, save in a few areas such as parasitology, needed to gain familiarity with American standards and equipment. Drivers and mechanics had to be employed, and since few individuals in Japan could drive at the time, many came from the former Japanese Army. While recruitment of properly trained Japanese staff was difficult, the recruitment of American personnel proved to be an even more frustrating task. Anti-Japanese sentiment in the United States had not completely subsided, and not every suitably qualified American investigator was enthusiastic about a position of uncertain tenure in Japan. Few were prepared to invest more than two years of their lives in the program, and the constant turnover of American scientific and administrative personnel compromised the continuity so essential to success. For many of those recruited, life in Japan and the work environment would prove unusual, if not strange, and separation from one's own culture and isolation in another required an adjustment.[5] The community of investigators, even at its largest, was relatively small, and as a result, it was difficult to separate one's work from one's social life. Thrown together for many of the day's waking hours, there was an opportunity for a clash of personalities. However, the work environment has been congenial, despite its unusualness from the American perspective.

One of the more pressing needs to further recruitment was the establishment of some mechanism to assure that suitable recognition was given to those investigators, Japanese and American, who had seen the value of specific studies, had initiated them, but could not bring them to fruition in the term of their appointment. Advancement of their careers often hinged on their research productivity, and this was commonly judged by the number and quality of their scientific publications. Although this need is inherent in all long-term studies, it was especially acute in the Commission's instance, since the institution had no strong academic ties in either Japan or the United States. The problem proved to be one more readily recognized than fairly resolved, and still plagues the Radiation Effects Research Foundation (the Commission's successor) to some degree, particularly in respect of its American employees who, unlike many of their Japanese colleagues, do not expect to make a career of their involvement with the Foundation. A solution to this issue equitable to all remains elusive.

Many unnecessary administrative barriers impinged on recruitment too. In the early years, for example, a security clearance was a prerequisite to employment, although newly employed individuals had no access to classified material, even of the most lowly order. Perhaps this merely reflected the paranoia of the McCarthy years, but it aroused unfounded suspicion in the minds of one's friends, neighbors and employer, since it was not known that the clearance investigations were routine. Worse still, the implied secrecy troubled acceptance of the findings by some members of the scientific and lay communities. It was presumed that the Atomic Energy Commission (AEC), which had been established in 1946 and provided the funds for the research, would suppress those discoveries it found prejudicial to its program, which then focused heavily upon the development of nuclear weapons to meet the nation's presumed defense needs. This was, in fact, not the case. Never to my knowledge has the AEC or any of its successor agencies tried to suppress or intrude on the freedom to publish any of the findings of the scientists of the Commission or the Radiation Effects Research Foundation. Publication has always hinged on the quality of the science and not upon the political acceptability of the findings to either the Academy or the funding agencies.

Once investigators were recruited, there was still housing to be procured and the tools acquired to carry out their work. The economic circumstances in Japan were severe. Housing was scarce; food and clothing were rationed; transportation, public or private, was limited. The Commission had to be self-sufficient—its self-sufficiency achieved in the context of an occupied, impoverished nation. It not only needed its own fleet of trucks, jeeps, buses and other vehicles, but the means to maintain and repair these. Even bicycles—several dozen, in fact—had to be purchased for those employees who would arrange patient visits. Medical record forms had to be printed, and printing facilities established to accomplish this. Clinical laboratories had to be built and staffed. Data processing equipment had to be bought or leased, and even laundry services established. Diagnostic tools, such as X-ray machines, and photographic facilities, virtually unobtainable in a devastated Japan, had to be brought from the United States.

As the program and staff in the two cities grew, difficulties in the coordination of the activities in Nagasaki with those in Hiroshima multiplied; telephone service was erratic and travel between the cities excruciatingly slow and dirty. Neither the main rail line in western Japan nor the spur from Tosu to Nagasaki was electrified. Diesel engines were not used, only coal-burning ones that deposited soot impartially on people, animals, and the passing terrain. Urgent instructions were often sent through the Japanese special postal delivery system which, although less rapid than the telephone, did assure overnight delivery. Most of the staff with programmatic responsibilities, although usually living in the Hiroshima area, visited Nagasaki regularly to ensure that the activities there did not differ from those in Hiroshima.

But nagging questions remained: Could an adequate research program be maintained in both cities? Was a program in Nagasaki necessary? Since the number of survivors presumed to have received significant exposures in Hiroshima was much larger, should not all the resources be focused on the studies in this city? In March 1951, the Commission's Executive Research Committee did, in fact, recommend terminating the genetic study in Nagasaki, which, since it was by far the largest undertaking there, was tantamount to a closure of most of the research program. This did not occur once it was pointed out that a study in Hiroshima alone could not possibly provide enough information to estimate the nature and extent of the genetic effects, and the surveillance of pregnancy terminations continued for another three years. This would not be the last threat to the continuation of the studies, however, nor did it resolve the more basic question about the scope of the Nagasaki effort. Defining this scope has remained a troublesome scientific and administrative matter to the present.

Other issues, often more pervasive and pernicious, plagued the early administrative and scientific planning. As growth occurred, the administrative structure of the institution became increasingly convoluted. There was simultaneously a greater centralization of authority in some areas, and an increasing dispersion in others. A variety of advisory committees came into existence—to supervise and systematize research, to administer the library, and to encourage recruitment of personnel and study participants. Effective communication of local needs and the programmatic aspirations of the Commission through many levels of decision making and different agencies within Japan, including those of SCAP, and in the United States was particularly vexing. Even the availability of housing for the recruited staff rested on potentially capricious decisions by the Occupation authorities. The area of western Honshu where Hiroshima is situated was in the zone of occupation assigned to the British Commonwealth Forces, which were under the command of an Australian, Lieutenant General Horace C. H. Robertson. Australia at that time maintained an exclusionary policy toward Asians and this fact, coupled with the rules restricting fraternization with the Japanese, understandably annoyed some of the Japanese-American employees of the Commission, who could not freely fraternize with their own Japanese-born relatives or reside in the housing under Australian supervision.[6]

Matters were not made easier by a succession of different heads of the Division of Medical Sciences of the NRC, under whose aegis we worked. First, there was Lewis Weed, an anatomist, and then Milton Winternitz, a pathologist. Both had come to the NRC after terms as Deans of two distinguished medical schools—Johns Hopkins and Yale—and were perceptive advocates of the studies in Japan. Indeed, from the outset of his appointment, Winternitz had argued forcefully for the need to introduce some basic pathology into the ABCC studies. However, greater improvement in the managerial issues came with the arrival of R. Keith Cannan, a biochemist with a deep sense of the worth of the studies in which the Commission was engaged, who recognized early that the work in Japan was essentially epidemiologic. At a particularly critical period in 1955, when the studies had not only lost momentum but direction, through his efforts the Commission was reinvigorated and set on a scientific course that has served the studies well in the decades that have followed.

Over all of this hung the uncertainty of ABCC's future. Only a few years after its establishment, sharp differences arose between the AEC and the NRC on the cost and management of the studies in Hiroshima and Nagasaki. These differences culminated in a statement by the AEC's Division of Biology and Medicine that it was unable to provide more than a certain fixed amount of money for the support of the studies; this amount was much less than that which the National Research Council saw as necessary to do a creditable job. Rather than sacrifice scientific standards, the Academy's Committee on Atomic Casualties recommended the closure of the program. A party of two, Ernest Goodpasture, a member of the AEC's Advisory Committee, and Willard Machle of the NRC, was sent to Japan to explore this and other administrative issues that had arisen. They were accompanied by Merril Eisenbud, the director of the Health and Safety Laboratory of the AEC's New York Operations Office and the administrative officer responsible for the contract under which the studies proceeded. MacArthur, however, opposed the proposed action; apparently he feared that terminating ABCC would create a void soon to be filled by others and that the quality of later observations would be poor, or worse, self-serving.[7] While his position was undoubtedly largely political, other defenders of the program arose who believed the decision to close the studies was scientifically indefensible and strongly opposed the decision. The Advisory Committee reversed its recommendation, and the work of the Commission continued, but the frailty of its mandate had been exposed. It existed then and even now through the sufferance of agencies and individuals who were poorly informed, often uninterested, and generally had other, more immediate, administrative, political, or scientific concerns.

Intrusions from Washington into the orderly management of the program, however well-intentioned, occurred repeatedly. In February 1950, for example, Carl Tessmer, then the Commission's director, and Grant Taylor, its associate director, were summoned to the United States to report on the status of the studies. Once in the capital, to their surprise they found that decisions about the size and the implementation of the program had been made by others, in particular the Academy's Committee on Atomic Casualties, most of whom were only marginally familiar with what was actually occurring in Japan. The program in Kure was to be abandoned, and the contemplated one in Sasebo not to begin. These external comparison groups were to be replaced by internal ones, drawn from the cities themselves. This was a development that was welcomed, since it had been repeatedly urged by the investigators in Japan, most forcefully perhaps by James Neel who, once familiar with the composition of the populations of the burgeoning cities, had argued that programs in Sasebo and Kure were not essential to the genetics studies and, by implication, to the other studies as well. But the decision regarding the relative magnitude of the activities in Hiroshima and Nagasaki, though defensible, appeared to be the outcome of political and economic rather than scientific considerations.

Obviously, what members of the Commission saw as intrusions were often seen differently by the AEC and the NRC. One of the latter's consultants, Willard Machle, lamented that the Council through its Committee on Atomic Casualties was not providing enough scientific and managerial guidance to the field operations, and that too much responsibility was delegated to the Commission's staff. He deplored, too, the failure of the AEC to communicate its concerns about the prosecution of the science to the Commission through the Council. He observed "More recently AEC has taken more active interest in the details of the field operation and the manner of prosecuting the research. These interests have not always been expressed through the medium of NRC and the consequences have been somewhat unfortunate, since direct contact and review of field activities by AEC personnel, independent of NRC assistance, has led to uncertainties in the minds of field personnel as to just who is to advise them."[8] How widely his views were shared may be arguable, but there was an obvious need to clarify the individual responsibilities of the three agencies involved—the AEC, the NRC, and the Commission—and to define better the short- as well as the long-term goals of the research.

Some of these difficulties or differences in perspective could possibly have been avoided if the Commission had had a more permanent American directorial staff. This was not the case; the organization had no less than six directors in the first 10 years of its existence. Fortunately, there was greater continuity within the agencies of local government and the city and prefectural medical associations, since the Japanese government, through its Ministry of Health and Welfare, had established branch laboratories of the Japanese National Institute of Health (JNIH) in Hiroshima and Nagasaki once ABCC was formed. JNIH assigned personnel to participate in the studies at the administrative and research levels, and local contacts were maintained largely by the directors of the branch laboratories, who also served as associate directors of ABCC. The first of these to join the Commission was Dr. Hiroshi Maki, who was appointed director of the Hiroshima Branch Laboratory of the JNIH in 1948. He was joined in 1956 by Dr. Masanori Nakaidzumi, who had been professor of radiology and dean of the University of Tokyo's Medical Faculty, and finally, in 1957, the triumvirate was completed with the appointment of Dr. Isamu Nagai as director of JNIH's Nagasaki Branch Laboratory. Each of these men contributed greatly to the conduct of the studies and the establishment of the Commission's administrative policies. They could and did articulate the concerns of the survivors and the local governments, and offered solutions to vexing financial problems when these deepened in the mid and late 1960s. It was Dr. Nakaidzumi, in fact, who at that time suggested the Commission's reorganization as a *tokushu hôjin* (a public corporation enacted by law) under the aegis of the Ministry of Health and Welfare. When the reorganization to be described later did occur, the negotiations led to a status similar to but not legally identical with that which he had suggested.

Disgruntlement was sown in other ways. The early relationship with the Medical School of Hiroshima University was burdened by distance—the town of Aga, where it was then located, is 20 or so miles from Hiroshima. Nor was this relationship helped by the brief history of the school. It had come into existence in response to a directive of SCAP that each prefecture was to have a national university, and to meet this requirement Hiroshima Prefectural Medical College, a school then of low standing and few years existence, was incorporated into Hiroshima Bunri Daigaku, a national college of arts and sciences. As a result, the new University's medical school and its faculty, some of whom were repatriates from lesser Japanese medical institutions in Taiwan and Korea, had little influence in medical circles within the prefecture. This was in sharp contrast to the situation in Nagasaki, where the medical school was not only the oldest in Japan but had great local prestige. When ABCC sought help from the former Imperial Universities in Tokyo and Kyoto, this was seen as a slight that strained the relationship with Hiroshima University for many years. Undoubtedly, some envy colored these relationships as well, since the resources at the disposal of the Commission were much greater than those available at the University. It is debatable, however, whether this appeal to the prestigious universities bettered recruitment opportunities. ABCC's association with the Japanese Ministry of Health and Welfare (*Kôseishô*), largely forced by SCAP's Public Health Section, made employment of Japanese scientists from the national universities difficult, since they were employees of another agency, the Ministry of Education (*Monbushô*). This difficulty, when combined with a position of uncertain duration, made ABCC attractive only to the academically footloose, who had few, if any, better employment opportunities.

Finally, about two years after the beginning of the Commission's studies, another wholly unanticipated complication arose. In June 1950, North Korean forces invaded the southern half of the politically divided Korean peninsula. They rapidly overran the defenses mounted by the numerically smaller, largely unseasoned American and South Korean troops, and a debacle was in the making. Eventually, the front stabilized, but not before the defenders had been pushed into a small portion of Korea surrounding the port city of Pusan. While these events did not appear to threaten Japan, doubts were raised about the continuation of the studies, at least in the minds of those who were in Hiroshima at the time. The Commission was still logistically dependent upon the Occupation for many things—for instance, gasoline and parts for its numerous former military vehicles—and the situation in Korea could impinge on the meeting of these needs. The latter were certainly less important in the eyes of the military than the defense of South Korea. Moreover, some of ABCC's professional personnel, including the Director, Carl Tessmer, were still in uniform, and others held reserve commissions either in the Army or the Navy and could be called to active service again. If this occurred they would have to be replaced, and as has been said, recruitment was not a simple matter. None were called directly from the Commission and the logistic requirements continued to be met. Nevertheless, several members of the staff, the surgeon Warner Wells for one, did serve in a consultative capacity at the military hospitals in western Japan. And Grant Taylor, the Associate Director, was sent to Korea briefly to assess the prevalence in that country of Japanese encephalitis, a mosquito-borne viral disease.[9] But there was a compensation. As the hospital of the British Commonwealth Forces in Kure began to expand, some of its more research oriented professionals sought to participate in the studies in Hiroshima when time allowed. This was an unexpected boon. Two of these, Colonel John Menzies and Captain James Renwick, proved especially helpful. Menzies became an informal consultant to the Department of Medicine, and Renwick, a medical geneticist, made important contributions to the genetic studies, most significantly to the review of the program on radiation-related spontaneous abortion that was then under way.

A year after the beginning of the Korean War the hostilities with Japan that began with the bombing of Pearl Harbor ceased formally with the signing of a Peace Treaty in San Francisco. The end of the Occupation, however, made the Commission's legal status in Japan unclear, and this remained so until an exchange of Notes Verbale occurred between the U.S. Embassy and the Japanese Ministry of Foreign Affairs in October 1952. This exchange established the Commission as an agency attached to the U.S. Embassy, granting it quasidiplomatic status, a situation that would continue until April 1973, when another exchange occurred. The latter terminated ABCC's special status as an agency attached to the Embassy but continued to recognize the Commission as a U.S. government research facility in Japan.

Throughout the years of the Occupation, the Commission's employees were not unionized. In retrospect, this was odd, since the policy of SCAP had been to encourage the cause of organized labor. However, soon after the termination of the Occupation, while Grant Taylor was director, unions affiliated with the generally anti-American General Council of Trade Unions of Japan (Sôhyô) were formed in both cities and began, or at least attempted, to influence the Commission's activities. These unions did not then nor do they now include either professional or administrative employees—the latter at the level of a section chief or higher—and as a result have frequently been led by individuals without administrative experience or professional training. Many have been distinctly left of center in their personal political allegiance. As a consequence, economic issues were frequently secondary to political statements. The recurrent negotiating themes were "job security" and "Yankee go home," but how the one was to be achieved if the other came to pass was never discussed. Moreover, the unions did little to further a rational attitude toward improving employee working conditions. Frequently, the leaders refused to serve on Commission-established committees that sought improvements and even forbade other members such service. It seemed to be the position of the union leadership that administration-suggested improvements in working conditions would be accepted, but constructive union-sponsored ideas would have to be obtained in classic Marxist tradition—some form of "struggle" was necessary to legitimize any change. Perhaps this only reflected the inexperience of a leadership unpracticed in labor–management relationships, or a perceived need to establish their position within the union itself.

Fortunately, over the years, only two significant strikes occurred, largely as a result of wage disputes. But the policy the union followed on these occasions was calculatedly truculent and unusually political in character. Pickets blocked entrances to the facilities in both cities, clinical examinations were necessarily suspended, and the nonunion staff were prevented from entering their offices or laboratories. The walls of the research facility were festooned with bits of paper containing derogatory, personal remarks that soured relationships between the American and Japanese employees for some time. All of this was unnecessary, since the legitimate aspirations of the union could have been achieved in other ways, but the union's leadership has often been manipulative and not, as a rule, sensitive to the employees it presumably represents.

Still other problems arose that struck at the purpose of the organization. Should it, indeed could it, legally provide medical treatment to the survivors? And if so, what should the treatment be? Moreover, if treatment were undertaken, what would be the long-term impact not only on the health care of the survivors but on the practice of medicine in these communities? Would it, through the special attention paid to one large group of inhabitants of Hiroshima and Nagasaki, stultify the development of adequate health resources for all? Would it leave the survivors lacking other means of care once the studies ended? Eventually, largely at the urging of the two governments and the local medical communities, it was decided that ABCC would not extend treatment. It was argued that for language reasons, if no other, Japanese physicians were better prepared than their American counterparts to treat the survivors. Many of the Commission's directors and physicians were not pleased by this arrangement, contending that it was not only out of keeping with the traditions of their profession but politically unwise as well, and they ignored the policy, distributing penicillin and other medications to needy families.[10] Retrospectively, their position was the wiser, since this decision, more than any other, has impaired public understanding of the aims of the studies and has haunted the research effort since it was initially made. It provided a needless basis for continued Japanese criticism of the Commission's activities by the laity, professionals, clergy, and political groups both pro- and anti-American, and inhibited staff morale. It was further argued that it compromised the quality of the clinical findings, since "If a critical illness should occur we are unlikely to know about it, and if death ensues our chance of obtaining an autopsy is greatly lessened due to loss of contact and also the lack of sympathetic rapport with the family, which is of equal, or even greater import."[11] But matters of this nature transcended individual persuasions. They involved two governments and two societies more disparate then than now. Answers were important, nonetheless, since upon these could easily depend local participation in the studies, a matter of moment to the Commission as it has subsequently been to the Foundation. Without a sufficiently high proportion of the survivors participating, the findings would be suspect—one could not be confident that those who were motivated to participate were representative, healthwise, of the possibly larger group who were not so motivated.

Individual participation has been voluntary since the outset of the studies but in the early years, while Japan was occupied, this may not have seemed so to the survivors themselves. Although ABCC was not a formal part of the Occupation, it was perceived as such, and not without reason. Many of our vehicles, although differently painted, were obviously military surplus, and we were housed and supported logistically by the Occupation. Japanese attitudes toward the Occupation have been understandably ambivalent. Not all that was done under SCAP's direction was viewed as constructive. But members of the Occupation did contribute to the rebuilding of these cities, and in the case of Hiroshima figured prominently in its long-term city planning. Indeed, the city's attractive and justly famed central boulevard (now known as Heiwa Dôri, but originally called the *Hyaku Metoru Dôro*, or hundred meter road), which was constructed during the war as a firebreak, was retained at the urging of an Australian consultant, a member of the British Commonwealth Forces.

ABCC's situation was somewhat different. It carried an onus that was not of its choosing. Although few, if any, survivors held individual members of the Commission directly responsible for their tragedies, as representatives of the country that had perpetrated the bombing, it was to be expected that the relationship between the survivors and the American staff would be ambiguous. Japanese social customs, however, would not have countenanced an open display of animosity; this would have been seen as rudeness and would have reflected poorly upon them regardless of their personal attitude. As individuals we were undoubtedly judged by our understanding of the survivors' needs, the sensitivity with which interpersonal relationships were managed, and the degree to which our conduct conformed to their expectations. But this was not enough. It was important that the survivors saw some immediate benefit to themselves stemming from participation in the Commission's studies. This was not easily achieved when treatment was not offered.

The early years were unsettled too by the recriminatory charges repeatedly hurled at ABCC by the print media, the leaders of its own unions, and some of the more politically active survivors themselves. Most of these focused on four issues: the Commission's failure to treat, its location overlooking the city, its alleged failure to inform the Japanese citizenry and scientific community of its findings in easily comprehensible words, and the arrogance of its employees. The latter is an especially ironic charge, since 90% or more of the employees have always been Japanese, locally hired. But it was argued they had been Americanized; chameleon-like, they had taken on the supposed attributes of the victors. Whether these charges rested on a misunderstanding of ABCC's mission on the part of the survivors or a lack of understanding by the Commission of the survivors' expectations is debatable, but the charges became more vitriolic in the spring of 1954 when the United States conducted its first test of a hydrogen bomb. The test, known as Bravo, scattered fallout over some of the participants, the inhabitants of several of the nearby atolls, and the men on a Japanese fishing vessel, the Lucky Dragon (*Fukuryû Maru*). This event focused still more press attention on ABCC and its activities.[12] Matters came to a head when physicians from the Commission sought to examine the fishermen upon their return to Yaezu, the vessel's home port. Eventually they were able to examine a number, albeit not all, but only after a brouhaha that seemed likely to involve the two governments themselves. The Japanese physicians charged with the care of the fishermen could and did see the presence of American physicians as a lack of confidence in their own skills and knowledge; whereas the Commission's physicians were responding to a request from their own government and were sincerely seeking to contribute to the management of the patients whose health problems were unclear. It was an awkward situation at best, and certainly did not need provocative newspaper articles to fan the flames.

These difficulties notwithstanding, the studies initiated in 1948 have been binational in their implementation since soon after their inception. At the outset, as was fitting, considering the immediate postwar economic plight of Japan, the United States was the greater financial contributor through contractual arrangements between the National Academy of Sciences and the AEC, and subsequently the Energy Research and Development Agency (ERDA). As time progressed and Japan's economic situation improved while America's worsened relatively, these initial administrative and financial arrangements grew less equitable. In the United States, concern mounted over the sharply rising cost of the studies. Inflation in Japan, especially in the 1960s and early 1970s, and the reevaluation of the yen in 1971 had dramatically increased the annual expenditure in dollars. Despite a systematic reduction in staff that saw the number of personnel fall from over 1,000 in 1950 to somewhat more than 600 in 1974, the annual operating budget (the budget exclusive of US and other foreign personnel costs) during this same time rose from $270,000 to about $1.6 million when the Unified Study Program (to be described shortly) began in 1956, and to over $6.7 million in fiscal 1974.[13] Means were sought to stem this flow of dollars and to make more equitable the financial and administrative participation of the two nations in the studies.

Creation of the Radiation Effects Research Foundation

On March 31 1975, the Atomic Bomb Casualty Commission was dissolved to be replaced by a private juridical foundation, a *zaidan hōjin*, organized under Japanese law, known as the Radiation Effects Research Foundation (RERF). The negotiations that preceded the creation of this new institution were protracted and the representatives of the Japanese and American governments wrestled with many issues. An Act of Endowment had to be drawn specifying the institution's governance and administrative structure, cost sharing, and the relationship of the Foundation to other Japanese supported research institutions. Even the choice of a name proved nettlesome. While the expression Radiation Effects Research Foundation would seem simple enough, when rendered into Japanese it becomes *Hoshasen eikyo kenkyusho* which is painfully similar to the name of the Japanese National Institute of Radiological Sciences, established in 1957, namely, *Hoshasen igaku sogo kenkyusho*. And since acronyms are commonly used to describe both institutions, the potential confusion was even worse. One is called *Hoeken* and the other *Hoiken*.

This new organization, in which the two nations would participate on an equal administrative and financial footing, offered continued employment to ABCC and JNIH staff and inherited not only ABCC's physical facilities—its buildings, equipment, library and the like—but all of the accumulated data, which were irreplaceable and whose worth was incalculable. Immediately prior to this transition, however, a major scientific review of the Commission's research program occurred under the chairmanship of James F. Crow, a distinguished geneticist and member of the National Academy of Sciences. This Committee made a number of recommendations, but the overriding and most important one was "that the basic elements of the ABCC program continue under the Foundation."

Administrative authority for this research institution rests in a 12-member board of directors, six members from Japan and six from the United States. Normally, six of these directors, the so-called "permanent" ones, live in Hiroshima or Nagasaki and constitute the Executive Committee responsible for the day-to-day activities of the Foundation.[14] Although the Act of Endowment provides for rotation of the chairmanship, the chairman is customarily Japanese and the vice-chairman American; the head of the Foundation's Secretariat, its office of general administration, is Japanese and the chief of research is American; the other two permanent directors have designated areas of research responsibility. A 10-member science council, with equal representation from the two countries, exists to provide scientific advice and research counsel, and meets at least annually to discharge its responsibilities. Oversight of the Foundation's financial affairs rests in the hands of two persons, one Japanese, nominated by the Ministry of Health and Welfare and the other an American, recommended by the National Academy of Sciences. The terms of office and provisions for their rotation, the responsibilities of these various administrative bodies, as well as the principles guiding the Foundation and its research are specified in the "Act of Endowment" that created the Foundation or in subsequent revisions of the Act.

The bulk of the Foundation's employees are Japanese, as was true of ABCC, but there continues to be a contingent of American professionals, primarily in computing, epidemiology, and statistics, where Japanese competence has lagged until recently. Funding from the United States is through contractual arrangements between the National Academy of Sciences and the Department of Energy. Japan's contribution continues to be through the Ministry of Health and Welfare. Money matters remain an administrative concern, nonetheless, since the total annual budget in 1990 exceeded $35 million. This does not represent an exorbitant growth in the budget of the Foundation in the years of its existence, the budget in yen having doubled in approximately 18 years (an increase of only about 4% per year, compounded annually), but the dollars required to meet the commitment of the United States have increased almost five-fold due to the precipitous fall of the dollar against the yen. Fiscal problems have been further complicated by differences in the budgetary processes in Japan and the United States, with two different fiscal years. While these are not important problems to the Japanese, since the budgetary process proceeds largely under the aegis of the Ministry of Health and Welfare and the funds needed are projected in yen, they can be especially vexing to the American side, particularly in years where the fluctuation in the currencies is large and to the detriment of the value of the dollar. Over the years of its existence, the Foundation has seen the yen appreciate from 270 to the dollar to less than 100 to the dollar as of this writing. This is tantamount to an almost three-fold increase in the annual contribution from the United States. While some budgetary economies could undoubtedly be achieved if the means for long-range fiscal planning existed, the financing of the Foundation's activities on an annual basis largely precludes this planning. The absence of such planning, however, increases the likelihood of administrative blundering, particularly in Washington, where the direction of the Department of Energy hinges on the appointment of individuals selected less for their competence than their political affiliation. Their limited sense of the worth of long-term studies such as this one can foster dubious short-term economies and a proclivity toward micromanagement. These tendencies could compromise the program and wound it mortally.

This administrative reorganization and the changes it spurred have not significantly altered the basic research strategy initiated under the Commission. Most of the studies continue to involve the fixed samples defined either as an outgrowth of the recommendations made in 1955 by an earlier special review group, known as the Francis Committee, or by the genetics studies launched in 1948. A larger laboratory component has been added to further investigations at the molecular and cellular levels that were not practical while ABCC existed. The Foundation, perhaps even more than its predecessor, has also been involved in collaborative studies with the faculties at the universities in Hiroshima and Nagasaki as well as the special radiobiological research institutes established in 1961 at these institutions. Moreover, as recommended by the Crow Committee, where fitting, the Foundation has financially supported research at other institutions in Japan when this research contributed directly to its mission and the specific research competence needed was not represented within the Foundation's own staff and facilities. No less importantly, RERF has continued, indeed broadened, the educational program initiated under ABCC's aegis. Its scientific personnel, through a program of summer internships and temporary foreign staff appointments, has figured prominently in the training of scientists from a variety of countries. And with the occurrence of the nuclear accident at Chernobyl in April 1986, the Foundation has served as a source of advice and counsel in the design and support of the studies of the exposed populations in the Soviet republics of Belarus, Russia, and Ukraine initiated in the Commonwealth of Independent States (the former USSR). The Foundation has also assisted in the studies in the Chelyabinsk oblast in Russia (the site of another nuclear accident).

Both ABCC and the Foundation have been actively involved in efforts to better the lot of the survivors. The Commission, for example, was instrumental in obtaining the funds from the United States Agency for International Development that built the first of the hospitals in Hiroshima and Nagasaki dedicated to the health care of the survivors, and utilized the money it received under the provisions of the Atomic Bomb Sufferers Medical Treatment Law to establish a fund to support other needs of the survivors. The Foundation has, in turn, worked closely with local authorities in extending this care and these facilities. To further these objectives, liaison committees comprised of local persons, exist in both cities and meet periodically to discuss social and scientific matters of concern to the survivors and to recommend policies to achieve a better mutual understanding of the aspirations of the survivors and the mission of the Foundation.

With the reorganization, labor–management relationships have become more tranquil and a greater spirit of cooperation has emerged. Even such potentially thorny problems as salaries and fringe benefits are generally amicably resolved through the use of the recommendations of the National Personnel Authority, which defines the salaries and benefits of Japanese governmental employees. Not all intemperance has disappeared, however, nor has the union's leadership ceased to further its own political agenda wherever possible.

Scientific and administrative changes of this nature do not come easily, rapidly, or without compromise, and it would be naive to suggest that they have been trouble-free. Maintenance of administrative equality in an organization as complex and visible as the Foundation is a demanding, continuing task. Administrative and scientific policy, for example, is to be established by the Executive Committee acting on behalf of the entire Board of Directors, but it is not always obvious when the Executive Committee *must* be consulted, since policy can be established inadvertently through the most perfunctory decision. This places a heavy burden on the administrative representatives of the two countries since, on the one hand, they must defend their stewardship before their own national agencies but, on the other hand, eschew a parochial attitude that would compromise the integrity of the science or the binational character of the organization. Unfortunately, the changed legal status of the institution has tempted some local interest groups, as well as national ones, to exert an unhealthy pressure on the Ministry of Health and Welfare and its representatives. Among these groups, there has been an inclination to see the Foundation as a Japanese enterprise reluctantly suffering American participation. Understandably, the lot of the survivors remains a strong and emotional issue locally, one that can be and is used to serve a number of political ends. As a result, pressure is intermittently exerted on the Ministry to alter the Foundation's mission to include a major treatment component. The treatment issue, although politically attractive, is now a specious one, however, since care is already available to the survivors through a variety of institutions, including a fine, well-equipped and -staffed Comprehensive Health Center managed by the Hiroshima Atomic Bomb Casualty Council. This Center offers not only programs of health examinations and treatment, but health promotion programs too. A similar facility exists in Nagasaki. Thus far, this change of mission has been resisted and hopefully will continue to be since the institution's hard-won scientific credibility would be quickly compromised if the research program became a puppet dancing to political machinations, American or Japanese. Ultimately, the Foundation can best serve the cause of the survivors through a strong research program that carefully delineates the risks of exposure to ionizing radiation and identifies the molecular and cellular processes that radiation impairs.

As just stated, achieving, indeed maintaining, a balance of views, has been and will continue to be one of the more challenging issues confronting the Foundation. Nevertheless, there has been a healthy spirit of accommodation of often disparate points of view as to the purpose of the institution, and the role and management of science in attaining this goal. Out of such understandings have emerged lessons pertinent to other large long-term epidemiological studies, national and international. Among these is the need for a strong scientific interest in the issue under study that will ensure an equally strong and lasting financial commitment of the agencies involved, the necessity for a core scientific staff prepared to dedicate a major portion of its working lifetime to the project, and lastly, a socially responsible research strategy that is sufficiently flexible to incorporate new technologies and points of view as they arise.[15]

Finally, it warrants noting that it is doubtful whether in the immediate postwar period Japanese scientists could have mounted a study of the scope undertaken by the Commission and now pursued by the Foundation. Given the myriad bureaucratic agencies in Japan that could have claimed an interest, there was no mechanism in the early years to ensure the centralized program that was patently needed. This is not solely a personal opinion. At a Joint Symposium on the Late Effects of Atomic Bomb Injuries held at the University of Tokyo Medical School (*Kōjin-Kai*) on the 5th and 6th of February 1954, Taku Komai, an emeritus professor at Kyoto Imperial University and the "father" of human genetics in Japan, commented, albeit in specific reference to the genetic studies, that he and his colleagues could not have achieved what had been done. He said the resources and trained personnel at their disposal would have been inadequate. His statement had broader implications, however, since many of the disciplines essential to the other studies of the Commission were not then well developed in Japan, and some, such as biostatistics, are still not yet well served. And Dr. Yoshimasa Matsuzaka in his otherwise critical "Notes on a Study of the Former ABCC" admits frankly that without ABCC and the resources of the United States, the medical effects of this tragedy would not have been scientifically documented for posterity.[16] Whatever their other faults, the Occupation and the authority of the Allied Supreme Commander when coupled with the prestige and position of the National Academy of Sciences in the United States and the Japanese National Institute of Health could and did provide the means to establish a research program with a common purpose, a common study design, and a centralized administration in the two cities.

3

The First Decade

Out of necessity, in the first years of the Commission's activities, scientific efforts were directed by the staff's clinical judgment and intuition, and retrospectively often appear opportunistic or ill-focused. They were guided by no clearly defined research strategy, few epidemiological clues, and no well-organized, extensive body of experimental data. However, a somewhat amorphous literature did suggest that ionizing radiation was potentially capable of producing cancer, new mutations, and developmental malformations. The applicability to the human of much of this research was questionable, since it had often been conducted on experimental animals given whole-body doses, usually of X-rays, much higher than the human will tolerate, and no less frequently on poorly calibrated machines so that the actual doses were less reliable than would have been desired. Moreover, radiation biology had not as yet become a well-recognized discipline, and potentially relevant studies were scattered through a vast biomedical scientific literature and not easily identified. It remained, therefore, for the Commission to determine what was applicable to its mission, and with limited library facilities, lacking the computer-assisted means of searching the scientific literature that now exist, this was difficult. But certain investigative directions, largely based on experimental animals, were clear, specifically the impact of atomic radiation on the occurrence of mutations and hematologic changes.

Hematological Investigations

Accordingly, among the first studies to begin was a hematological examination of a series of presumably "heavily" exposed survivors in Hiroshima that began in March 1947 and ended in April 1948.[1] The primary purpose of this study was to determine whether the blood-forming systems of a group of survivors, who had received a dose sufficiently high to have experienced a radiation-related depression in bone marrow activity immediately following the bombing, had recovered. Although this survey was seen as exploratory by the investigators, approximately 1,000 individuals were examined, all of whom had experienced epilation due to their exposure, and a like number of non-exposed individuals living in Kure who provided a basis for comparison to the Hiroshima findings.

These examinations suggested that the peripheral blood of the epilated individuals had almost completely recovered from whatever damage had been experienced. To the extent that immune competence is reflected in the peripheral blood, and in particular the number of circulating white blood cells, this finding suggested that their ability to ward off disease was now normal or nearly so. However, there was a small, but statistically significant, lowering of the average red blood cell count, the hemoglobin concentration, and the hematocrit (the volume percentage of red cells in whole blood) among the survivors. At face value, there was also some diminution of the white cell count, but this was not statistically significant. These lesser average values were not associated with gender or any particular age group, including the very young. Interestingly, the different hematological parameters were consistently more variable among the survivors than the comparison group, suggesting differences between individuals in the rapidity of recovery of the blood-forming apparatus following exposure, or variation in the doses received, since individual-specific estimates were not available. Retrospectively, the latter now seems the more likely explanation.

Somewhat more than a year later, an effort was made to reexamine as many of these individuals as could be located, and to enlarge the comparison group in Kure. The results of this second examination were similar to the first, although the total white cell counts that had been somewhat lower on the initial examination were no longer so.[2] As these investigators noted, the small disparities that did exist between the two studies could be interpreted either as evidence that a delayed, but complete hematological recovery had now occurred, or that the less significant earlier findings were due to unspecified nonradiation-related differences between the exposed and comparison groups, such as diet, that had lessened or ameliorated with time.

The Surveillance of the Outcome of Pregnancy

Almost simultaneously with the initiation of the hematological studies, a continuous surveillance of the children born in Hiroshima and Nagasaki to the survivors and to unexposed parents was begun. The aims of this surveillance were two-fold, first, to estimate, if possible, the number of newly arisen radiation-induced mutations, and second, to determine their long-term public health impact on the populations of these two cities. To these ends, in 1948, James Neel, in consultation with Ikuzo Matsubayashi, the chief of the Health Section in the Hiroshima City Office, had designed a survey strategy that hinged on a special provision in Japan's postwar rationing system. In the course of the war, Japanese authorities had found it necessary to promulgate a law making it possible for women, upon registering their pregnancies with the proper local governmental office, to obtain rationed food to sustain themselves and their unborn offspring through gestation, and clothing for the infant once the child was born.[3] Under the Child Welfare Law enacted on 12 December 1947, this registration could occur at any time after the mother was aware that she had conceived, but the supplemental rice ration (50 grams daily, or about one additional individual bowl of cooked rice) was only authorized after the 20th week of pregnancy. As a result, it was the practice to report pregnancies after the onset of the 5th lunar month and not before. Aware of this fact, Neel initiated a program wherein these mothers-to-be enrolled their pregnancies with the Commission's offices when they registered with the municipal authorities. To make this dual registration as simple and convenient as possible, ABCC's clerks were stationed at the city offices, where ration registration occurred. In addition to identifying the mother and father, these clerks obtained data on parental exposure, previous reproductive performance, and the expected date of confinement of the mother for the current pregnancy. The form on which this information was recorded was completed in duplicate; one copy was given to the pregnant woman or her representative and the other ABCC retained. The pregnant woman was encouraged to give her copy to the attendant at her delivery who then recorded the infant's life status, birth weight, and some particulars associated with the delivery itself.

Implementation of this program required personnel, American and Japanese, and Neel assembled the nucleus of an American staff—a vital statistician (Richard Brewer), a cytogeneticist (Masuo Kodani), and a pediatrician (Ray Anderson)—and set into motion the steps necessary to recruit local personnel in Hiroshima and Nagasaki—clerks to interview the prospective parents and to manage the records, physicians to perform the examinations—and to establish the transportation and facilities needed to support their efforts. Some space was already available in the Hiroshima Red Cross Hospital but more was needed. As stated earlier, this was obtained in a former military assembly hall, the Gaisenkan, in Ujina in the port area. The latter building, about 4,000 m from the hypocenter, had not been damaged by the bombing. Fortuitously, the search for medical personnel produced two bilingual, American-born Japanese-trained physicians—Koji Takeshima (in Hiroshima), then a member of the staff of the Red Cross Hospital, and Robert Kurata (in Nagasaki)—who proved invaluable as the study unfolded.

Most pregnancies at that time terminated at home in the presence of a midwife, and newly delivered mothers were reluctant to take their infants out of the house until they were a month or so old, when the child was taken to a shrine or temple to be introduced to the gods. As a result, the surveillance program depended heavily upon the involvement of the midwives and home visits.[4] To encourage midwife participation, each was remunerated for the births she reported to the Commission.[5] In retrospect, the stipends were small, albeit not disproportionate, considering the economic circumstances at the time, and the additional work imposed on the individual midwife was not unduly burdensome. Beyond weighing the infant with the simple scales the Commission provided, the midwife was not expected to recognize and carefully describe all of the possible congenital malformations that might arise. It was essential, therefore, that the genetics program develop a means whereby each newborn infant was seen by a physician as soon as possible after birth. To achieve this, the Commission turned to the many young unemployed doctors in Japan.

The education of physicians had been accelerated during the war to meet the country's military needs. To many of the 20 or so medical schools that then existed in Japan, another medical school had been added to make possible an increase in enrollment, and the curriculum was shortened from 7 years to 5. With defeat and the economic stringencies it brought, many of these young physicians were unemployed and were too impoverished to launch a private practice. But they did provide a reservoir for the recruitment of physicians to meet the Commission's needs.

Once a pregnancy terminated, if the infant was stillborn, grossly abnormal, died shortly after birth, or was handicapped in some other manner, the attending midwife informed us immediately by phone or by visiting ABCC's offices, whichever was more convenient. If the newborn was apparently normal, as was generally the case, notification was on a more leisurely schedule, usually occurring within a week. In either event, a physician, accompanied by a public health nurse, was sent to the home as quickly as practical to examine the child and to describe the findings. Their charge was to record all abnormalities, minor ones as well as those that were potentially handicapping or life-threatening. This led, of course, to the recording of a great many minor anomalies, such as a nuchal hemangioma (a reddish vascular birthmark at the nape of the neck) or auricular appendages (fleshy tabs on the ear lobes), as well as some abnormalities that would disappear as the child grew older, for instance, a "mongolian spot" (a bluish birthmark at the base of the spine) or a small umbilical hernia (a slight protrusion of the bowel through the umbilicus or "belly button").

This initial examination was the typical physical assessment of the health of a newly born infant; it included auscultation and neurological evaluation. Not all major malformations—those that are incompatible with life, are life-threatening, or that seriously compromise an infant's ability to grow and flourish—can be recognized shortly after birth through such an examination. Many heart defects or failures of motor or mental development are not recognizable until later. If these were to be diagnosed, another examination at an older age was needed. This second examination, when the infant could be brought to the clinical facilities of the Commission, was initiated in 1950, and provided not only the opportunity for new diagnoses, but also the means to confirm those that had been made at the time of the home visit.[6] It generally occurred between 8 and 10 months after birth; this age, chosen in consultation with the pediatricians who supported the genetics program, was selected for three reasons. First, it was about the earliest age when the bulk of potentially severe motor and mental defects would be recognizable; second, at this age, the infants were more readily manageable and less fearful of strange adults than they would be later in life, and finally, it would still be possible to institute corrective measures for some of the defects, such as the congenital dislocation of a hip, before irreparable damage was done through inattention. At this examination, which proved exceptionally popular among the mothers, American pediatricians were available to supervise and consult with their young Japanese associates on the findings, to reassure the mothers about the health of their children, or to suggest courses of therapy where necessary. But even before this program of systematic reexaminations was begun, a consultative clinic had been set up to aid in the confirmation of certain diagnoses such as congenital heart disease or dislocation of the hip and to afford a means through photographs, X-rays, and laboratory procedures to better document unusual clinical findings.

The home visits, although essential, were tiresome, and to the examining physicians undoubtedly never-ending. At the peak of the genetics program, 15 to 20 Japanese physicians were employed in these examinations in each of the two cities. Although most served on a full-time basis, each physician spent half of his time in the field, and the remainder in continued training and rotation through clinical services, either the Commission's or more commonly one of the local hospitals. Since 6,000 to 7,000 infants were born annually in Hiroshima and in Nagasaki, six examination teams were generally in the field 7 days a week. Although infants were not normally examined over the holidays associated with the New Year, often considered the most important annual holiday in Japan, arrangements were made for emergencies that might arise at that time.

Most of the time consumed in a home visit was spent in search of the child to be examined. Though the designation of streets and home addresses has since changed, particularly in the larger cities, in those years streets in Japan were generally not named, and consequently house addresses described an area rather than a fixed location on a street. Within this region, houses were not numbered consecutively along a road, but in the order in which they were built. An address was merely an invitation to search in a particular region of the city. Our strategy was either to locate a grocer within the area, who might know the family of the infant, or, better still, the local police box. At each such box was a map that identified every house and its occupants within the jurisdiction of the policemen who staffed the box. Their knowledge of the families living in their area and their map simplified enormously the search for a particular home.

The examination rarely required more than 15 or 20 minutes in the summer, when the infant could be quickly and safely disrobed. In the winter, in an unheated home, it took somewhat longer. The mother and the physician were loath to expose the baby to any more of the chill in the air than necessary, and, moreover, it was usually swaddled in layer after layer of clothing, which took time to remove. While the examinations proceeded, the mothers were encouraged to ask questions about their babies, their care, and whatever else concerned them. Generally, the physician responded to these questions as he or she recorded their observations on the child, and made recommendations for care when appropriate. Upon completion of the examination, as the physician and nurse left, they gave a bar of face soap, usually Ivory, to the mother to use on the baby. Gentle soaps were difficult to obtain in Japan in the early postwar years, and skin rashes from harsh ones were common. This little gift was in the tradition of reciprocity that governs gift-giving in Japan, and acknowledged ABCC's appreciation for the gift of participation of the mother and her infant.

When a pregnancy ended with a stillborn infant, or the child's death shortly after birth, or a physical abnormality, major or minor, a more detailed record form was completed, and blood was drawn from the mother to test for venereal disease, which was common and can cause certain congenital defects.[7] This form identified the nature of any abnormality, usually with drawings, and recorded more information on the household and the course of the pregnancy. To provide comparative observations, a similar form was completed on every infant, normal or abnormal, whose registration number ended in zero. Each of these longer forms was carefully reviewed to ensure that all of the questions were answered and the responses were clear. If the latter was not true, either the examining physician was consulted or another visit to the home was made, frequently by Dr. Takeshima in Hiroshima or Dr. Kurata in Nagasaki.

Typically, the visiting physician and nurse were not aware of the exposure status of the parents, and did not inquire into this, since the information had been obtained at the time the mother or her designated representative registered her pregnancy. This "blinding" was done to forestall possible biases in response; it is common for mothers of abnormal children to search their recollections for events they think may have caused the abnormality. This search tends to be self-fulfilling, and has been the basis of errors in estimating risk in the past. Epidemiologists try to avoid such possibilities, when practical, through prospective studies; that is, through careful inquiries that precede the events of interest so that the persons interviewed do not unwittingly attempt to "explain" the events. Thus, the data on radiation exposure were obtained before the pregnancy in question ended, and before the prospective parents knew whether their child was normal or abnormal. Infrequently, but occasionally, it was impractical to keep the physician in ignorance of the mother's or father's exposure, since now and then it was necessary to verify an observation recorded at the time of the registration or to resolve an inconsistency in a mother's or father's statement at the births of two different registered children.

To make more complete the recognition and description of life-threatening congenital defects, a program of autopsies was begun shortly after the beginning of the genetics surveillance. In the years 1948–1953, about 750 infants were autopsied. In Hiroshima, most of these examinations were conducted by a young military pathologist, William Wedemeyer, who had been assigned to the Commission, and in Nagasaki by Naomasa Okamoto, a member of one of the pathology departments at the University of Nagasaki Medical School, who was particularly interested in congenital anomalies. Although not generally an active participant in these autopsies, the physician who had been initially designated to examine the child was encouraged to attend the post mortem whenever possible so as to be fully aware of the nature of the findings. It was hoped too that this program, and their clinical experiences under the tutelage of American pediatricians, would sharpen their diagnostic skills and make their employment more intellectually rewarding.

The results of these autopsies were routinely reported, orally and in writing, to the midwife or physician who had referred the body, but the decision as to whether these results would be communicated to the parents was left to their discretion. This course of action was taken for two reasons. First, the autopsy occasionally revealed the presence of congenital syphilis and it seemed likely that this information could be more gently and sensitively presented to the family, and treatment urged, if it came from either their midwife or physician, rather than one of the Commission's doctors. Second, birth injuries sometimes occurred through mismanagement of the delivery and sometimes for reasons beyond the attendant's control. Whatever the basis, however, it seemed best to allow the midwife or physician the opportunity to inform the family, or to keep the information to themselves if they deemed this the wiser course of action. But the visiting physicians were available and prepared to explain the findings to the family if the midwife or their physician requested, as occasionally occurred.

Completeness and Accuracy of Information

Completeness of this system of registration and the accuracy of the clinical and other information obtained were constant concerns. To assess completeness, ABCC's data were periodically compared with the birth records of the city, and the midwives were encouraged to report all terminations, including those of unregistered mothers. Often these unregistered births involved infants born out of wedlock, or occurred to mothers who were registered in other communities or who simply "forgot" to register. Accuracy of the general information obtained at the time of pregnancy registration or the home visit was more difficult to evaluate, and could be done only by the comparison of successive registrations, where this occurred, or through comparison of the information obtained on the parents with that obtained through other studies of the Commission.

Accuracy of the clinical information and the diagnostic acumen of the young Japanese physicians could be assessed through comparison of their findings at the time of the home visit with those made at the follow-up clinic. One such comparison in 1951 in Nagasaki, for example, involved the findings of three physicians—the one who had conducted the home visit, a second Japanese physician who examined the child at the time of the follow-up in ABCC's clinics, and finally, the supervising American pediatrician. Among 100 children in whom an abnormality, major or minor, had been diagnosed at the time of the home visit, confirmation occurred in 91 instances. And among the 9 abnormalities not confirmed, 8 involved either a simple hemangioma, an umbilical hernia, a hydrocele (a collection of fluid about the testis), or an inguinal hernia (a protrusion of the bowel into the canal through which the testis descends into the scrotum). Since these defects could have been spontaneously resorbed between the time of the two examinations, it was questionable whether they actually represented incorrect diagnoses. Even if they were assumed to be errors, our young colleagues were apparently correctly reporting and diagnosing the vast majority of the malformations they saw.

Two other sources of evidence existed to evaluate the accuracy of the diagnoses of congenital defects. First, the findings on the children born to nonexposed parents living in Hiroshima and Nagasaki could be and were compared to other studies on the occurrence of malformations among the Japanese. The largest and most extensive such study had been reported by Drs. Mitani and Kuji of the Tokyo Red Cross Maternity Hospital. It involved 49,645 births in the years from 1922 through 1940. The overall frequency of malformed individuals in this series was 0.92%, whereas in the children of nonexposed parents in Hiroshima and Nagasaki it was 0.85%. The correspondence was surprisingly good when one notes that a hospital series in Japan at that time could not be considered a random sample, but probably included a disproportionate representation of complications of pregnancy. Moreover, certain of the malformations, such as hydrocele, congenital teeth, and partial albinism, which were included as major malformations in the Tokyo series, were seen as minor by the Commission and are not represented in the figure on the children of nonexposed parents. When these diagnoses are excluded from the Tokyo series, the correspondence is even better: 0.88% versus 0.85%. Second, for those children seen in a hospital either in Hiroshima or Nagasaki the findings of the visiting physician could be compared with those of the hospital physician examining the child. One such comparison in Nagasaki in 1951, involving the Commission's findings and those of the Nagasaki Medical School Hospital and the Mitsubishi Hospital, revealed 12 major "malformations" diagnosed at the two hospitals, of which the visiting physician had seen five. The remaining seven involved congenital dislocation of the hip, congenital heart disease, idiocy with funnel chest, and hydrocephalus. Since hydrocephalus, congenital heart disease, and congenital dislocation of the hip are often not recognizable until some months after birth, these were not considered "missed" diagnoses, and, furthermore, they would undoubtedly have been identified at the second clinical examination. The only diagnosis that should have been made at the birth examination but was not was the case of funnel chest (pectus excavatum), which is a hollow at the lower part of the chest caused by a displacement backwards of the breast bone.

Inevitably, as the genetics study progressed, other problems arose. One particularly disturbing matter was the changing attitude and policies of the national government toward limiting the growth of the Japanese population. The repatriation of so many Japanese, civilian and military, from Korea, Manchuria, and elsewhere, and the booming postwar birth rate were taxing the nation's already strained economic resources. Families were encouraged to have fewer children, and as a further inducement, in 1948 the government liberalized the legal basis for the artificial termination of a pregnancy. This act allowed a pregnancy to be interrupted if its continuance was likely to lead to a severely handicapped member of society (the so-called eugenics clause), posed a serious threat to the mother's health, or created an intolerable economic burden on the family. Although use of the provisions of the law was voluntary, to avoid capricious implementation the circumstances under which each of these alternatives could be invoked were defined. To abort a pregnancy for economic reasons, the family had to be currently receiving governmental assistance. To utilize the eugenics clause, the likelihood of an untoward pregnancy had to be evaluated by a committee of physicians. However, the threat a pregnancy constituted to the mother's health was a matter of the clinical judgment of her personal physician. Since any pregnancy entails some risk, most of the abortions performed were authorized under this last provision, the least intrusive and easiest to satisfy.

Pregnancies were usually terminated between the 4th and 8th week after fertilization by dilating the neck of the uterus and scraping the developing embryo from the uterine wall. It was a process that could be completed in a doctor's office without necessarily entailing hospitalization. Under the law, these interruptions were to be reported to the local health offices, but no systematic enforcement of this requirement was discernible. It was very difficult, therefore, to obtain a reliable estimate of how many pregnancies were being interrupted. The women involved were generally, and understandably, reluctant to state whether they had terminated a pregnancy, and the physicians did not want to discuss the matter since many did not report the fee they received for performing the abortion to the local tax office. Finally, after strenuous efforts to assure a sample of physicians known to be performing abortions that the information would be held in strict confidence, a few consented to provide the number of pregnancies they had interrupted over a period of several months in 1950. Their numbers, when extrapolated to the city, suggested that almost as many pregnancies were being interrupted in Hiroshima each month as came to term, and presumably the same held true in Nagasaki.

This had awesome implications for the study—a fall in the birth rate could be precipitous enough to compromise collecting sufficient data to evaluate the radiation hazard or, worse still, could introduce biases that would be difficult to manage, if pregnancies were more likely to be interrupted when one or both parents had been exposed. There was also, of course, the possibility that an induced abortion could alter ovarian function and make it difficult for the woman to conceive later. However, the studies of Masamichi Suzuki and Teruo Watanabe at the Commission suggested the latter was unlikely, since they found that ovarian function was not permanently impaired, and normality was achieved within two or three menstrual cycles.[8]

Nonetheless, it was obvious that this development had to be monitored carefully and continuously to appraise its possible effect on the study and the conclusions to be drawn from the data. Fortunately, as was soon discovered, economic considerations, and not the fact of exposure, were the major determiners of a family's decision to abort a pregnancy. Nevertheless, a precipitous drop in the birth rate did occur. Largely through the interruption of pregnancies, Japan's crude birth rate (the number of births per 1,000 persons in the population) fell from 30 or so in 1950 to about 16 in a period of roughly 5 years and has continued to decline, although more slowly. Or stated somewhat differently, whereas the average woman of child-bearing age in Japan at the beginning of the genetics studies in 1947 had 4.54 births, by 1973 this number had fallen to 2.14, and in 1989 it was only 1.57.

This clinical surveillance ended in the early spring of 1954, but only after the interim findings had been presented to an ad hoc committee of senior American geneticists and statisticians—including Nobel laureate George W. Beadle, Donald R. Charles, Cecil C. Craig, Laurence H. Snyder, and Curt Stern—that met in Ann Arbor, Michigan on 10-11 July 1953. This committee concurred with the opinion of Dr. Neel and myself that the surveillance had reached a logical end, since the survivors who were adults at the time of their exposure were completing their fertility. It was recommended, however, that further data be collected on the survival of newly born infants and on the sex ratio—the proportion of pregnancies terminating in a male infant. As Neel has subsequently written, "These two observations were continued because, on the one hand, a sex ratio effect of borderline significance had been observed and, on the other hand, it could not be precluded that an effect of parental exposure on a child's survival might become apparent during childhood."[9]

To implement these recommendations, however, the genetics program had to be reoriented to some degree. Food was becoming more readily available with the passage of time, and the need to ration staples less pressing. Sooner or later rationing would cease, and it would then be necessary to identify births through some other means. The offices of vital statistics in the two cities provided an alternative; with their cooperation, the Commission was permitted to abstract the information needed from the official birth certificates. Accordingly, the registration offices that had been maintained throughout the period of the clinical examinations were closed, the clerks were transferred to other units within ABCC, and henceforth all additions to the roster of children of the survivors were to be ascertained through the national system of birth

The First Decade

registration, which in these cities had now fully recovered from the trauma of the bombings.

Abortions and Miscarriages

Ideally it would have been desirable to have pregnancy registration occur at the earliest practical moment, but as earlier stated, the provisions of the rationing act did not encourage a woman to register her pregnancy before the 20th week of gestation. It was a concern, therefore, whether exposure to atomic radiation could have increased the frequency of miscarriage or spontaneous abortion among conceptions occurring after the bombing. If it had, the surveillance system would not identify this fact, since the pregnancies would have been lost before registration could take place. To remedy this, an effort began in September 1949, under the supervision of Dr. Saburo Kitamura, a University of Pennsylvania trained obstetrician, to identify all of those women experiencing a premature spontaneous termination of their pregnancy. Identification of these women hinged on the cooperation of the medical community, since the local physicians were the persons most likely to be consulted following a spontaneous abortion or miscarriage. At that time some 98–99% of the spontaneous abortions in Hiroshima were handled by physicians. To obtain their support, Dr. Kitamura and a young medical colleague, Hisao Sawada, began a series of systematic visits to the various local physicians and hospitals involved in the care of pregnant women to ask for their cooperation and to solicit their views as to how the abortions might be reported to the Commission. After consultations with these physicians, it was agreed that the abortions could be reported in any one of three ways—through a personal interview with the attending physician, if this was preferred, through a telephone call to either Drs. Kitamura or Sawada, or through a letter addressed to the Commission. Most practitioners chose the first of these methods since it required less of them, assured more privacy to the patient and her physician, and was more in keeping with the notion of a physician-to-physician exchange of privileged information.

Two complications arose. First, given the increased frequency of induced terminations previously mentioned, it was not always clear whether a woman would be candid in acknowledging that the premature termination of a pregnancy was not spontaneous but artificially induced. Some were embarrassed by the circumstances that made an artificial interruption necessary, and others might have experienced a sense of guilt. Conceivably, the artificial terminations could be identified through the members of the Eugenics Committee in Hiroshima, since legally they were the only individuals permitted to interrupt pregnancies. This committee then consisted of all the city's privately practicing obstetricians–gynecologists as well as those connected with hospitals, both private and public—some 35 individuals. But it seemed probable, given the economic circumstances that prevailed, that other physicians were also involved, and to visit every physician in the city was impractical with the resources available. Second, tuberculosis was widespread in Japan, including latent tuberculosis of the urogenital tract in women.[10] The latter disease can contribute to infertility, and possibly to the premature termination of a pregnancy, since the fallopian tube(s) through which the fertilized egg normally descends into the womb is often blocked or severely narrowed. However, at that time the diagnosis of urogenital tuberculosis was not always easily made. If the body or neck of the womb was involved, the presence of tuberculosis could be established through clinical examination and biopsies. But if this was not so, diagnosis had to rest to a large extent upon demonstrating the presence of the tubercle bacillus in the cervicovaginal and menstrual discharges. This usually entailed the inoculation of an experimental rodent, generally a guinea pig, with the bacteria isolated from cervical fluids of the woman involved and the later demonstration of the pathogenicity of the bacterium in the guinea pig. This was a lengthy procedure, and moreover, "false negatives," that is, the failure to recover the bacillus when it was actually present, were common. Neither of these difficulties was ever satisfactorily resolved.

Matters worsened in late 1950 when Dr. Kitamura returned to Tokyo and the private practice of medicine. The burden of supervising the study fell on his young colleague, Dr. Sawada, who lacked the age and recognition in the local medical community needed to elicit the level of cooperation that Dr. Kitamura had obtained. As a result, some of the physicians who had previously been participating in the study withdrew and eventually this program was terminated. The data that were collected are inconclusive, even contradictory, differing between the period of Kitamura's supervision and subsequently. Careful scrutiny of the methods of obtaining the medical radiation histories from the women and their husbands revealed serious discrepancies between the 907 histories Kitamura had obtained and the 532 collected after his departure. Reluctantly, it was concluded that the two groups were not comparable and could not, therefore, be pooled, and singly neither provided enough information for rigorous analysis.[11]

Although exposure of pregnant women to atomic irradiation increased the frequency of loss of the exposed embryo or fetus, particularly when exposure occurred early in pregnancy, it is not clear whether conceptions occurring after exposure were more likely to be spontaneously aborted. Later studies of the fertility of exposed women suggest

The First Decade 51

no statistically significant impairment in their ability to conceive, or if an impairment did occur, it was too small to be recognized by the techniques available.

The Search for Chromosomal Abnormalities

At the outset of the genetics program, it was known that ionizing radiation could produce chromosomal abnormalities. Indeed, in the early 1920s, an American geneticist, James Mavor, had demonstrated that X-irradiation increased the frequency of chromosomal nondisjunction— the failure of homologous chromosomes to separate properly—during cell division in the common fruit fly, *Drosophila melanogaster*.[12] In addition, the occurrence of chromosomal fragments and "bridges" among testicular cells from prisoners who had "voluntarily" undergone X-irradiation suggested that deletion of parts of chromosomes might be common following exposure. Efforts were made, therefore, to determine whether these abnormalities were also to be seen among the atomic bomb survivors. These early studies came to naught, however, since the techniques then available were inadequate to the needs.

The earliest observations, those of Masuo Kodani, rested on cells obtained through voluntary needle biopsies of the testes, a painful procedure, which yielded tissue that is hard to study. Although the chromosomes can be uniquely stained, using aceto–carmine or the Feulgen reaction, and were discernible, the cells cannot be readily flattened and, as a result, the chromosomes do not all lie in a single visual plane. The microscopist must follow a chromosome through different, narrowly defined visual fields. It is difficult under these circumstances to be certain whether the investigator is looking at one, two, or more chromosomes, and as a consequence, there even existed a difference of opinion as to the number of chromosomes human beings had. The consensus number was 48, but some investigators maintained the number was 47, others 46, and some still a lesser number.

Progress would not come until the normal number of chromosomes was known unequivocally and better techniques were available to study changes in the fine structure of human chromosomes. Finally, two scientists, Joe Hin Tjio and Albert Levan, presented persuasive evidence that the number was 46 rather than the 48 commonly believed.[13] Their conclusions, first presented at an international congress of human geneticists in Copenhagen in 1956 and reiterated at an international symposium in Japan later in the same year, were based on simpler, more reproducible techniques than those Kodani employed.[14]

In a surprisingly short time after these better methods of study were available, it was possible to establish not only that a variety of chromosomal abnormalities occur spontaneously, often to the detriment of the individual, but that ionizing radiation increases the frequency of specific kinds of aberrations of chromosome number and structure. Still further progress came later with the introduction of new staining methods that revealed more of the detailed structure of the chromosome. For example, the technique termed G-banding (it derives its name from the Giemsa stain that is used) can readily define some 400 or so different chromosomal regions, and disclose changes in these regions. Even finer chromosomal definition is possible, but there is a limit to the detail that a cytogeneticist can carefully evaluate or that can be assessed by current automated methods.

Growth and Development of Exposed Children

Another early concern of the Commission was the possible impact of atomic radiation on the physical growth and development of the children who were exposed. Assessing this was contingent on the existence of standards of normality—growth and development without exposure. Some normative data on Japanese children were available from a nutrition survey that SCAP had conducted in 1946–1947 and earlier work of Japanese pediatricians interested in growth, but still other comparison groups were essential to the assessment of any growth retardation that might have occurred following radiation exposure in these cities. In the summer of 1947, under the auspices of the Committee on Atomic Casualties and with the aid of the Public Health and Welfare Section of SCAP and the Japanese National Institute of Health, William W. Greulich, a professor at Stanford University Medical School and an internationally recognized authority on childhood growth and development, began an anthropometric evaluation of physical development among exposed children. The subjects were chosen with the advice of Japanese physicians who had been assigned to assist him by the National Institute of Health. The children were all in the first 6 years of school at the time of their examination; in Hiroshima 145 were seen, in Kure 253, in Nagasaki 295, and in Sasebo an undisclosed number. The participants in Kure and Sasebo were to provide comparison groups for those studied in Hiroshima and Nagasaki, respectively.

In the autumn of 1948, Greulich returned to Japan and added another 597 children (339 of whom had been exposed) to the Nagasaki sample. His observations, when coupled with those the Commission made in 1949 and 1950, revealed a retardation in height, weight, and skeletal development among the exposed children. Generally, boys tended to be more growth retarded than girls in the same age group. Greulich, in his interpretation of the study findings, cautiously noted that these adverse effects were probably attributable not only to exposure to ionizing radiation but to physical injuries as well as psychological trauma and widespread postwar undernutrition.[15] The origin of the apparent difference, however, obviously needed resolution, and the study was continued by ABCC under the supervision of Wataru Sutow, a pediatrician and former student of Greulich's at Stanford.

To provide guidance to this continued activity, in June 1951, Earle Reynolds, a physical anthropologist, was brought to Japan as a consultant to the program. He had been involved in the highly regarded longitudinal studies of childhood growth and development at the Fels Research Institute at Antioch College in Yellow Springs, Ohio.[16] His responsibility was to assess the program, the measurement techniques it was employing, and to recommend changes, if these were needed. At his urging, photographs, which had been taken but were then abandoned for the older girls, were to be reinstituted, and better standardized. Their purpose was to evaluate sexual maturation using a classificatory system developed at Fels, which Reynolds had modified to be applicable to Japanese children, and to afford a basis for assessment of body build, using the somatotypes William H. Sheldon had pioneered. To make these photographs acceptable to the participants, since they were to occur in the nude, the girls were to be photographed by a woman. Aware of the probable objections of the teen-aged girls, Reynolds strongly stated "The entire photographic procedure must be impersonal and professional, without extraneous remarks or behavior on the part of the operators. A definite effort should be made to tie in the photographic procedures with the medical examination, so that the patient will associate the two operations."[17] Despite these admonitions, participation of the older girls in the study fell dramatically. Eventually, the whole procedure came to haunt the Commission. At best it was seen as unnecessary by the children's parents, and at worst a form of voyeurism.

Nonetheless, in each of the years 1951, 1952, and 1953, somewhat more than 4,000 children, ranging in age from five through nineteen, were examined in Hiroshima. On every child, twelve different measurements of the body were made, and seven further ones were obtained from a standardized X-ray of the calf of the leg. The latter measurements included the absolute and relative thicknesses of fat, muscle, and bone in this area. To insure accuracy, every measurement was made in duplicate and carefully checked before the data were actually entered into the retrieval system. Analysis of the information accumulated in these three years led Reynolds to conclude

> [T]here does appear to be an association between exposure to atomic radiation and growth status in Hiroshima children. If

body size and maturation can be accepted as indicators of physical well-being, this association is in the direction of better physical status in the control children, and in children who were exposed, but presumably to lesser degree than other children.[18]

He noted, however, that there were many complicating factors, especially nutrition, that prevented an entirely satisfactory analysis, and urged that the children continue to be studied.

Once adequate clinical facilities were available in Nagasaki, Sutow and Emory West began a roentgenographic survey to measure skeletal growth in the children exposed prenatally. Chronological age is often a poor indicator of biological or developmental age. The latter can be better evaluated through changes in the skeleton, the eruption of teeth, and the appearance and maturation of sexual characteristics such as the emergence of axillary and pubic hair, the development of the breasts or penis, or the onset of menstruation. Skeletal growth and development is most commonly estimated from X-rays of the wrist and hand, since there exists an orderly progression in the appearance of centers of bone formation and the ultimate conversion of the cartilaginous forerunners of the wrist bones into bone. There are objective roentgenographic criteria for this progression, known as the Greulich–Pyle standards, and it was these, suitably modified for Japanese children, that Sutow and West used to evaluate possible radiation-related growth and developmental retardation. Their survey failed to disclose differences in the incidence of skeletal abnormalities among children exposed before birth, but did contribute to establishing standards for evaluating skeletal development among Japanese children that are still used.[19] Today, roentgenographic studies of this nature would be viewed with some misgivings, since there is a desire to reduce exposures to ionizing radiation whenever possible, but in the context of the time and the state of knowledge, the risks were thought to be minimal.

Continued anthropometric evaluation of the growth and development of the children exposed to the atomic bombs has shown growth to be impaired, seemingly in proportion to the extent of the irradiation received. This impairment is more than just a delay, as measurements of these individuals as young adults (when 18 years old) have established, and it is generally more pronounced the younger the individual was at the time of exposure to the atomic bomb.[20]

Radiation-Related Changes in the Lens

During these early years, three health effects of exposure to the atomic bombing were identified—the first cases of radiation-related opacities of the lens of the eye were seen; leukemia, especially the acute types, was noted to be markedly elevated among the survivors; and evidence emerged of the damage to the developing brain of those survivors exposed prenatally. Each of these discoveries arose in a somewhat different manner, but each almost fortuitously.

Even in 1945, it was known, largely as an outgrowth of the use of X-irradiation in treating cancer of the brain, that if the eye is within the radiation beam and receives a high enough dose, perhaps 200 roentgens or so, a characteristic change will occur in the lens that leads to a loss in its translucency. The cells involved in this process originate in the layers known as the germinative epithelium, found on the front surface of the lens. It is believed that the radiation-damaged cells are unable to develop into lens fibers as would normally occur, and that their remnants are gradually pushed to the back of the lens under the pressure of the remaining undamaged cells that continue to divide and develop throughout life, albeit more slowly with age. The opacification or change in translucency arises from the breaking down of the normally transparent proteins in the damaged cells.

These changes are first seen not in the front of the lens, as might be expected, since it was closer to the radiation, but in the back, just beneath the covering (the capsule) of the lens. As a result, they are said to be posterior (back), subcapsular (beneath the capsule) opacities. Some investigators have referred to these changes as cataracts but the word cataract connotes to most individuals a defect that impairs vision, although it is also commonly used to describe any detectable change in translucency in the lens. To avoid confusion, we shall refer to them as lenticular opacities, or opacities of the lens, since the cases to be described are not restricted to those that impair vision measurably.

The events just detailed, when combined with two further observations, made it probable that radiation-related opacities would be found in Hiroshima and Nagasaki. First, lesions at the back of the eye, in the area known as the fundus, had already been reported in survivors with acute radiation sickness, although these injuries did not involve the lens and were thought to be due to the blood disorders produced by whole-body radiation. A variety of acute effects had also been seen, ranging from thermal burns of the eyelids, dislocation of the lens, hemorrhages in the retina and vitreous, and separation of the retina. These findings suggested that the eye might be especially vulnerable to damage from ionizing radiation. Second, radiation-related opacities had also been seen among the physicists employed on the Manhattan Project.[21] Some 10 persons, all individuals who worked extensively with cyclotrons—machines designed to alter atoms—developed radiation opacities. It was customary among the physicists, when the cyclotron was being brought to power, to peer into the machine through windows in the apparatus to determine whether the beam of energy was on its target. This could be determined visually, since wherever the beam impinged, a peculiar but characteristic glow arose. It is said, possibly apocryphally, that in one instance the physicist found himself staring directly at the beam itself. Whether, in the other cases, the cataracts arose through exposure of their eyes to atomic radiation—primarily fast, highly energetic neutrons—while orienting the beam or merely through carelessness in the course of their work, is debatable, although the latter seems more likely.

To determine whether similar changes had occurred among the survivors in Hiroshima and Nagasaki in the late summer of 1949, a team of three ophthalmologists was sent to Japan—David Cogan and Samuel Martin from Boston, and Samuel Kimura from the University of California Medical School in San Francisco. Unknown to them, radiation opacities had already been seen in Hiroshima and Nagasaki, and rightly diagnosed by three Japanese ophthalmologists, Kinnosuke Hirose, Tadashi Fujino, and Hiroshi Ikui. Hirose and Fujino had seen their case, a 14-year-old boy exposed some 750 m to the northeast of the hypocenter in Nagasaki, in June 1949, and Ikui had seen two cases, both of whom had been exposed within 1 km of the hypocenter in Hiroshima, in September of the same year.[22]

One of the first affected individuals Cogan, Kimura, and Martin saw came to their attention serendipitously. At lunch in the Commission's dining room, one of the waitresses complained to Dr. Kimura of blurred vision and mentioned that she had been exposed. Ophthalmoscopic examination of her lenses revealed the tell-tale signs described above. Her situation proved especially tragic. Her vision grew steadily worse as time progressed, and eventually she died as an indirect consequence, possibly, of the progressive deterioration of her sight. In 1977, she was struck and fatally injured by an automobile while crossing one of Hiroshima's streets.

Subsequently, as the study proceeded, other cases were found. Among the 1,000 survivors Cogan and his colleagues examined—231 of whom were exposed within 1,000 m—10 cases of presumed radiation opacities were observed. All were exposed within 550–950 m of the hypocenter and all had epilated, suggesting that their doses had been large, possibly 1 Gy or more.[23] They estimated the frequency of radiation cataracts among survivors exposed within 1 km to be about 2.5%, but thought this figure might increase with time. A second, still larger survey involving 3,700 survivors conducted in 1951–1953 revealed 154 individuals with posterior subcapsular plaques large enough to be visible with an ophthalmoscope.[24]

Inevitably, as these surveys continued, new questions surfaced. Would the opacities be progressive? And if so, would the progression be slow or fast? Would they necessarily impair sight? To answer these questions, and in particular the issue of progression, means had to be found to quantify the degree of involvement of the lens seen at each successive ophthalmic examination of an individual. Drawings were made of the plaques by the examining ophthalmologist and by a medical illustrator, Geoffrey Day. While these could and did exhibit the nature and location of the opacity, they were too subjective to be useful quantitatively. Photographs would obviously be more objective, but conventional ones would not have had the needed depth of perspective. An effort was made, however, to photograph the changes stereoscopically, a technique still under development at that time. David Donaldson who was involved in the design and construction of the equipment necessary for ophthalmic stereophotographs and a colleague of Cogan's at Harvard, was sent to Japan to see what could be done. He seemed optimistic as he set up his twin cameras, adjusted their angles, the stroboscopic lighting, and ancillary paraphernalia to begin the photographing. While the stereoscopic photographs were impressive, and clearly revealed blast-embedded glass and other debris in the lens, minor changes in the posterior area remained elusive, and no further effort was made to photograph the opacities routinely.

Today it is believed that the opacities are generally not progressive and do not always impair vision. Indeed, most affected individuals are unaware of their presence. These are clinical judgments, however, and there now exist more objective methods of recording changes, such as stereolaminography or Scheimpflug imaging, which would better disclose any progression in these opacities, particularly if that progression is slow and might be inapparent to the examining ophthalmologist. The former technique is similar in its aims to other methods of imaging living tissue, such as magnetic resonance imaging or computed tomography. It produces a series of computer-reconstructed small slices, as it were, through the lens, which can be studied visually. Neither it nor Scheimpflug imaging has been used as yet to document the lenticular changes in the survivors with opacities, but both appear to be promising technologies.

Studies of the exposure of the atomic bomb survivors, as well as those of persons whose opacities stem from the therapeutic use of irradiation, suggest there is an exposure threshold—a dose below which these changes do not occur.[25] This value is not known precisely, although clinical studies following X-irradiation suggest that it may be in the neighborhood of 2 Gy. The threshold value may, of course, differ for various qualities of irradiation. Based on the atomic bomb survivors, the threshold for gamma-irradiation, measured in terms of the dose absorbed by the eye, appears to be about 0.73 Gy, a unit of dose to be described later; whereas the threshold for neutron-irradiation is much lower, under 0.10 Gy. Superficially, these values, particularly the one for gamma-irradiation, would seem lower than estimates from the clinical studies. But if the neutron dose is weighted to account for its greater biological effectiveness, the estimated minimal dose of radiation is about 1.5 Sv. It must be noted, however, that a threshold could be spurious, arising solely because of our inability to detect very small changes, and may not reflect an actual threshold in their occurrence.

Surveys conducted in the 1960s, and even more recently, have not revealed any new cases of radiation-related opacities, and thus a survivor now free of them will presumably continue to be so.[26] But this is a supposition, since experimental studies on rabbits, dogs, and monkeys exposed to highly energetic charged particles suggest the possibility that seemingly stationary opacities may suddenly begin to progress and that some opacities may not become apparent until long after exposure.[27,28] If the interval of time between exposure and this progression seen in experimental animals is extrapolated to the human situation, only now would one begin to see similar changes in the survivors. This fact argues for the need for continuing studies of lenticular changes in Hiroshima and Nagasaki.

The larger survey in 1951–53 to which reference has been made revealed a second type of ocular damage—polychromatic changes in the lens.[29] They are of two types. The least conspicuous is a faint multicolored sheen located at the posterior pole of the lens and is associated with an increase in markings of the lens covering. The more readily visible change involves a distinct grainy sheen usually seen in the central zone of the lens near its rear, where a diffuse reflection of light normally occurs as a result of small irregularities in the surface of the lens. These irregularities make the surface itself visible from any angle with an ophthalmoscope; the zone in which they occur is known as the zone of specular reflection. If these irregularities were not present, the posterior surface of the lens would be more difficult to visualize and the changes described elusive, if recognizable at all. The plaques themselves are best seen with a device known as a slit lamp (an ophthalmic biomicroscope) with the pupil fully dilated, which is an inconvenience to the person examined, particularly among individuals with darker pupils, such as the Japanese.

Neither of these two changes impairs vision nor are they necessarily indicative of exposure since they are also seen in unexposed individuals; however, both increase in frequency as dose increases, and within exposure groups as age increases. Later ophthalmic surveys of the survivors, using the same individuals insofar as practical, have confirmed these findings.[26,30]

The Appearance of Leukemia among the Atomic Bomb Survivors

The first intimation that leukemia, a frequently fatal disorder involving the blood-forming organs of the body, was elevated among the survivors arose through the perceptiveness of a young Japanese physician, Takuso Yamawaki. As early as 1949, he believed that he was seeing more cases of leukemia in his clinical practice than he would have expected, but since he was not a hematologist, nor an epidemiologist, he was uncertain of his diagnoses and how his perception of an increase could be validated. He sought the advice of Wayne Borges, a Harvard-trained pediatric hematologist, who had recently joined the Commission. Borges reviewed the blood slides on which Yamawaki based his clinical assessments, and agreed with all of the diagnoses. Neither of these investigators were apparently aware, however, that as early as August 1947 John Lawrence, the professor of hematology at the Medical School of the University of Rochester, in a consultant's report to the Committee on Atomic Casualties, had recommended "the incidence of leukemia or other blood dyscrasias (in Hiroshima and Nagasaki) should be determined and compared with the normal" or that leukemia had been identified in the planning document as one of the diseases that should be studied.

To understand fully the importance of Yamawaki's astute clinical observation and its implications, some familiarity is needed with the origins and nature of leukemia itself. Briefly, our blood is divisible into a fluid and a cellular part, consisting of red and white cells. The red cells are biconcave disks, called erythrocytes, which are red because they contain the oxygen-transporting protein, hemoglobin, whereas the white cells, or leukocytes, are irregularly shaped and largely colorless. The largest percentage of white cells in the peripheral blood have segmented nuclei and specifically staining granules; these cells are known as polymorphonuclear leukocytes. A small percentage of cells are known as monocytes. These too are morphologically distinguishable since they have indented nuclei with scarce, fine granules. About 20–30% of peripheral blood cells are lymphocytes that contain round or oval nuclei and no cytoplasmic granules.

An individual with too few red cells is said to be anemic; with too many, polycythemic. Insofar as white cells are concerned, if there are too few, the individual is leukopenic and prone to infections, often with serious consequences. Most survivors who received moderate-to-large doses of radiation to their bone marrow experienced a period of leukopenia of varied duration, and some succumbed to infections during this time when their natural blood defense mechanisms were compromised. A person with too many white cells is likely to have an infection or leukemia. Any pus-forming bacterial infection can raise the white cell count; more rarely, other disorders such as infectious mononucleosis, a viral disease, will do so also. Extremely high white counts can occur too in patients with cancer, Hodgkin's disease, and other disorders.

The term leukemia, however, does not describe a single disease, but a group of diseases with certain common findings, notably a progressive anemia, internal bleeding, exhaustion, and a marked increase in the number of white cells (and generally their immature forms) in the circulating blood. Most all of the body's white cells arise in the bone marrow through a process known as hematopoiesis. This process begins with primitive precursors, called hematopoietic stem cells, which can either self-renew or differentiate into one or the other of two lines of specialized cells, known as the lymphoid and myeloid lineages. The former lineage leads to the development of B- and T-lymphocytes, about which more will be said later, whereas the myeloid lineage is the source of the polymorphonuclear cells and monocytes described above. These cells will, in turn, produce white cells that possess the property of being able to ingest bacteria, foreign proteins, and even other cells. A variety of growth factors and other substances are involved in this transition from primitive to more specialized cells. Some of these agents enhance cellular proliferation and maturation and others appear to exert an inhibitory effect on the process. However, if one or more steps in the transition are impaired, intermediate-stage cells not normally found in the circulating blood will usually appear.

Traditionally, the leukemias have been distinguished clinically on the basis of their duration and character—whether acute or chronic—and the type and tissue origin of the white cells involved. As mentioned, white cells can be myelocytic in origin, that is, arise and mature in the bone marrow, or lymphocytic, in which case their final maturation (and differentiation) is in the lymphatic tissue, such as the spleen, thymus, tonsils, and lymph nodes. If the leukemia involves the former cells, most of which contain microscopically visible cytoplasmic granules, it is said to be myelocytic or myelogenous; whereas if it is the latter, with few or no visible cytoplasmic granules, it is lymphocytic or lymphatic. Leukemia-involving monocytes are monocytic in type, and since monocytes are in the myeloid cell line, they may merge with myeloid cells to form myelomonocytic leukemias. The term agranulocytic leukemia is usually reserved to indicate only the presence of a few granulocytes in the bone marrow. Acute leukemias that are not myeloid in type usually are referred to as acute nonlymphocytic leukemia (often abbreviated ANLL).

Today, leukemias are classified somewhat differently than was true in the early years of the studies in Japan. Better understanding of the cellular origin, differentiation, and maturity of leukemic cells has given rise to a system of classification known as the French–American–British (FAB) method, which recognizes many more subtypes of this disease and provides a better distinction between acute myelogenous and acute lymphatic leukemia than was previously practical. Moreover, it has been recognized that specific chromosomal abnormalities are associated with some specific FAB subtypes and that one type of lymphatic leukemia, adult T-cell leukemia, is related to infection with a human virus (known as HTLV-I) that is preferentially incorporated into those lymphocytes, designated as T-cells, that reach maturity in the thymus or other lymphoidal tissue.

To progress from Yamawaki's clinical intuition to a statistical statement about the frequency of leukemia among the survivors was more difficult than confirming his diagnoses. Yamawaki had a series of cases, but did not know how many individuals had been exposed and were at risk. To compute the number of cases one might expect to see in the absence of a radiation-related effect to compare with the number actually seen requires an estimate of the number of individuals exposed, and although a city-wide census had begun, the results were not yet available. He did know the distances at which the individuals with leukemia had been exposed, and these distances could be used to determine whether his perception of an increase was real, if certain assumptions about the number of individuals who may have been exposed were true.

The argument went somewhat as follows:[31] The evidence available to the Joint Commission in the autumn of 1945 had shown that the population in Hiroshima was fairly uniform in distribution within 2,000 m or so of the hypocenter when the bombing occurred. This implied that the number of individuals within a given area at that time depended largely upon the size of the area. Thus, the population exposed within 1,000 m should be only one-third as large as that exposed between 1,000 and 2,000 m, since the area within a radius of 1,000 m is only one-third the size of that within a concentric zone beginning at 1,000 m and ending at 2,000 m. If radiation had no affect upon the occurrence of leukemia and mortality was independent of distance, there should be three times as many cases in the 1,000–2,000 m zone as within the 1,000 m zone.[32] However, it was known that comparatively fewer individuals survived in the latter instance than the former and, therefore, the proportion should be even greater than 3 to 1.

In 1949, Yamawaki and Borges had seen nine cases of leukemia among the Hiroshima survivors. All had been exposed within 1,500 m of the hypocenter. The probability of this occurring, if the frequency of leukemia was not related to exposure, would be very small, and while it could not be concluded that the excess was due to exposure, radiation seemed the most likely cause. This finding, the first evidence of a possible increase in any cancer among the survivors, immediately prompted an effort to confirm and extend what was apparently being seen.

The task was made difficult, however, by the absence of individual dose estimates, the lack of a systematic case-finding mechanism, and uncertainties about the size of the population at risk. Moreover, the clinical facilities that were available before construction of those the

Commission occupied late in 1950 and early 1951 were limited, but once the latter could be used, the number of individuals examined rose rapidly, reaching almost 1,000 a month. Through these examinations, physician referrals, and death certificates, Jarrett Folley, who was then the chief of the Commission's Department of Medicine, Borges, and Yamawaki were able to identify 19 individuals who had either died from or had the onset of leukemia in the years 1948-1950 among an estimated 98,265 survivors in Hiroshima based on the city's 1949 census of survivors (and 10 cases among 96,962 survivors in Nagasaki). When these cases and the survivors were distributed by distance from the hypocenter, a significant increase in the number of cases was seen within 2 km.[33] And by 1953, when 50 cases of leukemia had accumulated in Hiroshima, William Moloney, a hematologist who had replaced Folley as the chief of the Department of Medicine, and Marvin Kastenbaum, a statistician, were able to show that cases increased in frequency from about 1 in 12,625 individuals exposed at 2,500 m or beyond to 1 in 80 if exposure occurred within 1 km. Thus, the risk within 1 km, where the radiation was intense, was more than 150-fold greater than that at a distance where the dose was presumably very low.[34] When the cases and individuals at risk were divided into those with severe radiation complaints (a history of epilation, oropharyngeal lesions, or purpura, or some combination of these) and those without symptoms, the increase in incidence with declining distance from the hypocenter was more striking among those individuals with complaints, but as Fig. 10 illustrates, the effect of distance could be seen in both groups of survivors. Statistically, the data were consistent with a straight-line relationship between the logarithm of the distance of the survivors from the hypocenter and the logarithm of the incidence of leukemia, a fact suggested but not established in the earlier study. To phrase their findings in simpler terms, the decline in frequency of leukemia with increasing distance is rapid initially, and then slows as the risk approaches the background rate. Interestingly, since the logarithm of the average dose of the survivors is roughly proportional to distance, except at the very close distances, where shielding cannot be ignored, this implies a curvilinear relationship between dose and the occurrence of leukemia that is similar to the one now generally accepted.

As the number of cases continued to grow, it was possible to examine the relationship of the type of leukemia to exposure, and the gender and age distribution of the affected individuals as well as their distance from the hypocenter. Acute forms were the most common, followed by the chronic myelocytic types, but a paucity of cases of chronic lymphocytic leukemia was also noted.[35] Although age was recorded in terms of the apparent onset of symptoms, rather than age at exposure, the risk of leukemia was clearly greater at younger ages. Finally, since some of

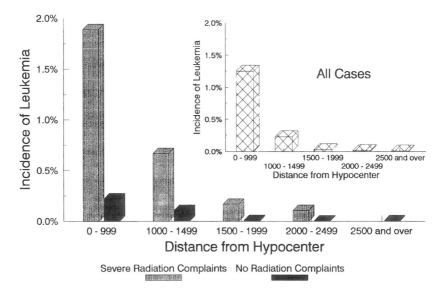

Fig. 10. The distribution of cases of leukemia in Hiroshima by distance from the hypocenter and the presence or absence of symptoms associated with acute radiation sickness. The inserted graph gives the distribution of cases by distance without regard to symptoms of acute radiation illness. (After Moloney and Kastenbaum, 1954)

these affected individuals had been repeatedly seen in the hematological surveys that had occurred immediately before or shortly after the establishment of ABCC, it was possible to gain some insight into the length of time intervening between exposure and the onset of leukemia. It appeared to lie between 2 and 8 years.

These observations led Moloney and Robert Lange to search for preclinical signs of leukemia. They reasoned that early recognition might enhance the survival of an individual with this disease, which was almost invariably fatal, and that such knowledge would broaden an understanding of its natural history. They believed that the acute forms of leukemia were poor candidates for their intended investigation, since the preclinical course in the acute disease was too short to be effectively studied; virtually every acute case that had come to their attention already exhibited the full array of symptoms of the disease. Chronic myelogenous leukemia was more promising, since the preclinical course was known to be very long. Their work was hampered, however, by the absence of suitable study techniques. Modern cytogenetics had yet to develop, the role of some viruses in the etiology of leukemia was unknown, and none of the tools of contemporary molecular biology were available. It had been shown, however, that the activity of a cellular enzyme, known as alkaline phosphatase, was markedly depressed among individuals with leukemia,[36] and Moloney and Lange reasoned that measurement of this enzyme might provide an early clue.

At that time, the Commission maintained a small diagnostic ward, and potential cases of leukemia could be admitted, examined, and followed thoroughly. Moloney and Lange were able to show that the presence of even a small number of immature bone marrow cells, when coupled with an increased absolute count of basophils (white cells that contain granules with an affinity for basic dyes) and low alkaline phosphatase activity in separated white cells, was highly suggestive of early myelogenous leukemia. Moreover, the low alkaline phosphatase activity appeared to precede cytological evidence of white cell immaturity by many months.[37]

While the leukemia story was unfolding, another possible radiation-related hematological disorder emerged—refractory or aplastic anemia, although the evidence that it arises more frequently following exposure is still far from clear. Aplastic anemia is a relatively rare disorder, resulting in a chronic, marked diminution in the number of circulating red cells, that does not respond to any current form of therapy. Patients with this disease have low red cell, white cell, and platelet counts, that is, they have a pancytopenia. The diagnosis can be established unequivocally only by bone marrow biopsy, which reveals greatly diminished to absent marrow cellularity. The disease is typically fatal and its occurrence, as a possible occupationally related hazard of radiologists, had been reported as early as 1924.

Six cases of this disease had been seen in Nagasaki through 1951. Of these, 4 were exposed within 1,500 m and all had had acute radiation sickness including epilation.[38] The other 2 cases involved individuals who were present in the city but at substantially greater distances (one at 3,800 m, and one at 4,400 m) and had experienced none of the symptoms of acute radiation sickness. Five cases were stated to have been seen in Hiroshima, but these were apparently never fully described. These early studies, like those on leukemia, were flawed by diagnostic difficulties, incomplete case ascertainment, and the absence of estimates of individual doses. Finally, in 1970, when 156 cases of definite or probable aplastic anemia had been seen in Hiroshima and Nagasaki, a dose-related study could be done.[39] Forty of these individuals were atomic bomb survivors (20 each in Hiroshima and Nagasaki); the remainder were not present in the city at the time of the bombing, or were born after 1945. Twelve of these 40 were members of the fixed sample known as the Life Span Study (to be described shortly) and eight had received doses then estimated to be 0.01 Gy or more. The risk of occurrence of aplastic anemia among these latter individuals, as contrasted with those survivors receiving less than 0.01 Gy, was 1.8-fold greater, but given the small number of cases, this apparent increased risk was not statistically significant. Until a similar study is conducted using the new doses and the better diagnostic criteria now available, aplastic anemia will remain an uncertain but possible late effect of radiation exposure.

The Effects of Prenatal Irradiation

Studies of the effects of X-ray and radium exposure on embryonic development began soon after the isolation of radium and the discovery of X-rays. As early as 1903, Bohn had reported the maldevelopment of the sea urchin following exposure of its eggs to radium, and still other investigations established the fact that ionizing radiation could produce developmental malformations in experimental mammals. Retrospectively, it is surprising that such defects were not also seen among human beings equally early, since the use of radium and X-rays for diagnostic and therapeutic purposes was widespread. At that time, in fact, exposure to ionizing radiation was even being used as a means to interrupt pregnancies. However, soon after the introduction of radium therapy, case reports began to appear suggesting that when this therapy was used on pregnant women, generally to treat a malignancy of the cervix of the uterus (the constricted portion of the womb), the developing fetus was often seriously affected.

Finally, in 1929, two obstetricians from the University of Pennsylvania, Douglas Murphy and Leopold Goldstein, attempted a more systematic study. Questionnaires were sent to some 1,700 gynecologists and radiologists in the United States seeking to identify women who had received radiation therapy during pregnancy. One hundred and six women were identified, and of these 74 were delivered of full-term children. Thirty-eight of these children had more or less serious disturbances of health or development, and 16 of the 38 were described by these investigators as "microcephalic idiotic children." Fifteen of these were born to women who had received either radium or X-ray treatments early in pregnancy.[40] This was a number far greater than would have been expected if no causality were involved.

These observations stimulated an investigation of the children exposed in utero to atomic radiation. Soon after his arrival in Japan in the fall of 1949, James Yamazaki and his colleagues in Nagasaki began a study of the outcome of pregnancies among women who were pregnant at the time of their exposure. Their findings suggested that a disproportionate number of these pregnancies terminated in a live-born child who was mentally retarded. Since the number of cases they had seen was small, a second survey was initiated in Hiroshima under the supervision of Jane Borges, who was the head of the Commission's Department of Pediatrics. Two hundred and five living children were identified as having been exposed in the first half of gestation. Among these, 11 had been exposed in utero within 1,200 m of the hypocenter and 194 beyond. Strikingly, seven of the 11 had mental retardation and an abnormally small head, but none of the 194 were affected.[41] While the sample was ill-defined, the findings were so dramatic that a more systematic effort was made to obtain a study group that would permit better estimates of the risk to the embryo or fetus.

This effort led to the establishment of at least three overlapping samples of individuals exposed prenatally to the atomic bombings of Hiroshima and Nagasaki. These have been termed the original PE-86 sample,[42] the revised or clinical sample, and the in-utero mortality cohort. Differences between these samples originated in the different purposes for which the participating individuals were chosen, e.g., as the bases for clinical examinations, or for mortality surveillance. The samples themselves are the culmination of a series of actions that occurred between 1950 and 1960 (see note 43 for details). Briefly, four avenues of ascertaining information about prenatally exposed survivors were used: the birth records required by Japanese law and maintained by the appropriate city offices in Hiroshima and Nagasaki; the supplementary schedules of the Japanese National Censuses of 1950 and 1960, which sought to identify all survivors then alive; the 1950 Sample Census conducted by the Commission; and finally, the fortuitous recognition of a prenatally exposed individual through the Commission's Master File (to be described shortly), or some other chance encounter. The nature and time of availability of these sources of data defined and limited their utility. For example, individuals ascertained *only* by means of the 1960 census can contribute to mortality data for the years subsequent to 1960, but have never been a part of either the initial clinical sample or its subsequent revision.

Mental development was measured in a number of different ways, varying with the ages of the children at the time of each examination. While they were still young, assessment of the children's intelligence rested largely on the examining pediatrician's clinical impression of their intellectual status relative to their peers, their ability to count and perform simple arithmetic problems, and their capacity to form coherent sentences. Once they were older, however, structured intelligence tests were used. These included two Japanese intelligence tests, the Tanaka B and the Koga,[44] the Bender–Gestalt, and the Goodenough "Draw-a-Man" test (about which much more will be said later) (see Chapter 7).

Most of the mentally retarded, prenatally exposed survivors are still alive. Although sociologically handicapped, they are not witless. Some exhibit a surprising degree of knowledge in particular areas but an inability to integrate what they know. For example, one, a male, can recite the names and uniform numbers of all of the members of the local baseball team, the Hiroshima Carps, and many of those of the Tokyo Giants, the Japanese team with the greatest national following. Moreover, through the institution where he has resided for the past 20 or more years, one that helps to support itself through farming, he is familiar with and can identify a wide range of commercially important plants and farm machinery. He cannot, however, do simple arithmetic problems or read or write, save at the most elementary level. His speech is slurred, most probably as a result of the disproportionate development of his head and the subsequent severe malalignment of his teeth, but he has a healthy curiosity. If encouraged, he can be quite gregarious, chatting about the things he knows and intrigued by strangers, particularly foreign ones. Another, a small, muscularly built male shows an unusual, indeed unexpected degree of independence. He is employed in simple tasks in a family-owned catering service, and has traveled widely in Japan by himself, but he has no sense of the value of money and cannot count the change he receives in simple transactions. He is married to a nonradiation-related retarded individual, but childless. His concern for his wife, who is more handicapped than himself, is touching and illuminating for the concern he exhibits toward others.

Collectively, the behavior of these two individuals is an insight into the intricacy of the normal human brain, so marvelously set forth in Richard Restak's books, *The Brain* and *The Mind*.[45] Patently, portions of their brains function adequately, but other portions do so poorly, or not at all. As individuals, they are not unique, since a number of the retarded, prenatally exposed survivors are equally lively, with a childlike curiosity and a ready smile. Others are withdrawn and apprehensive, frightened by a world that shares too little of its love. Most appear acutely aware of their reluctant acceptance as human beings, and are exquisitely sensitive to slights. When their curiosity and openness are rebuffed, they inch further into a world they are not obliged to share, one free of hurt. As their parents age and die, more and more of these retarded individuals, unable to guide their own lives, are committed to the care of others, usually institutions. Although the latter frequently combine a policy of work at simple tasks with attention to bodily needs, there is a concern for self-worth, an encouragement often reluctantly extended by others, and within the limits of my experience, love is not niggardly given by those charged with their welfare. However, no single consequence of exposure to the bombing is more disturbing, more heartrending than the fate of these individuals.

Minoru Omuta, a journalist with the *Chugoku Shinbun*, Hiroshima's newspaper, has written movingly of their travail—shunned, frequently incapable of simple work, and often totally dependent upon others for their existence, they eke out each day of their lives. Selflessly, he has fought for recognition of their plight and their needs. Through his manifest concern, he has become a source of counsel and encouragement to these survivors and their families. Troubled by their lot, he asks searchingly "Why ... should these children have been condemned to such wretched lives—alive, yet with their mental faculties and speech hopelessly impaired? As we watched them innocently romping or sitting, gazing fixedly into space, we found ourselves searching deep in our hearts for the answer."[46]

The Master File

As ABCC's studies grew in number and the data on exposed and unexposed parents increased, it was obvious that some central registry was or would be needed in which to record the different ways of ascertaining the subjects selected for clinical and field programs, and to compare various sources of information. To fulfill this need, a file, known as the Master File, was created, and a strategy initiated for assigning to each individual, when first coming to the Commission's attention, a unique number. This file had, in effect, two aims: first, to determine whether an individual had been earlier identified, and if so, to collate the different sources of information. The second was a more hopeful one, and assumed that if all possible sources of information were investigated, every survivor still residing in these cities or their environs would become known through at least one or possibly more of these sources. Slowly but unceasingly, this file has grown until in 1991 it involves more than three-quarters of a million individuals, both exposed and unexposed, or their children. They have come to the notice of the Commission or Foundation through many mechanisms—the genetic studies, ad hoc censuses, registration for rice rations in the immediate postwar years, the National Censuses of 1950 and 1960, unsolicited visits to the institution's clinical facilities, and the like.

Initially, management of the accumulating data was largely manual. But as quickly as possible, machine-oriented retrieval techniques were introduced. Computers as they are now known did not exist, and the first system used punched cards, a technique devised by Herman Hollerith, an American inventor, for the U.S. Bureau of the Census. This system, whose commercial manufacturer would become the International Business Machines Company (IBM), allowed the encoding and recovery of numeric and alphabetic data on rectangular cards through holes punched in fixed locations in the card. Information was retrieved by stacking the cards in hoppers holding 1,000 or so, and mechanically passing these through a machine where they were counted and sorted into separate bins through the electrical sensing of the locations of the punched holes.

The Commission's questionnaires and clinical forms were designed so as to expedite the coding and machine recording of the data, assuming a punched-card system of retrieval was to be used. Considerable thought went into each step in the card's preparation, since the aim was to be able to recover as much of the basic data as possible through the punched-card equipment without having to refer constantly to the original record. Punched-card machines were far less flexible than contemporary computers, however, and much slower. Each card was limited to 80 entries or columns of information. To record the observations on one questionnaire, often more than one card was needed, and thus the order of each of the subsequent cards had to be identified also. Once a card was punched, to minimize errors in data entry, it had to be verified—a process very similar to a second punching of the original information. If the keystrokes of the verifier did not match those on the original card, the card was rejected, pending manual comparison of the discrepant entry. Once verified, a hundred or so cards could be fed through the sensor each minute, but at this speed cards frequently jammed if they were slightly bent in handling or storage. When this occurred, the machine had to be turned off, and the card, or the remnants of it, removed. If the obstructing card was damaged, as was often the case, a new one had to be made and the process repeated. Humid weather frequently led to warping of the cards, making sorting and counting a tedious affair.

As the data grew, hundreds of thousands of cards had to be sorted and counted. To obtain a specific piece of information on a particular survivor or group of survivors, two or more steps were often needed. First, the cards had to be sorted to isolate all of those with an entry in a given column, and then these were passed through the sorting machine again to count the various alternative punches in that column or possibly in another column. Storing the massive number of cards that quickly accumulated required space that could have been used more productively had there been an alternative system for managing the data.

Most of the statistical processing of the information was done on mechanical calculating machines such as those once manufactured by Monroe, Marchant, or Friden, since multiplication and division could not be readily performed with the punched-card equipment. The latter could sort, count, add, and subtract, and through an ingenious use of a process known as progressive digiting—based on the principle that multiplication is merely successive additions, and division is successive subtractions—multiplication and division could be laboriously done, but only on a very limited number of variables at one time. Statistical use of the data had to be simple, and would be considered inelegant by contemporary standards. Complex analyses, requiring tiny fractions of a second on present computers, such as those entailing the simultaneous scrutiny of six or seven variables, required hours when done manually. However, the limited computational flexibility available had a beneficial effect—it forced the investigator to think seriously about the biological issues involved and to plan the analysis carefully. It was virtually impossible to browse through the data, trying one method or another. Moreover, once planned, there remained the arduous task of writing detailed, extremely specific instructions to the machine operators who managed the card files.

Eventually, in the 1960s, this mechanical system of data management was replaced by an electronic one, using a machine known as an IBM 1401, and the punched card information was transferred to magnetic tapes, which could store greater amounts of data and be read rapidly. Although the conversion to this system was more belated than warranted, largely as a result of administrative decisions made outside the Commission's control, this was a healthy development, and once accomplished it permitted prompter changes to newer computers, first an IBM 1440 and subsequently a series of Nippon Electric Company (NEC) machines. Eventually, new computing techniques evolved that led to a greatly broadened set of analytical possibilities. Finally, with the advent of desktop computers and interactive computing software, data analysis has become still more convenient. Searches of the data are more readily made, and the investigator can manage the data directly, without the intervention of others. This frees one to explore alternative methods of analysis to examine their impact on the findings, and to represent the data graphically in ways that may be more revealing than merely examining an array of numbers. This greater flexibility has also encouraged the development of methods of analysis and computing programs uniquely suited to the data at the Foundation's disposal, but potentially applicable to other large epidemiological studies elsewhere, especially those focusing on the biological effects of ionizing radiation. For example, using punched card equipment, so-called longitudinal analyses, which seek to study changes in the health status of an individual as revealed by successive clinical examinations, were virtually impossible. The results of each successive examination were studied separately as if independent of what had preceded or would follow. Obviously, this was not only inefficient but sacrificed information that could be indicative of incipient disease. Longitudinal analyses, although they remain difficult, are now practical. Similarly, the use of complex regression models that strive to take into account modifiers of radiation-related risk, for example, gender and age at exposure, which would have been unimaginable in the early years, are now routinely used in the assessment of the data. The importance of these developments cannot be undervalued. They have permitted much better characterization of the findings than were possible heretofore and have contributed to a sharper delineation of the risks of exposure.

4

A New Research Strategy

Fortuitous findings of radiation effects, such as those just described, are weak scaffolds on which to erect a major scientific study, particularly one of this importance. If the Commission's activities were to be truly long-term, as it was evident they must be, it was essential that the scientific program reflect this orientation. Opportunism might suffice briefly, but the piecemeal approach that prevailed through the early 1950s could not continue indefinitely nor could the investigative program resonate to the whim of the ever-changing staff. A new research strategy had to be found, one that would provide a coherent basis for all of the studies then underway or contemplated. This need was sharply focused by an investigation of deaths among the survivors and their relationship to exposure that was initiated in 1951 in Hiroshima. The initial results of this study suggested that there was a sizable increase in mortality among young exposed adults, particularly males.

The design of this inquiry was simple. Death rates were to be determined through an effort, first, to collect all of the city death certificates, second, to estimate the city's population presumably at risk at that time, and third, to divide the number of deaths by the population to obtain the mortality rates to relate to exposure status. However, this simplicity was the study's undoing. It tacitly presumed, for example, that the definition of residence for the population was the same as that for the deaths recorded in the city. This was not true. The deaths the city recorded included not only those of individuals actually residing there, but elsewhere in Japan, if the individual's legal address (and hence family record) was in Hiroshima. Moreover, it assumed that the loss of population through migration was independent of the different radiation exposures, but migration varied with the origin and place of birth of segments of the population under scrutiny. The question this study raised was sufficiently important, nonetheless, that it demanded an answer. The only acceptable strategy, one that would avoid these erroneous assumptions, was a study based on a fixed or "closed" sample of survivors who could be followed to death wherever they might reside.

In the autumn of 1955, the Francis Committee, named after its chairman, Thomas Francis Jr., a distinguished epidemiologist and member of the National Academy of Sciences, was sent to Japan at the instigation of Keith Cannan, the chairman of the Division of Medical Sciences of the National Research Council, to evaluate the scientific program, including the death certificate study, and to recommend changes, if these seemed warranted, as Cannan undoubtedly suspected was true. Francis, a virologist known scientifically for his work with influenza but more popularly for the large Salk vaccine trial in the early 1950s, was accompanied by Felix E. Moore, who was then at the National Heart Institute but would become the head of the Department of Biostatistics in the School of Public Health at the University of Michigan, and Seymour Jablon, a mathematical statistician and a member of the Veterans' Follow-up Agency of the NRC. The program they suggested and which was adopted has sustained the activities of the Commission and Foundation for more than three decades, but it did not provide for the continuing need for information on the genetic effects of exposure to ionizing radiation nor did it stress the importance accurate radiation dosimetry would assume with time.

Critical of what was perceived to be a lack of continuity in direction, investigative leadership and stimulating support, and cognizant of the need to increase programmatic stability, integration of purpose, and effort, the Francis Committee recommended some sweeping changes. They argued that the individual studies had to serve the whole, and this could only be done by unifying them through a focus on a common set of survivors. Accordingly, the new research strategy they proposed, termed the "Unified Central Program," included a mortality surveillance, a clinical study to assess health and morbidity, and a program of autopsies.[1] The former two would center upon fixed samples of survivors and suitably age-, gender-, and city-matched comparison persons, whereas the last would entail the pathological study of as many deceased individuals from the mortality surveillance sample as practical.[2] Their recommendations had many merits, not the least of which was the anchoring of the studies on fixed samples, rather than the "open," constantly changing populations of these two cities as had previously been done. Thus, migration into and out of the cities, which plagued earlier assessments, would loom less large as a possible source of confusion.

To support the clinical program, they suggested establishing an epidemiological detection network to provide current information on the health status of members of the clinical study sample through a system of weekly health reports on the status of sample members or immediate notification of the Commission in the event of significant illness or death. They contended this network would serve two other purposes as well; it would yield information regarding the migration of individuals within the study groups and provide "a continuing and close relationship between ABCC and the members of the study groups." These ends were to be achieved through the designation of a series of lay observers, or "monitors," members themselves of the study group who would have under their supervision another 15 to 20 sample members, all living in the neighborhood of the monitor. If a member of a group moved to another location within the city, he or she was to be assigned to the monitor in the new neighborhood. The monitors were to be paid for supplying the information desired.

This was an interesting but unusual suggestion for at least two reasons. First, health networks of this nature were nonexistent in Japan and certainly rare elsewhere, and their cost-effectiveness was poorly known. Moreover, their acceptability to the survivors was questionable. They could easily be seen as an intrusion on the limited privacy of the average Japanese. Second, the proposed network was unwittingly tantamount to the recreating of the neighborhood associations, the *chonaikai* or *tonari-gumi*, that had existed before and during the war and had been used so effectively by the military oligarchy to stifle dissent and to mobilize the population. These uses had led the Occupation authorities to view the associations as vestiges of coercion, and one of their early acts was to disband them. Whether the members of the Francis Committee were aware of this fact is uncertain.

The Commission staff voiced other concerns about the value and feasibility of the network. One of these involved cost. It was estimated that in Hiroshima alone implementation would require over $200,000 a year, and it was questioned whether the quality of the information that would be obtained warranted an expenditure of this magnitude. Another issue raised revolved about the use of lay monitors. It was suggested that permanent employees of the Commission might be better, and their use would place control of the collection of the data directly in the hands of the Commission. But a further alternative suggested was the subcontracting of the work to a local institution such as Hiroshima University Medical School. This would have certain obvious advantages from the standpoint of public relations, labor union problems, working space, which was already limited, and the like. The disadvantage of a subcontractual arrangement was that it would give the Commission looser control of the collection of the data than would be true with permanent employees, but it was believed that this limitation could be overcome by a properly organized system for inspecting and checking the information. A more troubling question, however, was the one of the incentives or compensation that could be offered to the patients to ensure their cooperation. It was argued that it would be useless to expect every patient to spend 15–30 minutes each week detailing his or her aches, pains, illnesses, and absences from work without some incentive or compensation. These concerns proved premature, since the Francis Committee's suggestion in this regard was never implemented, and in its stead was substituted a history of illness obtained by interview at the time of the periodic physical examination.

Acceptance of the other recommendations came, but grudgingly so. The emphasis on the importance of an epidemiological approach rankled some of the professional medical staff in Japan who believed it denigrated the role of the clinician and was not responsive to the health needs of the survivors. It was argued that the death certificate analysis was "simply a tedious job for a group of statisticians." As the ABCC director himself opined, "I must state unequivocally that the determination of the late effects of radiation is a strict medical problem and the determination of why such effects should occur is also a most subtle medical problem.... The statistician can only count what has been done. He can only add up the cases of leukemia; he can't look through the microscope and diagnose the disease."[3] There was annoyance too at the comments of the Committee on teaching and training. Francis and his fellow committee members had seen these efforts as peripheral to the activities of ABCC, as indeed they were in a sense, and urged that they either be abandoned or held to a minimum, whereas some of the American staff in Japan saw their Japanese associates as poorly trained and their clinical observations untrustworthy as a consequence. One went so far as to say "Without good data the clinical program is lost and analysis of data not only becomes fruitless but pointless.... The data must be good data! And such data are provided by good doctors!"[4] The inference was clear, but whether this view of the quality of Japanese medical education at that time was warranted is debatable. Medical practices were not the equal of those in the United States nor much of Europe, but there was an earnest striving to do better, standards were steadily improving, and most Japanese physicians were aware of the shortcomings of their own clinical training.

Acceptance was undoubtedly delayed too by two other reports written more or less concurrently, which saw some aspects of the programmatic needs in a different light or suggested alternative strategies. The first of these was the assessment of Charles Burnett, a member of the AEC's Advisory Committee on Biology and Medicine and Chairman of the University of North Carolina Medical School's Department of Medicine. At the invitation of the Division of Biology and Medicine, he had accompanied the Francis Committee, although not formally a part of it.[5]

He had been assigned the task of reviewing ABCC's program in internal medicine. While he concurred with most of the general remarks of the Committee, he looked more favorably on the training activities, stating that it was these programs that would attract and hold promising young Japanese clinicians who could, in time, exercise more authority and responsibility in the Commission's work.

The second report was written by Thomas Parran, then the Dean of the University of Pittsburgh's School of Public Health but formerly the Surgeon General of the U.S. Public Health Service.[6] Parran had accompanied Drs. Shields Warren and John Bugher, both radiation pathologists and members of the Academy's Committee on Atomic Casualties, and Elbert DeCoursey, who was a pathologist also and had been head of the Army Group within Oughterson's Joint Commission, to Japan in February 1956 to learn more about the Commission's research. His report reiterates many of the problems identified by the Francis Committee. He was sympathetic to the need for stronger leadership and the unified program the Committee advocated. However, he took exception to the proposed morbidity detection network using lay observers. Parran was doubtful that this was the best means for determining the precursors of chronic disease generally, or those that might be associated with atomic bomb exposure specifically. He contended that household morbidity surveys usually elicit complaints so vaguely expressed as to defy diagnostic categorization and that such surveys considerably underestimate the prevalence of disease. He suggested instead a more focused approach utilizing employees of ABCC, rather than lay observers, one that would attempt to obtain information on those signs or symptoms relevant to conditions associated with radiation exposure and with other diseases of general health import.

Save for the epidemiological detection network, on 13 March 1956 the Academy's Committee on Atomic Casualties formally endorsed the recommendations of Francis and his colleagues, but before this endorsement was forthcoming an *ad hoc* conference had been held at the Academy on Sunday, 27 November 1955. The participants in this conference made a series of administrative, organizational, and scientific suggestions to further the work of the Commission but the salient ones were approval of the Unified Program, with its focus on closed populations, and an urging that the shielding studies that were then under way be pressed forward rapidly. The conferees also recommended that "the existing clinical and pathological programs be reoriented progressively and systematically toward conformity with the principles of the Unified Study Program."

Execution of the Francis Committee recommendations began in 1957 under Keith Cannan, who was serving temporarily as director of the Commission, and were continued vigorously by George B. Darling, who assumed the directorship later in 1957 and retained this position for 15 years. Wisely, Darling saw the orientation of the research program as not only a means to provide a firmer basis for future studies, but an opportunity to strengthen the relationship of the Commission with the local communities and to enlarge the responsibility of Japanese investigators and institutions for the research design. Of particular importance in the latter regard was the Japanese National Institute of Health, which, although involved in the studies from the outset, would now assume a larger role. He revitalized the Japanese Advisory Council to the Commission (the *Kyôgi-kai*), the counterpart to the Academy's Committee on Atomic Casualties established in 1955 but which met desultorily. Panels of Japanese professionals chosen by the Institute were asked to review carefully the selection of people to be studied as well as the nature of their examinations. To sustain this interaction of agencies and their professionals, renewable five-year enabling agreements were written to codify the roles of the Commission and the JNIH in the new endeavor. Darling saw this widening of responsibility and involvement as essential to a long-term program, but a broadened commitment alone, he argued, would not suffice. There had to be a steadfastness of purpose over time that was unprecedented, and not only the assurance of the necessary funding, but a continuity in the recruitment of research and administrative personnel.

As perceptive as Darling's administrative vision was, if the new directions were to be achieved, there had to be an equally clear recognition of the scientific problems and how these were to be surmounted. The Commission was exceptionally fortunate, therefore, to have had Gilbert W. Beebe as its head of Epidemiology and Statistics in the years 1958 to 1960. It was his task to define the samples the Francis Committee envisaged, and to organize the methods to be used to collect and retrieve the data central to the implementing of the new strategy. These were awesome challenges, and it is difficult to believe that the program would have succeeded as admirably as it has without his clear-headed analysis of what was entailed and his perseverance.

It was more arduous than may be appreciated, or was probably foreseen by the Committee, to define a study sample of survivors of the required dimension and representativeness, and to choose an appropriately matched comparison group or groups. Sources of data had to be identified, and lists of survivors and possible comparison persons compiled and verified. Fortunately, in the early years, the Commission, often maligned for its seeming lack of direction, had the foresight not only to encourage the first country-wide effort to identify survivors of the atomic bombing of these cities through the use of supplementary schedules during the national census in October 1950 but to be instrumental in machine processing the information obtained through this census. As a result, the national government made available the names and addresses of the survivors that had been identified. There were earlier local censuses, of course, initiated under municipal auspices as well as those of ABCC. In 1946 in Hiroshima, for example, Ikuzo Matsubayashi, the head of the city's public health department previously mentioned in connection with the genetic surveillance, had conducted a census of atomic bomb casualties. And in 1949, Commission field investigators had called at the homes of all individuals identifying themselves as exposed in Hiroshima's 1948 Rice Ration Census to determine more about the details and place of their exposure. These undertakings were incomplete, however, since they were geographically restricted and did not include exposed individuals who did not return to the city until after the specific census had occurred, or never returned.

To evaluate the suitability of the nonexposed populations in Hiroshima and Nagasaki as comparison groups, in 1950 and 1951, ABCC undertook two special censuses. These were based on a technique known as stratified sampling. This entailed dividing the entire population of each of the cities into a series of distinct subpopulations, called strata. Within each stratum, a separate sample was then randomly selected. The strata in this instance were the census blocks, 10% of which were randomly selected. Within each selected block all exposed and nonexposed individuals were enumerated. And, on 3 June 1953, at the time of the Daytime Census of Hiroshima, the national government approved the addition of a question regarding exposure in this city at the request of the Commission. Finally, in 1958, an interim census of 11 cities within Hiroshima Prefecture, including the city of Hiroshima, occurred, and the survey schedule used included a question on presence in the city at the time of the bombing similar to the one employed in the 1950 census. Again, ABCC was given access to the various schedules as compensation for its role in planning, enumerating, coding and tabulating the data. These local surveys proved especially useful in identifying nonexposed comparison individuals, that is, persons who were not present in either city at the time of the bombing.

Once all of this information was at hand, selection of the individuals to be incorporated in the proposed morbidity and mortality surveillance could proceed, but the task remained formidable. Choice of the survivors focused on the supplementary schedules used in the 1950 National Census alluded to previously. This census had identified 159,000 persons who had been exposed in Hiroshima and 125,000 in Nagasaki. Of these 284,000 survivors, 98,000 were residing in Hiroshima on October 1, 1950, the date of the census, and 97,000 in Nagasaki. These 195,000 resident survivors and another 70,000 or so persons, who were not exposed and were identified through the special censuses to which reference has been made, formed the basis for the construction of a "master sample." But before this sample could be defined, since the supplementary schedules included only the survivor's name, sex, date of birth, city of exposure, and actual address, other information had to be obtained, notably the individual's distance from the hypocenter and location of their family register. This information could only be obtained through directly interviewing these individuals or a suitable surrogate if the person in question had died since the census. Once these interviews started, steps to choose the master sample could begin.

Eligibility for inclusion in this sample depended on certain other criteria, however, such as confirmation of residence in one or the other of the cities or in a defined, immediately surrounding area at the time of the national (or special) censuses. And in Nagasaki, where the number of individuals exposed at distances of 3,500 m or more was so large, logistic considerations dictated a random elimination of some of these potentially qualifying individuals. As finally selected, the master sample consisted of about 163,500 individuals divided into two groups: a Master Sample Proper (126,000) and a Master Sample Reserve (37,500). These groups differed only in the location of their family register. In the first instance, these records were in one or the other of the two cities (or the defined adjacent areas); whereas in the second, they were located elsewhere. It was from these two groups of individuals that the surveillance samples were chosen in a series of three separate "selections." The first selection defined the subjects for the morbidity surveillance, the Adult Health Study, to be described more fully shortly, whereas the mortality surveillance sample included not only these subjects but the second and third selections, which numbered about 70,000 and 10,000 individuals, respectively. The latter surveillance sample included all individuals exposed within 2,500 m, but the survivors within 2,000 m functioned as the core group on which matching occurred. Equal numbers of individuals exposed at 2,500–9,999 m or not in the city at the time of the bombing were sought to match the core group by sex and age. This stepwise selection process was prompted by the need to launch the new clinical studies as rapidly as practical and the availability of the information necessary to select the much larger mortality sample.

The Life Span Study

Death is an inevitable consequence of life, and exposure to ionizing radiation cannot increase its frequency. Ionizing radiation can alter the cause, however, and the time at which death occurs. The purposes of the mortality surveillance—the Life Span Study (LSS) as it is commonly called—are many, but, among these is determining those causes of death that are increased as dose increases, the time at which this increase becomes manifest, and its subsequent course of expression. A large sample is needed to achieve these aims, and initially the group chosen for study embraced approximately 100,000 individuals of whom about 83,000 were survivors.[7] This sample or cohort was subsequently extended, first in 1967 and then again in 1979, and now numbers about 120,000 persons of whom 91,000 were exposed and have potentially estimable doses (some 2,000 more individuals are known to have been exposed, but the information on their exposure is so incomplete or complex as to defy estimation of their dose). Collectively, these exposed individuals represent somewhat less than one-third of the 284,000 survivors identified in the 1950 census. About two-thirds of the 91,000 survivors in the present LSS sample were exposed in Hiroshima and resided in that city in the census year, and the remainder were in Nagasaki.

A mortality study of this scope is possible in Japan because of the existence of a unique record resource. In 1871, a law, the *Koseki-ho*, was promulgated establishing a system of obligatory family registration or household censuses. The original intent of the law was twofold—to provide a continuing, as distinct from a periodic, census of Japan's population, and to define the cohort of males available for compulsory military service.[8] Nationwide compilation of these family records began in 1872. In 1898, enactment of a civil code formally established the *ie* (family or household) system requiring a register for each family and the designation of a household head who was to be responsible for the veneration of the family's ancestors, the continuity of the household, and the preservation of property. To achieve these ends the head of the household had considerable legal authority, including the right to refuse entry into the register of any marriage of which he did not approve. The force of this right rests in the fact that vital events become official in Japan only when they are entered into the family register. Legal marriage was, therefore, entry into the register, and a de facto marriage left no record for legal purposes. The *Koseki-ho*, with modifications, remained in effect until 1947 and the inauguration of the new constitution. The present law, dated December 22, 1947, retains many of the features of its predecessor and differs primarily with respect to items reflecting the changing Japanese conception of a family and intrafamily relationships. It abolishes, for example, the *ie* system as a legal entity but retains the designation of a household head whose authority, however, is sharply curtailed.

These family registers, known as the *koseki*, are under the jurisdiction of the Ministry of Justice. Each such family record is identified through the lawful head of the family, the *hittosha*, whose name appears first in the record. On this individual falls the legal responsibility of seeing that all vital events affecting the composition of the family—adoptions, births, deaths, and marriages—are reported to the office, the *koseki-yakuba* (or more briefly, *koseki-ka*), having custody of the family's register. Aside from the provisions of the law itself, further impetus to prompt and complete death reporting is to be found in the regulations concerning burial and cremation permits. And in the modern era, the commercialization of the economy and the development of individual property rights has provided additional incentive for families to see that the names of deceased members are removed from the *koseki*.[8]

An integral part of this system of record-keeping is the *honseki*, one's legal address, which identifies the location of the family register in which one's name is inscribed. Changes in this address can only occur through notification of the *koseki-ka* having custody of the family's record; thus, if at any time in life the *honseki* of an individual is known, it is possible to locate the appropriate family register and determine his or her vital status, wherever the individual may reside or have resided in Japan.[9]

Under the Life Span Study, on a cyclic basis (commonly three or four years), the *koseki* of every individual alive at the end of the last cycle is perused anew to determine whether he or she is still alive. If a person has died since the last cycle, the fact and place of death are recorded in the register and, through the Ministry of Health and Welfare, a copy of the death schedule filed at the regional Health Center can be obtained to learn the presumed cause of death and contributory factors. Follow-up is virtually complete, only very rarely is an individual lost to the study, generally as a result of permanent migration out of Japan. It should be noted, however, that resident aliens do not have a *koseki* and since most of the Koreans present in Hiroshima and Nagasaki at the time of the bombings were there as conscripted labor, this means of surveillance was not practical in their instance. Korean survivors are not, therefore, numbered within the mortality surveillance sample, nor are American-born Japanese who may have been in Hiroshima or Nagasaki at the time of the bombings, if their parents had not maintained their family register and reported the individual's birth to the Japanese consular authorities in the United States.

Still one further group is difficult to follow and has been excluded from the mortality surveillance. Feudal Japan was a rigidly structured society. At the top of the hierarchy of social classes were the warriors, the *samurai*, followed by the farmers, artisans, and merchants, and then two outcast groups the *eta* and the *hinin*. The *eta* were individuals engaged in occupations considered "unclean" to orthodox Buddhists such as tanners, slaughterers, and buriers of the dead. The *hinin* (the name literally means "nonhuman") were prohibited from making a living by any other means than begging. They were, however, subject to compulsory labor, including caring for victims of contagious diseases and taking criminals to execution grounds. These two groups, often referred to in the records as the *kawatta seikatsu* (implying a "different living standard") or the *suiheisha* (the water-level people) were obliged to live in designated areas. These regions in Hiroshima and Nagasaki were well known, and in the instance of Hiroshima included Fukushima-cho to the west of the hypocenter and portions of Onaga to the northeast. Although discrimination on the basis of social class has been illegal in Japan since 1871, it has been difficult to suppress, since the earliest of the family registers contained information on an individual's social status and consequently membership in any of the outcast groups could be easily determined.[10] Since membership in these groups had been hereditary in the feudal era, some social stigmatization of their descendants has continued. To preclude any semblance of cultural insensitivity, the decision was made to exclude individuals in these designated regions in the surveillance, although descendants of these groups who might be residing elsewhere in the city were not excluded.

Until the early 1970s, the *koseki* were public records available for anyone's scrutiny, but as concerns about invasion of privacy and loss of confidentiality mounted, legal steps were taken to close the system. Much of this concern stemmed from the fact, previously stated, that the very early records often indicated the social class of an individual. It was the outcast groups,[11] who were instrumental in the public closure of the files, although access is still permitted with the explicit approval of the family concerned.

In the initial years of the mortality study, through arrangements with the local Bureau of Judicial Affairs, all clerical activities involved in the surveillance procedure were undertaken by ABCC staff members. Today, although direct access remains possible with the approval of the Ministry of Justice and the demonstration of adequate safeguards, the actual examination of the *koseki* of interest is done by employees of the various *koseki-ka*, on a fee-for-service basis. However, the task is so extensive and burdensome, given the other activities of the *koseki-ka* that the total sample of survivors is divided into five subsets. Three of these involve individuals whose family register is located in Hiroshima or Nagasaki and two involve individuals whose records are maintained in *koseki-ka* elsewhere. Insofar as possible these groups are matched by age and sex, and within the three subsets in Hiroshima and Nagasaki a different subset is examined each year in the cycle. Information on the vital status of those sample members whose records are outside these cities and their immediately adjacent areas is sought concurrently.

The results of these perusals of the *koseki* are reported to the Radiation Effects Research Foundation (RERF), which then obtains copies of the appropriate death schedules for analysis. The procedures for processing and reviewing the death certificate information, once it is available, have been codified and are periodically reviewed within the institution. The copies of the death certificates and other personal information are stored in locked files. Access to the latter is limited to those persons with a definable need, specifically the physicians or clerks involved in the coding and transfer of the information to computerized data files for analysis. These working files do not identify study participants by name, only by their Master File number; the aim is to avoid inadvertent breaches of confidentiality. No information is released that includes personal identifiers without the explicit, written approval of the individual involved, and then only after the endorsement of the directors of RERF. To ensure that these and other safeguards are adequate to maintain privacy and confidentiality, the procedures are reviewed annually with representatives of the government, generally the local Bureau of Judicial Affairs.

Establishment of Tumor and Tissue Registries

Death certificate data have their limitations; some causes of death are poorly recorded, and might be consistently under- or overestimated. To offset these shortcomings, the death certificate data are supported by information from special tumor and tissue registries. At the instigation of the Commission the tumor registries—the first continuous ones in Japan—were started in 1957 in Hiroshima and in 1958 in Nagasaki, but only after prolonged negotiation with the various community hospitals and the medical associations of the two cities. These registries, from their outset, have focused on the populations of these cities and not solely on the Life Span Study sample. Operational support came initially from ABCC and now its successor, RERF, but sponsorship resides in the local medical associations. Although it was not appreciated at the time, this seemingly late start of the registries does not limit their utility much, if at all, since it is known now that the time from exposure to the appearance of most solid tumors is at least 10 years and more probably 15, and the registries had been in place and functional several years before an increased risk was seen for any solid tumor.

Establishment of uniform principles to guide the collection, recording, and review of the information on tumors obtained in the two cities was a daunting task, since the data had to serve not only the purposes of the Commission but those of the local medical communities as well. The needs of the local groups were not as great as those of the Commission and, understandably, the local medical communities sought to achieve their purpose as simply as possible. It was essential to the Commission, however, that the data be obtained in precisely the same way in the two cities if radiation-related effects were to be compared, and this entailed greater systematization of data collection and review.

This meant that procedures had to be established for the pathological review of the slides, tissues, and other material that was submitted. Means had to be found to ensure consistency and accuracy of the diagnoses and to permit the review and editing of the records. Committees had to be appointed to supervise collection of the data and to manage access to the accumulating information, and a mechanism developed to provide periodic reports to the sponsoring agencies. Implementation of a common basic structure occurred more quickly in Nagasaki than in Hiroshima, where there were more large hospitals and two medical associations to deal with.

Initially, each physician was to submit a short summary of every malignant tumor seen to the registry, but this soon became burdensome. The only way to remove the burden, however, was through the use of non-professionals, but since medical records are privileged documents, in the beginning there was an understandable reluctance to open them to lay abstracters. Moreover, in some instances, it was unclear whether the records were, in a sense, the property of the hospital or the physician who made the diagnosis. This was an especially difficult problem in Hiroshima where the university hospital relied upon the medical school's pathologists to make the diagnoses, and they perceived the records as their own, since they were not customarily paid for their services. There were other new precedents that had to be established too. Most pathologists were unaccustomed to having their diagnoses examined by other pathologists who might or might not agree with the diagnosis made. There was a sense of self at stake. Eventually, largely through the support in Hiroshima of Drs. Tetsuo Monzen and Shoji Tokuoka, who were the leading pathologists at the prefectural hospital and the medical school, respectively, and of Drs. Ichiro Hayashi, Issei Nishimori, and, particularly, Takayoshi Ikeda in Nagasaki, these difficulties were resolved, and it was possible to overhaul the registry operations to meet the demanding standards that prevail in registries elsewhere.

Presently, incorporation of cases into these registries is an active process.[12] At appropriate intervals of time, representatives of the registries visit each of the large participating hospitals in Hiroshima and Nagasaki to identify all tumor cases, irrespective of whether the individual was or was not exposed, and to abstract from the medical records information regarding the identity of the individual, the type of tumor, its method of diagnosis, and other pertinent observations. This can be a wearisome duty, since many Japanese hospitals do not maintain a centralized medical record library and the relevant information must be sought in the records of the various clinical departments in which a malignancy might be diagnosed. Moreover, Japanese medical traditions and the payment provisions of the national health insurance plan encourage the establishment of small, private hospitals known as *iin*. These must by law have fewer than 20 beds, but they are so numerous in Hiroshima and Nagasaki (over 100 in each city) that it has proven impractical to review all of their records. Although the number of tumor cases seen in these institutions is small, the registries are dependent nonetheless upon the voluntary submission of these cases. But fortunately, most of the cancers seen at these small hospitals are eventually referred to larger ones where the records are periodically reviewed.

At the outset, case ascertainment was confined to the two cities, but recently, the registry in Nagasaki has been extended to be prefecture-wide. Since the registries record tumors occurring among all citizens of these communities and not merely members of the Life Span Study, as previously noted, they serve not only as a research tool, but also as a source of information on cancer incidence generally and changes in this incidence that may occur over time. The registries could also provide the bases for evaluating survival after different treatment modalities, although thus far they have not seen use in this regard.

The tissue registries, which are functionally a part of the tumor registries, are more recent and unique in many respects. Hiroshima's was established in 1973 and Nagasaki's in the following year. Again, these registries are managed by the Foundation under local medical direction, and each has a series of committees responsible for its activities, including the review of cases. The purpose of these registries is to provide pathological confirmation of the presence of a cancer, to determine whether the malignancy was a primary or a secondary one (a metastatic tumor), and where possible, to identify the type of cell involved. At their initiation, however, there was another motivation. The financial situation at ABCC in 1973 was difficult; costs kept rising, largely as a consequence of the devaluation of the dollar, but the budgetary support from the AEC did not increase. Funds had to be sought elsewhere.

Since the National Cancer Institute in the United States was interested in the occurrence of cancer in the various ethnic groups that comprise the U.S. population, and since occurrence can be influenced by environmental factors, comparative rates and studies are important. These registries could, therefore, serve not only ABCC's interests but those of the National Cancer Institute as well. Through the intercession of Robert W. Miller, the head of NCI's Clinical Cancer Epidemiology Branch, the Institute decided not only to underwrite the expenses of the Tissue Registries but to provide funds for the establishment of a visiting scientist program for young faculty members from Japanese universities who would spend a year at ABCC, including two months at the National Cancer Institute in Bethesda, improving their understanding of cancer epidemiology. As Miller has aptly written, these developments served "to encourage greater participation in ABCC by Japanese scientists, and the continuation of the cancer studies."[13]

The systematic autopsy program initiated as a result of the Francis Committee recommendations had a purpose similar to that of the registries but, in addition, it offered an avenue to study the accuracy and completeness of the reporting of causes of death through a comparison of the statements on the death certificate with the findings of pathologists at the time of postmortem examination. Although ABCC began an autopsy program much earlier, it had not been oriented toward a specific group of survivors nor had it systematically sought the help of local physicians. Now, to stimulate a higher level of cooperation, a small fee, 540 yen ($1.50 at that time), was offered for the procurement of autopsies referred to the Commission's Pathology Department.[14]

However, recruitment of participants into the autopsy program was never easy; it required great tact, patience, understanding, and the active support of the family physician, since, for many reasons, postmortem examinations are not common in Japan. Indeed, many individuals see them as a defilement of the dead. To further family involvement, funeral expenses were borne by ABCC. Nonetheless, the proportion of deceased individuals actually autopsied never exceeded 45%, although an effort to encourage autopsies continued for approximately 15 years. Active solicitation was abandoned about 1975 because of declining interest, but until quite recently some autopsies were performed at the request of a survivor's family.

Finally, in addition to the tumor and tissues registries, there exists a leukemia registry. It is difficult to assign a date to its origin since it developed over time. Efforts had begun as early as 1949 to identify all of the cases of leukemia that had occurred or were occurring in Hiroshima or Nagasaki, but initially there were no formal agreements with the local medical associations such as those associated with the tissue and tumor registeries nor was the process of case finding fully codified. However, by 1959, the ascertainment of cases of leukemia in these two cities had evolved into a de facto registration process that has been continuously refined, extended, and strengthened. The mechanics of the process are described in detail in a technical manual written by Stuart Finch and his associates.[15] Briefly, the screening begins with more than 100 clinical diagnoses and terminological variants occurring either on the death certificate or in hospital records that could conceivably represent leukemia. All of the clinical, hematological, and pathological records on individuals with these diagnoses are assembled and carefully reviewed. When a diagnosis of leukemia is made, it is based on the concurrence of at least two hematologists and is guided by standardized criteria. The diagnostic process includes specification of the type of leukemia, certainty of diagnosis, chronicity, and certainty of chronicity. New cases are periodically reviewed by the registry staff and a number of consulting hematologists, and in recent years efforts have been made and are being made to ensure consistency between this registry and the morbidity registries previously described. An important function of the leukemia registry is to collect and store hematological specimens, since diagnostic standards have been constantly changing and the opportunities to classify leukemias on their molecular and cellular bases have been growing steadily. The registry is generally believed to include information on essentially all of the leukemia deaths or cases occurring in Hiroshima and Nagasaki since 1948.

The Adult Health Study

The health assessment program, termed the Adult Health Study, originally involved some 20,000 persons, all of whom are also members of the mortality sample. Selection of these individuals occurred largely in 1958 through the efforts of Arthur J. McDowell, who would later organize and direct the Health Examination Survey of the U.S. National Center for Health Statistics. The choice of the sample members was a complicated one, since the number of individuals who could be examined was determined largely by the staff and clinical facilities available. The sample is not a simple random one, but was constructed (stratified) so as to ensure maximum representation of those individuals who received the higher doses and lived within areas accessible to the contactors who would have to arrange for their participation. Although the preliminary T57 doses (to be described later) were potentially available, McDowell, Gilbert Beebe, Seymour Jablon, and their coworkers chose wisely, given subsequent changes in the estimates of individual doses, to emphasize distance and the occurrence of epilation, bleeding, and oropharyngeal lesions—symptoms associated with acute radiation sickness—as the bases for sample selection.

All individuals exposed within 2,000 m with one or more of these symptoms were automatically selected.[16] The age and gender composition of this group in each city was then used to define three other study groups—individuals exposed within 2,000 m who were without symptoms, those exposed within 3,000 to 3,999 m, and finally, nonexposed individuals enumerated primarily in the first of the ABCC sample censuses, although some individuals were chosen from the second census to achieve the desired matching criteria.[7]

The labor involved in the selection process was considerable, since the proposed participants had to satisfy other criteria as well, such as Japanese ancestry and *honseki* in Hiroshima (or Nagasaki) or an immediately adjacent area. Furthermore, almost without exception, since the number of individuals in a given age–gender–*honseki* class in the three symptomless groups was greater than the number of symptomatic individuals, some further selection was needed. This was done through the use of random numbers. Selection did not depend, therefore, upon any attribute of the individual other than gender, age, and location of *honseki*.

Members of the Adult Health Study are invited to receive clinical examinations on a biennial basis. To ensure that the examinees in each successive month in the two-year examination cycle are representative of the entire sample, they are divided into 24 matched schedule groups. As a result, short-term studies involving individuals examined over intervals of only a few months can be undertaken with the knowledge that the examinees are representative of the full sample and the full range of exposure. Transportation is provided and the results of the examinations are communicated to the participants, and, where desired, to their physician. The first of these examinations began in 1958 under the direction of James W. Hollingsworth and the 18th cycle was completed in 1994.

Implementation of the Adult Health Study required more than the selection of a sample, however; there had to be a study plan. Development of the latter was not straightforward. The sample selected was large and the possible manifestations of radiation-related injury many. Moreover, most physicians, save those few with epidemiological training, are not accustomed to using rigidly defined examination protocols—their examination of a patient is generally informed by their own clinical experience and acumen. But this latitude is not acceptable when the aim is to compare the findings on a large number of individuals. Clinical and laboratory examination procedures must be standardized, and the examining staff made fully aware of the aim of each question and observation if idiosyncratic interpretations are to be avoided. Standard forms had to be designed on which to record observations. And supervision of data collection and collation was essential if the observations were to be comparable and trustworthy. Inevitably, ambiguous findings arise and this necessitates a process through which they can be easily identified and resolved. Finally, if the clinical experiences of the medical staff were to be rewarding, opportunities for individual research had to be created. Here the matched schedule groups were particularly helpful in providing samples for short-term studies that could be extrapolated to the entire Adult Health Study, since these groups were representative of the whole.

The construction of an epidemiological investigation is, however, more than an exercise in logic or even perseverance; it involves people whose participation must be sought and whose sensitivities and expectations must be recognized. Thus, all of the ends we have described had to be achieved in a context that was socially responsible and addressed the concerns of the survivors themselves. Much of the burden of these activities fell on the successive heads of the Department of Medicine in the years from 1958 through 1964, specifically, James W. Hollingsworth, Stuart C. Finch, Lawrence R. Freedman, and Kenneth G. Johnson, who found themselves not only the supervisors and clinical mentors of their young fellow physicians but administrative officers as well. The task of their successors was somewhat less onerous. They inherited a well-integrated program, but they too had the responsibility of constant supervision and the need to adapt the examinations to new findings on the survivors as well as in the biomedical sciences of relevance to the health study as these emerged.

In the beginning, the examinations were general assessments of physical well-being, but increasingly the focus shifted to early cancer detection, since this was the major risk survivors face. Participation in this series of clinical examinations, although voluntary, has remained high. Initially, over 90% of eligible individuals presented themselves for examination, but then participation fell somewhat, stabilizing at about 80%. Recently, the rate has declined a bit more. As the members of the study sample have aged, it has become increasingly difficult for some to come to the clinic and it is customary to visit them in their homes. Since the outset of the program, however, a relatively higher proportion of study subjects in Nagasaki have been examined than in Hiroshima and this difference is now more exaggerated.

The success of this ambitious program hinged on the continuity in scientific direction and staffing that George Darling foresaw. Arrangements were therefore made by Keith Cannan with three departments or agencies in the United States to provide the needed long-term support. The medical program was to fall under the supervision of the Department of Medicine at Yale University, the pathology program under the Department of Pathology at the University of California at Los Angeles, and epidemiology and statistics were to be the responsibilities of the Veterans' Follow-up Agency in the National Research Council.[17] These various groups undertook to provide the senior staff, and to contribute to the recruitment of the junior American clinical and scientific personnel as well. Enlistment of the latter, which had been no less vexing than the recruitment of established investigators, was ameliorated to some extent when, during the years of the physician draft, the U.S. Public Health Service agreed to offer commissions to young physicians of the Academy's choosing who elected to fulfill their military obligation in the Public Health Service working with ABCC rather than through a stint in the Armed Forces. Over an approximately 20 year period beginning in 1956 no less than 54 young physicians would be assigned to the Commission's Departments of Medicine, Pathology, and Clinical Laboratories through this mechanism. But when the "doctor draft" ended in the early 1970s, the old problems returned.

Over the years, these surveillance activities have been altered or broadened in a number of ways. Initially, survivors in the pediatric ages were studied under a separate program administered by the Department of Pediatrics aimed at evaluating their growth, development, and health status. However, since the youngest of these individuals was almost 15 years old at the introduction of the Adult Health Study, they were incorporated into the Unified Program, and a Department of Pediatrics, separate from that of Medicine, ceased to exist. As time has passed and information has accumulated on the health experiences of the nonexposed group, these experiences have been shown to differ in important ways from those of exposed individuals, and this group no longer figures prominently in analyses of the radiation effects. Comparisons are now generally made with those survivors who received doses of less than 0.01 Gy. Finally, two special mortality surveillance samples have been defined. One of these, mentioned previously, involves the survivors exposed prenatally and an age-, gender-, and city-matched comparison group; the other one, encompassing about 75,000 individuals, consists of the children of parents, one or both of whom were exposed either in Hiroshima or Nagasaki and an age-, gender-, and city-matched group of children whose parents were not exposed (the F_1 Study). The sizes of these various samples and the interrelationships among them are illustrated in Fig. 11.

These other surveillance programs, which will be described more fully later, are managed in the same manner as the Life Span Study sample, although the time between the initiation of successive review cycles differs. For example, the interval between each new ascertainment of the life status of the offspring of survivors has generally been

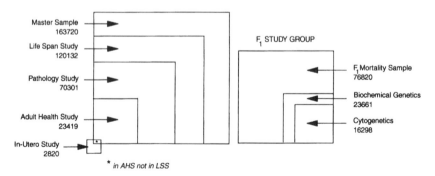

Fig. 11. The sizes and relationships among the major morbidity and mortality samples used in the studies of the atomic bomb survivors.

longer than the cycle in the Life Span Study for the simple reason that most of these individuals are at ages in life when few persons die. The oldest of the children of exposed parents was only 44 years old in 1990. At that age in Japan less than two individuals in every 1,000 will die in a year. Each surveillance cycle is so large an undertaking, dependent as it is upon the cooperation of the various *koseki-ka*, that it has not been practical, heretofore or now, to ascertain deaths in this cohort on a yearly basis.

After an examination cycle for any one of these mortality samples is concluded, the data are analyzed, and following a rigorous review by other scientists, both from outside as well as from within the organization, the results are described in technical reports and later in briefer articles in scientific journals. The technical reports—printed by the institution in English and Japanese—were begun in 1959 under George Darling's aegis and have served a variety of aims.[18] Initially, they addressed the unfounded, albeit prevalent, complaint of the Japanese press that the studies in Hiroshima and Nagasaki served a covert, unilateral scientific purpose, one designed to further the nuclear weapons program of the United States. Attempts to persuade the print media that this was not true, and that the information obtained in these studies was released in the open literature without restriction had been fruitless. Presumably this failure reflected in part the fact that most Japanese journalists and many Japanese scientists read English hesitantly or not at all, and did not have access to foreign publications. The bilingual nature of the technical documents and the summaries of these reports that appeared in the *Hiroshima Medical Journal* (*Hiroshima Igaku Zasshi*) made the findings available to a wider community of investigators. They have also been the means of fully describing the construction of the various samples used in the studies, or to report negative findings. Scientific journals are, in the main, not interested in such matters, but a permanent record is important to a program of this complexity and duration.

Recently, the practice of publishing a single, bilingual technical report has been abandoned, and in its place has been substituted separate English and Japanese versions. This change was instituted for two reasons. First, it allowed more rapid publication of the results of a study, since the report need not be delayed until both language versions were available, as had been previously true, and second, most Japanese physicians and scientists can and do read English more readily now than when the bilingual series began.

Most of the technical reports have been sufficiently detailed to permit other investigators to reproduce the findings that are discussed, and to analyze the data in alternative ways. However, the means to communicate scientific information continue to evolve, and to meet newer needs, RERF has begun publishing a Commentary and Review series that details general aspects of the program of studies or describes some of the statistical methods that have been developed to analyze the Foundation's data, but may have broader applicability. Finally, to this series of publications has been added a quarterly, known as *RERF Update*, that strives to provide not only a current summary of research at the Foundation but to do so in a historical context.

It has long been appreciated that this study and these data are more than a national or binational resource; they are also an international one. As such, it is imperative that the data and findings be shared as widely as possible. Copies of the technical reports are, therefore, available without charge to anyone who requests them, and for those individuals interested in the finer structure of the observations, the grouped, tabular material on which the analyses rest can be obtained on a floppy disk for a modest fee. These disks are available soon after RERF's own investigators have described their findings. Usually, this occurs after the pertinent technical report has been accepted for publication. Before the advent of personal computers, the data were made available to other investigators through extensive tables accompanying the technical reports. ABCC and the Foundation have also always attempted to respond to all reasonable tabulation requests from noninstitutional investigators. The aim has been to be as open with the data as possible without intruding on the privacy of the individual survivors, or the on-going or contemplated research of the Foundation's investigators, who have been instrumental in collecting or directing the collection of the basic information, and their freedom to describe the findings as they interpret them.

Over the years, however, the program initiated in response to the Francis Committee Report has undergone a subtle but important change, one only dimly foreseen, if foreseen at all in 1955. Emphasis has moved from the mere identification of the effects of ionizing radiation to their quantitative measurement. A greater concern now centers on numeric estimates of risk, on the best possible description of how radiation-related events increase as dose increases (the dose–response relationship), on the time of onset of an effect and its duration, and on the role that host as well as environmental factors may play in the manifestation of radiation-related damage. There is an increased interest too in the cellular and molecular mechanisms through which these effects arise. This shift of emphasis will undoubtedly continue, spurred by advances in molecular biology and by the development of techniques to analyze the accumulated data better. The fact that these studies can accommodate this changing emphasis attests to the perceptiveness of the Francis Committee and to the importance of the epidemiological approach it advocated so fervently.

However, the preoccupation with the fixed samples advocated by the Committee has to some degree stifled efforts to utilize more fully those observations collected in the earliest years of ABCC. This is lamentable, and particularly so insofar as leukemia is concerned, since the radiation-related increase in this malignancy was first noted in these years. Many of the initial cases of leukemia arose, and the survivors died, before the occurrence of the National Census in October 1950, the basis of both the Adult Health Study and the Life Span Study samples. The insight these cases can provide into the initial rise in frequency of leukemia will come only if these data are merged more effectively with the knowledge derived from the Life Span Study. To achieve this will require linking these cases to the population at risk when they were ascertained through a variety of sources of information, including the hematological surveys conducted in 1947–1954, and the pre-1950 local municipal censuses. Although some strides have been made in this direction recently (see note 18, for example), full reconciliation has not yet been achieved.

Other Studies of General Health Import

As the studies of radiation effects progressed, it grew increasingly obvious that the samples defined by the genetics program and by the studies of the survivors provided opportunities to examine other health issues of import to the populations of these two cities and to the world community. Some of these investigations were designed to characterize better the biology of the populations of Hiroshima and Nagasaki upon which radiation-related damage has been superimposed, and others were aimed at understanding some chance observation that might provide further insight into general health problems. One particularly interesting study involved a rare congenital abnormality known as acatalasemia. This disorder was first described in 1947 by Shigeo Takahara, an ear, nose, and throat specialist at Okayama University. He noted that among some of his patients with ulcerations of the mouth, when the mouth was swabbed with hydrogen peroxide (H_2O_2) the blood immediately turned a black–brown color.[19] Normally, when peroxide is used, the blood remains red, but foams exuberantly as a result of the action of an enzyme, catalase, in the erythrocytes. This enzyme reduces hydrogen peroxide to water and in the process liberates oxygen, which produces the foaming. Subsequent biochemical studies showed that the abnormal color stemmed from the production of a form of hemoglobin incapable of combining with oxygen, known as methemoglobin.

Once the number of recognized cases of acatalasemia was large enough, detailed investigations of their families became possible. These revealed the condition to be the result of the inheritance of two copies of a rare, recessive gene, one stemming from the individual's father and the other from the mother. Carriers of this gene—individuals with only one copy—were shown to have levels of catalase activity that were only half normal values.[20] Although it was first thought this enzymatic defect was confined to the red blood cells, this has proven not to be the case. As the means for culturing human cells developed, Robert Krooth and his colleagues demonstrated that fibroblasts (the cells that give rise to the fibrous tissues of the body) cultured from individuals with acatalasemia lacked enzymatic activity too, and that the cells of the carriers were deficient.[21] This discovery opened opportunities to study the nature of the biochemical defect more easily than was previously possible. Acatalasemia—or acatalasia as it is now commonly called, since the effect is not restricted to the blood—has been found to occur in many human populations, but for reasons still not clear, it is much more frequent among Orientals, particularly Japanese and Koreans.

Two larger and especially noteworthy investigations illustrative of this broader biological perspective are the studies of the life experiences of children of consanguineous marriages which were still common in Japan in the first decade or so following the war, and the studies of the risk factors associated with cardiovascular and cerebrovascular disease. These two studies, since their objectives were to provide information of a normative sort applicable to the widest possible community of humankind, focused primarily on individuals in Hiroshima and Nagasaki who were not exposed, or if exposed, received doses of less than 0.01 Gy.

Consanguineous Marriages and the Child Health Survey

Historically, the first of these studies—the Child Health Survey—centered on the life experiences of the children of consanguineous marriages, that is, marriages where the parents were biologically related. It was a collaborative undertaking involving not only the facilities and staff of ABCC but also members of the medical faculties of Hiroshima University, Kyoto University, Kyoto Prefectural University, Kyushu University, Tokyo Dental and Medical University, and the universities of Michigan and Indiana. Funds to support the study came from several private and public agencies in the United States—the Atomic Energy Commission, the Rockefeller Foundation, and the Association for the Aid to Crippled Children—each of which was interested in a particular aspect of the proposed health examinations.

When this study began, little was certain about the actual effects of consanguineous marriages. A few earlier surveys had occurred, indeed, one in the United States as early as 1858, but selection of the study subjects in these instances had been such as to lead to the possibility that the data might be seriously biased. There was, of course, anecdotal information attributing deleterious consequences to consanguineous marriages and numerous civil laws or religious prohibitions existed to prevent or limit their occurrence, suggesting a basis for concern. It was known that the children of related parents were more likely to be homozygous for a rare gene—to have two copies of that gene—than was true of children whose parents were not related to one another. If the gene's effects were harmful when homozygous, the children of related parents would be expected to exhibit these deleterious consequences more often than would the children of unrelated parents. This knowledge rested largely on studies of children selected because they were known to have a rare, inherited disease. It was not known, however, how common these deleterious effects would be if a group of children of related parents were studied who had not been selected with a view toward some specific health outcome.

The earlier genetic studies in Hiroshima and Nagasaki, through the previously described routine registration of all pregnancies, had identified several thousand children born in these cities whose parents were related. Analyses of the data collected at or shortly after the birth of these children had revealed that congenital defects were more common when the parents were related, and more of the children were prone to die in the first year of life than would be expected normally.[22] The purpose of the Child Health Survey was to extend these observations through providing more information, over a longer period of time, on the health status and growth and development of the children born to a random group of consanguineous marriages involving parents of differing genetic relationships. The findings on these children were to be compared with children whose parents were not known to be related and had also been seen in the early clinical studies. To ensure as much accuracy as practical in determining how the parents were related (if they were), their relationship was to be established in two ways—through personal interviews to obtain the family pedigree and through the examination of the obligatory family register of their parents and other antecedents, since these would identify a relative common to both the child's mother and father. Comparison of these two sources of information showed that they accorded well with one another.[23]

At its 13 March 1956 meeting, the Committee on Atomic Casualties enthusiastically endorsed this new initiative, asserting "the consanguinity study offered the opportunity to make the previous work in genetics at ABCC even more significant—an opportunity that should not be missed."[24] There remained, however, a need to demonstrate the feasibility of the proposed study, and to this end, in the summer of 1956, a pilot study was undertaken involving 1,500 children in each city ascertained through the original genetics program. Half of these children were the offspring of related parents and half were not.

This preliminary investigation revealed the loss of children to follow-up to be small, and not compromising to the proposed larger study. Moreover, it disclosed a significantly higher risk of death among the children of related parents than among the children whose parents were not related.[25] Accordingly, much of 1957 was spent planning the definitive study, and detailing procedures for identifying the sample of children to be examined. Three Japanese physician–geneticists, Norio Fujiki, Koji Ohkura, and Toshiyuki Yanase, spent the year at the University of Michigan participating in these activities and in the design of the bilingual forms to be used. Finally, in the summer of 1958, the first examinations began in Hiroshima and during the next two years 6,884 children, over 98% of those 6,969 selected and still living either in Hiroshima or Nagasaki would be seen again. This level of cooperation amazed the American investigators accustomed to participation rates of 70 to 80% in epidemiological investigations involving individuals who are not ill and are not, therefore, highly motivated to participate in health-oriented studies. It was even unexpected by our Japanese associates who had, nonetheless, assumed that participation would be high. When the examinations began, the children varied in age from 5 to 11 years, and the studies included clinical assessments of health status, stool examinations for parasites, dental evaluations, anthropometric measurements, neurophysiological tests, and in Hiroshima, psychometric studies.

Before the actual examinations could begin, however, there was a lengthy series of meetings with the local municipal and educational authorities, parent-teacher associations, and the medical community to seek approval of the study and understanding of its objectives. These groups quickly endorsed the study's aims, and the school authorities approved the release of the children from school on the day of their examination. But other logistic problems remained. A cadre of contactors had to be recruited and trained to solicit the participation of the study cases, and means found to transport the child (and parents, if they wished to accompany the child, as the mother often did) to the clinic. Transportation was provided by leasing local taxis, each of which displayed a small pennant identifying the study. At the conclusion of the examination, the examining physician informed the accompanying parent of the initial findings, and responded to whatever questions the parent might have about the child's health and development. Later, when the laboratory studies were completed, a letter was sent to the parents detailing all of the findings, and recommending medical attention where indicated. To achieve all of these ends no less than 19 professionals were involved in the examinations, along with a substantially larger support staff.

Out of this collective effort has grown the largest, most complete body of data on the biological consequences of inbreeding, that is, of being the child of consanguineously related parents, available in any nation, as well as a better appreciation of the relative magnitude of these effects.[23] Indeed, this study has become the standard against which all later studies have been judged. But it had one inherent shortcoming. Since ascertainment of the children who were studied was through the original ration registration process described earlier, it could not provide information on whether consanguineous marriages were more or less fertile than marriages involving unrelated spouses, or whether the pregnancies of related parents were more or less likely to terminate in spontaneous abortion. Eventually, in 1964–1965, this limitation would be offset by another study, the Hirado Health Study, funded by the U.S. Atomic Energy Commission and organized by the University of Michigan with the logistic support of the Commission and the cooperation of several Japanese universities, in particular Kyoto Prefectural University of Medicine, Kyushu University, and Tokyo Dental and Medical University.

The children who were examined in the Child Health Survey were a cooperative lot, oblivious to minor aches and pains. The bulk were generally in good health; however, intestinal parasites were common. Some 80% of them had either roundworms (*Ascaris*), hookworms (*Strongyloides*), or whipworms (*Trichuris*), and many had multiple infestations. Flea bites were equally frequent, and the parents found the fleas difficult to eradicate from infested floor mats. Scabies, a contagious skin disease caused by a mite that burrows into the skin, was prevalent, and dental care was poor. But insofar as the effects of consanguineous marriages are concerned, the study revealed that if one contrasts the child of a first-cousin marriage with one born to parents who are not biologically related, the following is found: The child of first cousins is 17% more likely to die early in childhood, is 37% more likely to have one or more major congenital malformations, is about 10% more likely to have a dental problem, is 1–4% smaller physically, has a somewhat poorer performance on neurophysiological tests (1–3%), and an IQ that is 3–5% lower. Moreover, the child of first cousins does 3–4% more poorly in school than a child of similar age and the same gender born to unrelated parents, but is less likely to exhibit an allergic disease. This is generally thought to reflect the fact that allergic disease often arises as a consequence of an immunological incompatibility between mother and child, inherited through the father. Such incompatibilities are less likely to occur when the mother and father are related, and are more genetically similar, on average.

Briefly summarized, these data reveal widespread, but small, biological effects that are explicable only on the basis of the relationship obtaining between the parents and not on socioeconomic or other differences among the consanguineously married and the nonconsanguineous. Whether these effects are sufficiently great to deter the marriage of relatives, given their cultural and socioeconomic advantages in some societies, is a matter of individual judgment.

Although important in their own right, these studies were not tangential to the charge before the genetics program. The aim was not merely to count newly arisen radiation-induced mutations, but to estimate their long-term impact on a population. Since it was known that mutations could lurk in a population for generations before manifesting themselves, it had to be determined how genetic variability was maintained through this period prior to manifestation. Herman Muller had alluded to this problem in his presidential address, entitled "Our Load of Mutations," delivered at the first annual meeting of the American Society of Human Genetics in 1949.[26] But there were several competing theories as to how genetic variability was maintained in populations, and little human data to provide guidance as to which of these was correct. Finally, in 1956, Newton Morton, James Crow, and Muller indicated how studies of the children of consanguineous marriages might contribute to the estimation of this "load," and to a resolution of the issue of the relative importance of these competing hypotheses.[27] One of the motivating purposes of the studies in Japan of the children of related parents was, therefore, to provide the data to further this estimation. Although time and further studies have shown that the method advocated gives less clearcut answers than originally surmised, the need to know how genetic variability is maintained stimulated a great deal of theoretical, experimental, and epidemiological research, and in this sense, their paper was seminal, a major contributor to the growth of human population genetics.

Cardiovascular Disease and the Ni-Hon-San Study

As the 1960s unfolded, evidence amassed rapidly that numerous factors, some intrinsic to the individual and some not (for instance, diet), contributed to the occurrence of cardiovascular and possibly cerebrovascular disease. This evidence included information from presumably genetically similar groups of individuals living in different environments with different life styles. The existence of these groups offered an opportunity to examine the effects of different environments on disease occurrence, which would presumably be relatively free of the added influence of genetic variability. Comparison of Japanese residing in Japan with those living elsewhere presented an especially attractive prospect, since the limited information available suggested that the frequency of coronary artery disease among Japanese in the United States was much higher than in Japan. Cerebrovascular disease, on the other hand, appeared to be less common among the migrants than in Japanese still living in their homeland. If this information were correct, and if one could presume that on average the genetic constitutions were the same in these groups, then something associated with the change in life style and environment affected their health. But what could this be? And would the changes that had occurred in the health experience of the migrants to the United States be seen in time in Japan as well?

The study that evolved to answer these questions, supported jointly by ABCC, the U.S. National Heart and Lung Institute, and the U.S. National Institute for Neurological Diseases and Stroke, has been known as the Ni–Hon–San Study. Initially it embraced three groups of Japanese—those living in Japan (Ni, for *Nihon,* the Japanese word for Japan), those in Honolulu (Hon), and those in the San Francisco area (San).[28] Three different collaborating centers were involved—the Commission, the Honolulu Heart Study, a National Heart and Lung Institute contractually supported field investigation; and the School of Public Health at the University of California at Berkeley. Work began in Hiroshima and Honolulu in 1965, and in San Francisco in 1969.

The examinations, which employed a common set of components in all three study areas, included a complete physical examination, a series of diet, health, and socioeconomic histories, an electrocardiogram, detailed laboratory studies to assess blood lipid levels, and where possible, pathological studies. The investigations in Hiroshima and Nagasaki focused on males born from 1895 through 1924, who were between the ages of 40 and 70 when the study began. There were 3,322 such individuals in the Adult Health Study sample, and 13,126 in the Life Span Study. Examinations on the former had begun in 1958, and mortality in the Life Span Study had been under surveillance since 1950, as previously noted. Thus, there already existed seven years of observation that could contribute to estimates of the prevalence of these diseases among the Japanese in Japan, and 15 years of mortality experience to examine trends in the occurrence of deaths attributable to cardiovascular and cerebrovascular disease over time.

In the years that have followed the initiation of these examinations, an unusual body of data has accumulated, one that captured the period when the prevalence of cerebrovascular disease in Japan began to decline and greater concern arose about possible changes in the prevalence of coronary heart disease stemming from the increasing "westernization" of the Japanese diet, particularly the rise in the consumption of fat. To avoid inadvertently confounding differences in medical care, culture, and language when comparisons were made of the incidence of coronary heart disease in Japan, Hawaii, and California, cases were restricted to those individuals in whom the occurrence of a myocardial infarction ("heart attack") could be documented electrocardiographically. These comparisons revealed the frequency of infarctions to be the lowest in Japan, where values are half those seen in Hawaii; the frequency in California is 50% higher than in Hawaii.[29] However, similar comparisons of the incidence of either strokes resulting from intracranial bleeding (hemorrhagic stroke) or a blood clot (a thromboembolic stroke) indicate that the incidence in Japan is about three times higher than that seen in Hawaii. Why this should be is not known; however, it has been suggested that this noteworthy difference may be attributable to differences in oriental and western diets. The traditional oriental diet has been low in animal protein and saturated fat; whereas in western diets consumption of these foodstuffs has been high. Experimental studies suggest that animal protein and saturated fat have an inhibitory effect on the occurrence of stroke, and during the years of this study both were more commonly consumed by Japanese residing in Hawaii than those living in Japan.

Cerebral infarction generally occurs either as a consequence of sclerotic changes (a hardening) in the small vessels of the brain or through the deposition of fatty materials in the walls of the system of arteries at the base of the brain known as the circle of Willis. Interestingly, pathological studies in Hiroshima, Nagasaki, and Honolulu suggest that small vessel sclerosis is much more common in Japan than in Hawaii, but atherosclerosis of the circle of Willis is more severe in Hawaii.[30] This led to the further conjecture that the westernized diet alluded to above may suppress the onset and development of small vessel sclerosis.

Although the studies in San Francisco have ceased, observations continue to be made in Honolulu, Hiroshima, and Nagasaki, and periodic comparison of the two sets of data still occur. As we shall later see, this early interest in atherosclerosis, which was motivated by geographic differences in cardiovascular and cerebrovascular disease among ethnically Japanese individuals, has taken on added value as data accumulate, suggesting that mortality attributable to causes of death other than cancer, and in particular cardiovascular disease, might be elevated among the atomic bomb survivors.

Other collaborative investigations of the kind described have occurred, such as the study of rheumatic disease, and others will undoubtedly occur in the future. Indeed, even now, a study of senile dementia involving the Foundation and investigators in Honolulu is underway, and others will soon be initiated involving investigators at the Mayo Clinic. The size of the sample and the duration of the follow-up of members of the Adult Health Study offer an unparalleled opportunity to examine the changes in physiological and health parameters that take place as the population ages. A better understanding of these changes is of importance to all nations, but especially the industrialized ones and Japan, in particular, since the number of elderly individuals is increasing so rapidly.

The import of these collaborative studies lies, however, in more than just the biological effects they have revealed, or may reveal in the future. The study of vascular diseases, for example, contributed to a surge of interest in Japan in the epidemiology of these disorders, and provided training to a number of Japanese cardiologists who have gone on to distinguished careers in cardiovascular disease research. It has also served to further subsequent binational studies of those factors, particularly in childhood, that contribute to cardiovascular disease, for example, the collaborative studies involving the Medical School in Shimane Prefecture and the School of Public Health of the University of Texas Health Science Center in Houston. The investigation of the effects of consanguineous marriages helped to stimulate the development of human genetics and biostatistics in Japan and prompted a number of other studies of the effects of inbreeding, such as those in Fukuoka and Shizuoka. The Japanese geneticists involved in the Child Health Survey have stood in the forefront of research and teaching in their discipline in the past several decades. They have done much to stimulate interest in human genetics in Japan and so too have the statisticians who participated in the study. Traditionally, Japanese statisticians have been mathematicians, little interested in the application of statistical methods to biological problems. Akio Kudo and Koichi Ito, who were primarily theoreticians initially but who made major contributions to the study, have retained a lively interest in applied statistics and have encouraged a generation of Japanese students to equip themselves to address biological problems. As mathematicians, they have been in a unique position to further developments of this nature, which profit binational relationships as well as international ones. Their interests legitimize a field of inquiry and contribute to a better quality of life everywhere through dissemination of knowledge pertinent to health and disease.

5

Exposure and Dose

Biologically meaningful studies of the atomic bomb survivors, more specifically studies that can be extrapolated to other exposed individuals, are possible only if the amount of radiation each survivor actually received is known. This need to express quantitatively the radiation experienced leads to the somewhat arcane topic of radiation dosimetry—the science of measuring (and calculating) the energy transfer between radiation and matter.[1] Dosimetric ideas and concepts are relatively simple but, unfortunately, are often clothed in a specialized vocabulary and unfamiliar units of measurement. To keep these ideas and concepts as simple as possible, the word "exposure" will be used only in its nontechnical sense—a person is exposed to ionizing radiation when present in a radiation field. The damage caused by this exposure increases with the amount of energy that was absorbed, and the latter depends upon the absorbing material. Since some materials are heavy and others light, it is customary to quantitate the average energy absorbed in terms of the person's (or object's) mass, that is, as the energy absorbed per kilogram. This ratio of energy to mass is the absorbed dose.

Doses are expressed in metric units. In the most recent international formulation, one joule—an amount of energy—per kilogram is the approved unit of absorbed dose. This unit is given a special name, a *gray* (abbreviated Gy). While the term "joule" may seem unfamiliar, it is the only metric unit commonly used in the United States. A more widely known measure of energy is the watt—the energy requirement of household electrical appliances and light bulbs are usually expressed in this unit. One watt is the transmission of energy at the rate of 1 joule per second. As an illustration, one kilowatt-hour on a utility bill means that 3.6 million joules of electrical energy were used (1,000 watts × 60 minutes × 60 joules per minute per watt).

A joule is actually a very small amount of energy. The average 60 watt light bulb consumes 60 joules of electrical energy a second. And a dose of 1 Gy (1 joule per kilogram) will raise the temperature of water only about one millionth of a degree fahrenheit, an almost immeasurably small amount. Yet in exposed populations deleterious health effects are observed after doses of 1 Gy or even less, as we shall see later. To appreciate fully how such small amounts of energy can cause biological damage, it is necessary to consider the deposition of energy on a microscopic scale. While a quantitative description is beyond the scope of this chapter, it should be noted that the actual energy transfers in tissue take place in extremely tiny volumes along the track of a charged particle. When these particles interact with specific biological molecules, such as DNA, the molecule can be damaged, and if the damage is not properly repaired serious consequences may follow. The dose in Gy is really an average per unit mass of the many localized discrete energy transfers that occur when radiation is absorbed.

While the absorbed dose is a useful concept and a measurable physical quantity, it is difficult to calculate. As a result, the dose estimates for the atomic bomb survivors are based on a more easily computed quantity called "kerma." The acronym kerma stands for *k*inetic *e*nergy *r*eleased in *ma*terial. It differs from dose in that it is the energy *released* by the radiation rather than the energy *absorbed*. Except in very special circumstances, the energy released within a material is almost totally absorbed, so that the absorbed dose and kerma are numerically nearly the same.[2] Both dose and kerma are expressed in Gy.

In calculating the amount of ionizing radiation received by the survivors in Hiroshima and Nagasaki, it is convenient to know the kerma in a variety of circumstances. One of these is the kerma in air at a given distance from the detonation where there were no buildings or other structures present to perturb the radiation field. This is often referred to as *kerma in free air*. Since most of the persons who survived the atomic explosion at Hiroshima or Nagasaki were inside a building usually their home it is also necessary to calculate the kerma within a structure. This is termed the *shielded kerma* or *kerma in house*. Finally, the energy per kilogram transferred to the organs of a survivor's body is the *organ kerma* (or often the *organ dose*). It is always less than the shielded or in free air kerma because the various parts of the body shield one another.

Radiation from the Bombs

For practical purposes, the primitive nuclear weapons dropped at Hiroshima and Nagasaki produced two kinds of ionizing radiation—gamma rays and neutrons. The former are like the X-rays from a physician's diagnostic X-ray machine but have much greater energy. So much more energy, in fact, that when gamma rays were first discovered, soon after Roentgen's discovery of X-rays, it was thought that they were a new kind of ionizing radiation and were given a distinctive name. Neutrons are a bit more exotic. Unlike the negatively charged electrons associated with atoms or the positively charged particles known as protons in an atom's nucleus, neutrons are electrically neutral. They are found in all atomic nuclei (except hydrogen), where they act as neutral spacers for the charged protons. The bigger and heavier the nucleus, the greater the number of neutrons required to hold it together. When a heavy atomic nucleus—such as uranium or plutonium—breaks apart (a process known as fission), high-energy neutrons are emitted as fragments. This breaking apart of uranium or plutonium was the source of the neutrons at Hiroshima and Nagasaki. The neutrons that arose did so either within the bomb itself through fission or, to a minor degree, from the decay of radioactive fission products in the fireball. The former event gave rise to the so-called prompt neutrons and the latter, the delayed ones. The liberated neutrons varied in initial energies from the highly energetic— the fast neutrons—with the ability to penetrate the atmosphere at great distances, to those neutrons at or near thermal energy levels. Thermal neutrons have little energy (approximately that associated with the random and continuous movement of the atoms in the air about us), and do not penetrate tissues deeply. They therefore do not contribute importantly to organ kerma except possibly to the skin. Since all neutrons lose energy as they traverse the atmosphere, the intensities of thermal neutrons at the ground result from the same neutrons that had greater penetrating ability when they were closer to the explosion. As a consequence, accurate estimation of the thermal neutrons associated with the two weapons is important not so much for the possible biological damage accruing from exposure to these neutrons as for the evidence they provide on exposure to more energetic particles at an earlier stage in their path.

The biological actions of neutrons and gamma rays are qualitatively similar, albeit not identical. Experimental data indicate that neutrons are much more effective for the same dose, and this difference in effectiveness is often more pronounced at lower doses. To allow for this, neutron doses (in gray) are usually multiplied by a weighting factor of 10 or more, so that the biological effects of neutrons per unit dose are comparable to that of gamma rays.[3] This weighting allows adding the doses from the two kinds of radiation to obtain a total biologically equivalent dose. A dose so adjusted is called the "equivalent dose" (or dose equivalent) and is expressed in a unit known as a *sievert* (Sv).

Once a unit of dose is defined, quantitative estimation of the energy absorbed by an atomic bomb survivor is still not a simple task. The dose will depend upon, first, the survivor's distance from the point in air where the nuclear weapon actually exploded, often called the *epicenter* or *burst point*; second, the nature of the materials that intervened between the epicenter and the individual and may have diminished his or her dose; and finally, the human body's own ability to restrict or alter the dose received by specific organs through absorption and the scattering of radiation. As an illustration, the testes are closer to the surface of the body than the ovaries; therefore, the absorbed gonadal dose for a fixed dose to the skin, will be higher in males than in females. But the degree of protection of the gonads will also vary depending upon whether the individual was standing at right angles to the line of sight to the bomb or facing toward or away from it, since the amount of intervening tissue will vary with orientation.

To arrive at an individual's dose, several sources of exposure must be evaluated separately. There is the radiation released directly as a consequence of the detonation, which spreads out more or less equally in all directions at or near the speed of light. This is termed the *prompt radiation*. But there are other components, such as the *delayed radiation*, previously mentioned, and the *fallout*. The latter stems from the deposition of radioactive materials that were either formed in the explosion or were sucked up into the maelstrom created by the bomb and subsequently, as the cloud cooled, fell to earth. This fallout, which also occurred following the Trinity test at Alamogordo,[4] has been called the "black rain" in Japan because of its appearance.[5] Finally, there is the *residual radiation* resulting from the induction of radioactivity in materials in the soil through their bombardment by neutrons released from the bomb. This latter induction of radioactivity is also known as "neutron activation."

No less than five different estimates of the amount of radiation in the air at various distances from the hypocenter, the kerma in free air, have been proposed for the two weapons. The first, and the most tentative, involved the calculations made by the Japanese Army in 1945. The second were the estimates of Robert Wilson in 1950, which were based on work by Herbert York. But Wilson did not estimate the neutron dose.[6] Neither of these were ever actually used as the bases for estimating the dose to individuals, but the subsequent three estimates have been employed to this end. The first of these formed the basis of what were called the T57 doses—"T" for tentative and "57" for 1957, the year the estimates were computed.[7] Next came the T65 doses (in 1965),[8] and more recently, a third has been used—the DS86, an acronym for Dosimetry System 1986.[9] Each of these successive assessments has been more sophisticated than its predecessors and has led to changes in the dose at a given distance from the hypocenter. Table 1 provides a chronology of these efforts and the estimates of kerma in free air at 1000 m measured in gray.[10] Note that the estimates of kerma in free air in Nagasaki have not

Table 1. Chronology of atomic-bomb kerma estimates at Hiroshima and Nagasaki in free air at 1,000 m, in gray (adapted from Ellett, 1991).

Year and dosimetry	Hiroshima		Nagasaki	
	Neutron (Gy)	Gamma (Gy)	Neutron (Gy)	Gamma (Gy)
1945 Japanese Army	1.25	2.00	—	—
1950 Wilson	11.40	6.80	0.80	6.80
1957 T57D	3.11	5.72	0.61	8.65
1965 T65D	1.92	2.56	0.36	8.89
1986 DS86	0.23	3.94	0.14	7.84

changed much over time as contrasted with those in Hiroshima and that the greatest relative difference between the cities involves the neutron component. Retrospectively, it is clear that in computing the T65 doses an error had been made in the allowance for the build-up of neutron scatter within the iron nose of the bomb assembly itself. This resulted in the assumption of a higher neutron kerma in air than actually occurred.

The latest effort to estimate individual doses, unquestionably the most ambitious, began in 1981 and largely ended in 1986, although work on a smaller scale, is continuing. It involved recompiling and examining all of the original data by dozens of physicists in Japan and the United States. This was a monumental undertaking, a search for records as much as 40 years old stored in numerous archives and warehouses distributed about the two countries, but essential if the reassessment were to be independent of previous ones, uninfluenced by the possibly erroneous interpretations of earlier investigators.

Numerous meetings of the physicists of the two countries occurred to share findings and to seek agreement in their interpretation. Often, new observations or new experiments or further analyses of the original data were required. Even new samples of soil, steel reinforcing bars, bricks, and roof tiles had to be obtained. These were carefully chosen from areas known to have been undisturbed since the bombings. And to provide more direct insight into the yield of the Hiroshima bomb, a replica of this weapon was assembled at the Los Alamos National Laboratory in New Mexico and used to obtain experimental readings on the radiation it released.[11] These were compared with theoretical calculations. This replica differed from the original bomb only insofar as the gun barrel was shortened and the fissile materials used were less than those that would produce an explosion.

To check the theoretical calculations, extensive use was made of a technique known as thermoluminescence dosimetry (TLD). Parenthetically, thermoluminescence was discovered over three centuries ago—it is generally credited to the 17th century English physicist, Robert Boyle, who had observed in 1663 that a diamond will emit visible light if it is heated—and has been used to estimate radiation doses for three decades or so, but its broadest application has been in the field of archaeology, where it has been employed to date ceramic artifacts. Indeed, much of the scientific basis for its use in the present instance hinges on these archaeological studies.

The procedure rests on the knowledge that a portion of the energy absorbed by some crystals, for example quartz (sand) crystals in bricks and roof tiles, during exposure to ionizing radiation remains stored or trapped over long periods of time within impurities or imperfections in the arrangement of atoms within the crystal lattice. Normally, the electrons within a crystal occupy fixed positions, but when the crystal is exposed to X- or gamma rays, their energy increases and they are freed to move about within the crystal until they are trapped. However, if the crystal is then heated, the electrons are liberated from the traps, and will return to their normal fixed positions, losing energy in the process. This lost energy is released as light. The amount of light emanating from the crystal can be measured and used as the basis for estimating the energy initially absorbed by the crystal. Quartz crystals extracted from bricks and tiles thus became another means of estimating the in free air gamma dose at various distances in Hiroshima and Nagasaki. However, the measurement process is destructive and the luminescence occurs only on the first heating. To forestall inadvertent systematic analytical errors, five different laboratories participated in the measurements; these included the National Institute of Radiological Sciences and Nara University in Japan, the University of Utah in the United States, and Oxford and Durham universities in England. Whenever practical, each laboratory received samples obtained from the same specimen for analysis and comparison. Unfortunately, thermoluminescent dosimetry does not provide information directly on the neutron doses in Hiroshima and Nagasaki, where the greatest ambiguity still exists. Neutrons have no charge and interact with nuclei primarily by scattering or by attaching themselves to a nucleus, depending on their energy. Neutrons have relatively little influence on the change in energy in the crystal lattice whose electrons provide the basis of the estimates of ionizing radiation to which the crystal has been subject.

Prompt Radiation

To estimate the various sources of exposure and, ultimately, individual doses in Hiroshima or Nagasaki it has been necessary to determine, as reliably as possible, the elevation and coordinates of the epicenter and precisely where a given survivor was at the time of the explosion. If the individual was in the open and in an unobstructed line directly to the epicenter, the exposure at the surface of the body would be the amount of radiation present in air at that distance, taking into account the absorption and scattering of the energy by soil and the air that had intervened essentially the kerma in free air at that distance. If, however, an individual was inside a building, the orientation of this structure to the burst point had to be determined, since gamma rays or neutrons coming obliquely through a thick wall could be reduced substantially in the case of neutrons when contrasted with those rays or particles entering at right angles. It is also necessary to know what materials were used to construct the building and how many walls were interposed along a straight line between the person and the epicenter. Eventually, measurements had to be made of the attenuation of the radiation through its absorption or scattering by the materials through which it passed.

Most survivors (80–85%) within a kilometer or two of the *hypocenter* or *ground zero* (the spot on the ground immediately beneath the epicenter), where the irradiation was most intense, were shielded to a greater or lesser degree. The most common shielding structure was a conventional Japanese wooden building, a single- or multiple-family dwelling.

The burst point of the bombs (and their hypocenters) were determined by triangulation using the direction of fallen trees, bent power poles or chimneys, but chiefly the orientation of shadows etched onto buildings or roads by thermal radiation. Another source of information was the presumed flight pattern of the Enola Gay, the bomber that released the Hiroshima weapon. Reassessments of the evidence have occurred in Hiroshima and Nagasaki over the years. There has been general agreement as to the burst point in Hiroshima but not in Nagasaki, where estimates have differed by about 50 m. The epicenter thought most probable in Nagasaki was not "finally" established until recently.[12] While this discrepancy might seem small, radiation in air diminishes roughly with the square of the distance, and differences in dose could be caused by an error of this amount, particularly for survivors within the first kilometer or so of the burst point, although generally, even at a few hundred meters, the absorption of energy is a more important determinant of dose than distance.

To ascertain the house shielding factors a series of more or less conventional Japanese homes was assembled at the Nevada Test Site in the United States and these were exposed to ionizing radiation. The design of these houses was based on careful studies of Japanese handbooks of construction and extensive measurements made in Nagasaki in 1954–1955 of the average dimensions of a house and the relative densities of all commonly encountered Japanese building materials—mud-plaster walls, slate, wood siding, subflooring, floor mats, and bedded roof tiles. Similar measurements were made in Hiroshima in 1955. The first two of these dwellings were constructed in 1957 at the time of a series of nuclear tests, known as Plumb bob, using materials imported from Japan, including the framing and sheathing, bamboo lathing, roof tiles and clay, and oyster shells and seaweed for the stucco material.

Two houses did not provide a firm basis for studying the effects of mutual shielding by adjacent buildings. This was an important consideration, since at the time of the bombing Japanese structures were generally placed more closely together than would have been true in the United States. Subsequently, in 1958, after an intensive study of the types and distribution of houses within 1 km of the hypocenter in Hiroshima and Nagasaki previously described, seven more Japanese buildings were built—two large two-storied houses, two middle-sized ones, and three small homes—using American materials believed, on the basis of experimental evidence, to have the same radiation absorption characteristics as conventional ones employed in Japanese construction. Figure 12 illustrates one such single-story, single family dwelling and its floor plan. These buildings, which were movable to allow assessment of the effects of variation in grouping, survived three nuclear detonations following suitable repairs after each test. However, before all of the necessary measurements could be completed, the moratorium on the atmospheric testing of nuclear weapons, the Limited Test Ban Treaty of 1962, took effect. To conclude the studies, a series of further experiments, known as Operation BREN, was initiated in the spring and summer of 1962 under the leadership of John Auxier and Fred Sanders of the Oak Ridge National Laboratory. To measure the neutron and gamma radiation fields at large distances from a fission source the small Health Physics Research Reactor (unshielded and unmoderated) at Oak Ridge was suspended on a hoist car mounted on a high tower (1,525 feet tall) at the Nevada Test Site, and to simulate the gamma radiation fields arising from the fireball at the end of the reactor studies a cobalt-60 source of about 1200 curies was substituted for the reactor. The final measurements were obtained directly with systematically situated dosimeters in the open, in Japanese houses, and in clusters of Japanese houses. These were placed at three elevations within each house corresponding to the position of an individual who might be lying prone, sitting, or standing. Auxier's account of this experimental program, code-named *Ichiban* ("Number One" or alternatively and less accurately, as "The Greatest"), carefully describes the problems encountered and how each was surmounted.[13]

To be useful in estimating risk, however, these physical parameters had to be applied to data on the location and shielding of individual survivors at the time of the bombings. To ascertain the latter, lengthy

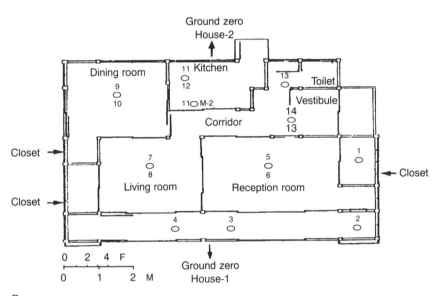

Fig. 12. One of the single-story, single family dwelling units constructed to evaluate building transmission factors in the *Ichiban* project (Panel A). Panel B indicates the interior floor plan of this dwelling and the location of the dosimeters (circles) used to measure the fluences within the house. (After Noble, 1967)

interviews of the survivors began in Hiroshima in 1951, initially under the direction of Richard Brewer and then of Lowell Woodbury and in Nagasaki in 1952 under the guidance of a bilingual Australian, Kenneth Noble. To conduct and process these interviews, groups of specially trained individuals were organized in the two cities under the supervision of Hiroaki Yamada (Hiroshima) and Yoshio Okamoto (Nagasaki), both of whom worked closely with physicists at the Oak Ridge National Laboratory in Tennessee. The task they confronted was enormous. Indeed, so much so that the original objective—to have a detailed history on every survivor in the study samples exposed within 2 km of the hypocenter in either Hiroshima or Nagasaki—had to be abandoned in Hiroshima, where many more survivors had resided within 2 km of the hypocenter. This meant that priorities had to be established, and these were based largely on two considerations. First, emphasis was to be placed upon those shielding categories having the largest numbers of individuals with the most complete and reliable associated medical records, and second, those categories where a team of consultants thought that good radiation measurements were possible through laboratory and full-scale field investigation.[14]

Through the use of several hundred aerial photographs taken shortly before and soon after August 6th and 9th, detailed city maps, block plots, and scale drawings (plane, front, and side elevations) of the interiors of individual houses, including surrounding structures such as other homes, wooden fences, and stone walls, the survivors exposed in the two cities were encouraged to locate themselves as accurately as possible. Figure 13 illustrates the results of one such interview of a survivor.

Preparation of the scale drawings was aided by Japanese home building practices which have been modular and, until recently, conservative over time.[15] Construction of an ordinary residence in Japan is still controlled by the traditional measure system, despite the fact that the metric system has been in use in other aspects of life since 1891. House sizes are customarily expressed in terms of *tsubo* (1 tsubo is approximately 36 square feet). The horizontal and vertical dimensions of a house are determined by the size of a *tatami*, a woven straw and reed floor mat about 3 by 6 feet in size. While room size can and does vary, customarily if the room is square, it is either 4.5 or 8 mats, or if rectangular, 6 mats. The width or length of a mat (or multiples thereof) determines the location of interior support columns as well as the height of sliding interior panels (known as *fusuma* or *shoji*), lintels, and ceilings. As a further aid to the reconstruction of the location of those survivors exposed in commercial or manufacturing structures, such as the dozens of factory and office buildings that made up the Mitsubishi arsenal complex in Nagasaki (where almost a fourth of the survivors

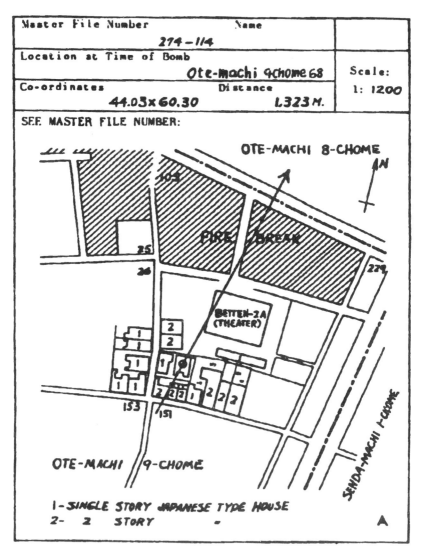

Fig. 13. A schematic representation of the results of a typical household shielding interview. Panel A indicates the location of the survivor's home in Otemachi and the direction (arrow) toward the hypocenter. Panel B indicates the floor plan of this home, the direction toward the hypocenter (arrow), and the location of the survivor within the home at the time of the bombing (circle). Panel C indicates the distance of the survivor from the epicenter, and the angle of the incoming radiation, as well as the number of walls and roofs intervening between the survivor and the epicenter.

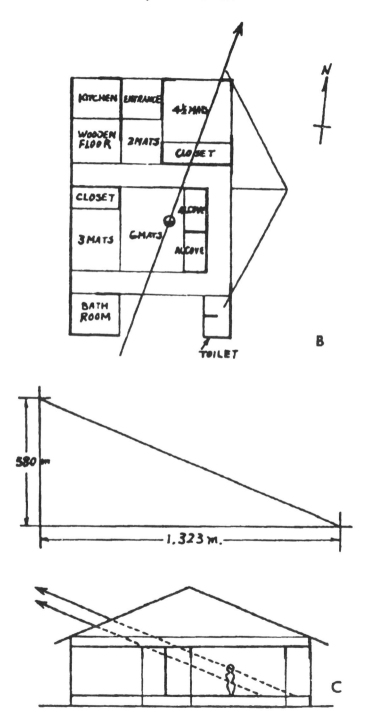

exposed within 2 km in Nagasaki were working), the cities were scoured for construction blueprints of these buildings.[16]

Collecting, evaluating, and verifying these shielding histories was intimidating, involving approximately 28,900 individuals (20,400 in Hiroshima and 8,500 in Nagasaki) and requiring almost a decade of careful interviewing and assessment. Eventually, these histories were available on virtually every survivor in the study sample in Hiroshima exposed within 1,600 m, a very substantial fraction within 1,600–1,800 m, but relatively few in the 1,800–2,000 m zone. Despite every effort to be as thorough and careful as possible, some of the histories do not lead to estimated doses that agree with biological evidence, and may be incorrect.[17]

The complexities do not end here. The two nuclear weapons that were detonated were different in the fissionable materials they contained and in their construction. The Hiroshima weapon, called "Little Boy," was a long, thin, more or less cigar-shaped bomb containing uranium-235 whereas the Nagasaki device, "Fat Man," an egg-like structure, used plutonium-239. Detonation of the Hiroshima bomb was brought about by a unique gun-barrel device through which one piece of fissile material was literally shot down a steel barrel onto another piece to achieve supercritical mass, the radiation from which produced a runaway chain reaction.[18] The Nagasaki bomb was detonated through implosion; the outer portion of the bomb's mass, through which plutonium-239 was distributed subcritically, was explosively collapsed onto a subcritical, centrally placed sphere of plutonium, making it supercritical. In both instances, conventional explosives were used to produce the detonation and supercritical masses (see Figure 14).

These differences in construction altered the yield and the spectrum of radiation emitted. Preliminary Japanese estimates of the yield of the uranium-235 bomb made soon after the bombing placed it at about 20 kilotons (kT), but it is now generally believed that it was closer to 15 kT—the energy released was equivalent to 15,000 tons or 30,000,000 pounds of dynamite.[19] The yield of the Nagasaki weapon is presently placed at 21 kT. More is known about the characteristics of this bomb than the Hiroshima one, since the initial nuclear explosion at Alamogordo on the 16th of July 1945, the Trinity Test, involved a Nagasaki-like device detonated on a tower. Weapons of this type were also studied in a series of experiments, called Crossroads (the Able and Baker Tests), in the Pacific in 1946 at Bikini Atoll and in many others at the Nevada Test Site, where their blast and radioactive characteristics were assessed directly. The Hiroshima bomb was the only one of its kind to be detonated. Others were built, but were subsequently dismantled, and the yield and radiation spectrum of the Hiroshima bomb have had to be reconstructed. Steps had been taken, based on the Trinity Test, to mea-

Fig. 14. Photographs of models of the two atomic weapons and schematic drawings of their firing mechanisms.

sure the yield photographically through the rate of growth in the fireball and the changes in the blast over pressures with time. However, the film of the Hiroshima detonation was accidentally destroyed, and only one pressure transducer gave even remotely usable readings. Subsequent assessments rest on this one pressure measurement, blast wave damage, charring of wood, radiochemical analyses of steel construction rods, and the like, and, most importantly, comparison with the Nagasaki explosion.

Every physical aspect of the shielding, whether by construction materials or the body, had to be carefully defined, and the consequences of variation in these examined. At the time of the bombing, most wooden Japanese structures had tile roofs, for example. The tiles were laid on a mud base an inch or so thick. This base was not generally uniformly applied, but laid down in strips. This meant that there were thick and thin areas that could differ in their transmission of radiation. The effects of these differences had to be scrutinized, as did the fact that tile size in Hiroshima was somewhat different from that in Nagasaki. This could affect the extent of the overlapping of adjacent tiles. Similarly the effects of variation in construction materials on transmitted energy had to be explored.

To understand how these factors as well as the tissues of the body affect the transmission of ionizing radiation it was necessary to know what happens when an enormous number of particles of varying energy bounce about in a geometrically complex medium that absorbs some and changes the energy and direction of others through elastic and inelastic collisions. This was done through a process known as Monte Carlo simulation, a method of obtaining a solution to a problem through computer modeling of the situation of interest. Literally, in the present instance, the computer creates a large number of imaginary photons or neutrons and the fate of each is studied as it courses its way to extinction. This produces a very large number of particle "histories," each one of which describes the changes the particle underwent as it traveled through the medium being modeled (the wall of a house or the tissues of the body). Included in the history is the energy of the particle, its direction of travel, the locations at which probabilistic interactions with the medium occurred, and the fates of the products formed by the interaction. These particle histories can be used to answer a variety of dosimetry questions, such as the energy deposited in a specific volume of body tissue or other material. In some instances, in the simulation of the Hiroshima and Nagasaki weapons, as many as 400,000 histories of the transmission of radiant energy through materials or the body were obtained. More commonly, to determine the effect of structural shielding, the fate of 40,000 neutrons or photons was followed as they entered (traversed) a given house or cluster of houses.

Each such step in the reconstruction, though time-consuming, was carefully examined and verified. Slowly, a consensus system emerged, one different in many particulars from the previous ones. The various steps necessary to develop the latest dose estimates are summarized in Table 2.

This reevaluation was stimulated by the seeming contradiction between some of the biological and physical data, particularly with respect to neutrons. It was recognized that neutron kerma had probably been

Table 2. Steps in the development of the Dosimetry System 1986. (Adapted from Ellett, 1991)

1. Reevaluation of weapon yield at Hiroshima and Nagasaki.
2. Calculation of source spectrum for both weapons. Comparison with field measurements using a replica of the Hiroshima bomb.
3. Calculations of air transport of prompt neutrons, prompt secondary gamma radiation, initial gamma radiation, and delayed gamma radiation from the fireball, and delayed neutrons.
4. Comparison of calculated gamma kerma in free air with modern thermoluminescent dosimeter measurements from building materials exposed at Hiroshima and Nagasaki.
5. Comparison of calculated neutron kerma with historical data on phosphorus-32 and cobalt-60 radioactivity.
6. Evaluation of doses due to soil activation and fallout.
7. Calculation of shielding by Japanese houses.
8. Calculation of body self-shielding.
9. Development of a computerized system to calculate organ doses for individuals.
10. Uncertainty analysis.

overestimated in earlier studies and that the mean free path length of a neutron (the average distance it will travel before it is scattered by an atom) was shorter than had been supposed. This distance depends upon the moisture in the air, and the initial experimental determinations of the mean free path were made in Nevada, where the air is significantly drier than in Hiroshima or Nagasaki, especially in the summer, when the bombs were detonated. Accordingly, to assess the neutron component in the radiation of each bomb, the meteorological conditions—temperature, humidity, and barometric pressure—that prevailed when each bombing occurred had to be studied anew. These conditions influence the density of the air and the moisture it contains at various elevations above the ground and vary to some degree with the time of day. The Hiroshima bomb was detonated at 8:15 AM, whereas the Nagasaki bombing occurred at 11:02 AM. The time of the detonation also affected the number of people who were at home, in school, at work, on the street, or downtown (closer to the hypocenter in the case of Hiroshima).

As a result of the recognition that the neutrons released by the bomb were captured in air more rapidly than had previously been estimated, the neutron kerma has diminished in the recent assessment. However, the gamma kerma, which is augmented by neutron capture, has increased in Hiroshima, notably at distances of 1,100 m or beyond, since the capture of a neutron by nitrogen in air results in the emission of high energy gamma radiation.

Delayed Radiation

Estimating the contribution to a survivor's dose made by delayed radiation from the fireball is particularly troublesome, since the exposure could have extended over 5 seconds or more, and the shielding of an individual could have changed within that time span. The blast wave emanating from the detonation of the bomb was reinforced by a secondary wave produced by the reflection of the blast by the earth and buildings sturdy enough to withstand the primary blast wave itself. This reflection, moving through a zone of diminished air pressure produced by the heat dissipated in the detonation, could more than double the pressure of the blast on a building and lead to its collapse, either exposing the building's occupants or burying them. These localized effects were so idiosyncratic to defy reconstruction in each individual instance, but loom large when one considers that the exposure to delayed radiation could equal or exceed the prompt component. At 1 km, for example, the cumulative exposure from delayed radiation in the open in Hiroshima in the first five seconds or so after detonation (virtually all of the relevant radiation from the fireball accumulated in this period of time) has been estimated to have been about 2.5 Gy. An individual inside a wooden building was shielded from approximately half of this amount, although the building undoubtedly began to collapse two seconds or so after the explosion when the blast wave reached the structure.[20] However, a survivor standing in the shadow of a wooden building was protected only for the first two seconds; once the structure began to collapse there was less material intervening between the individual and the rising fireball. As a result the individual's dose could have been higher than has been estimated, but to what degree is uncertain, since it would depend upon how rapidly the sheltering structure collapsed, and how the rubble was spatially distributed.

Fallout

Fallout occurred in both cities and could produce at least four different sorts of exposure—through proximity to deposited radioactive substances, absorption through the skin of radioactive dust and materials accumulated on the surface of the body, inhalation, or ingestion of radioactive substances. In Hiroshima the major fallout-contaminated area was about 3 km to the west of the hypocenter in portions of the city known as Koi-Takasu. In Nagasaki the prevailing winds are generally from the west and, consequently, the fallout was about 3 km to the east of the hypocenter in the region surrounding and including the Nishiyama Reservoir, one of Nagasaki's sources of drinking water.

Measurements made in Koi-Takasu in early October 1945 revealed the radiation stemming from fallout to be about 0.45 milliroentgens per hour, or a dose summed from one hour after the bombing to the expenditure of all radioactivity (the so-called *infinity dose*) of approximately 0.014 Gy. This is about the equivalent of naturally occurring background radiation in one year. Similar measurements made in the Nishiyama Reservoir area, where some 600 individuals were dwelling at the time of the bombing, were somewhat higher, about 1 milliroentgen per hour, or an infinity dose estimated to be 0.30 Gy.[21] These infinity estimates are, of course, upper limits; they are based on the supposition that an individual would have remained outside in the area, never entering their home or leaving the contaminated zone for the remainder of their lifetime. More realistic estimates are thought to be about a fourth of the infinity dose, or 0.0035 and 0.075 Gy in Hiroshima and Nagasaki, respectively. While these doses are not negligible, particularly in Nagasaki, both are below the level at which biological effects have been demonstrable epidemiologically. The newer reconstruction gives estimates very similar to those just stated—an infinity dose in Hiroshima of 0.01–0.03 Gy and in Nagasaki of about 0.20–0.40 Gy.[22]

Initially, it was speculated that these cities would be uninhabitable for decades after the bombing, but this has clearly not been true. Today, the zone of fallout in Hiroshima can no longer be readily identified through physical measurements—all parts of the city have a more or less uniform level of background radiation. But in Nagasaki, as recently as 1978, the area around the Nishiyama Reservoir was measurably more radioactive than other portions of the city, and presumably still is. However, it is believed that this increased radioactivity no longer reflects the 1945 explosion but rather subsequent atmospheric testing of nuclear weapons that injected radioactive particles into the upper air that fell to earth often hundreds and even thousands of miles from the actual test site. The region around the reservoir is a sump; it collects water, but it also collects and concentrates radioactive products that fall from the skies, whatever their source.

As previously said, fallout is generally thought to have contributed little to the exposure of most survivors. There is evidence, however, that in Nagasaki's Nishiyama Reservoir area, some of the long-term residents have an elevated burden of a particular radioactive form (an isotope) of the element cesium (cesium-137), a product of a nuclear explosion.[23] These burdens have presumably arisen from the consumption of food raised in contaminated soil, often in household gardens, since similar burdens are not seen elsewhere in the city in individuals who derive their drinking water from the reservoir. The average annual exposure in 1969—24 years after the bombing—from ingested cesium-137, one of the longer-lived radioactive substances, was 3 microsieverts (that is, 3 millionths of a sievert) per year for males and 2 microsieverts for females.[24] These internal doses are very much less than the current annual reference limits on intake (ARLI) of radioactive nuclides of 20,000 microsieverts for workers recommended by the National Council on Radiation Protection and Measurements, and substantially lower than the Council's annual limit of 1,000 microsieverts for continuous exposure to the public.[25] Indeed, these doses are even lower than the Council's annual Negligible Individual Dose (NID) of 10 microsieverts.

Similarly systematic studies of long-term residents of the fallout area in Hiroshima have not been conducted. Such measurements were seriously contemplated in late 1959 when it appeared that the whole-body counter used in the studies of the Marshall Islanders inadvertently exposed at the time of the Bravo test might be brought to Hiroshima.[26] This plan foundered on logistic difficulties. The whole-body counters in use at that time were very bulky and heavy, weighing as much as 21 tons as a consequence of the iron and lead shielding incorporated into their construction, and as a result the one available at Eniwetok would have to be transported to Japan on a naval LST (landing ship tank). This meant too that participants would have to go aboard the ship to have their body burden determined, and it was questionable whether they would be enthusiastic about this. However, without such a counter, which identifies specific radionuclides through scanning their energy spectrum, accurate measurements could not be made. Simpler measurements of total radioactivity would be virtually impossible to interpret and would be dominated, moreover, by one of the naturally occurring and long-lived radioactive isotopes of potassium, potassium-40. The latter is ubiquitous in the earth's crust and is present at much higher levels in everyone, regardless of where we live, than the radionuclide cesium-137, produced by an atomic bomb.

Residual Radiation

Residual radiation was largely restricted to the area immediately surrounding the hypocenter and dissipated rapidly, much of it in the first few hours following the detonation. Only those individuals who might have passed through this central zone shortly after the bombing, either to escape or on missions of mercy, would have been exposed to this source of radiation. The available evidence strongly suggests that it has added little to the total dose the survivors might have received or to that experienced by those persons who helped in the rescue operations but were not exposed to the actual bombing. This assertion rests first on the nature of the elements made radioactive secondarily and the rapidity of their decay, and second on the knowledge that extensive fires in the critical areas, particularly in Hiroshima, virtually precluded entry in the early hours when the level of residual radiation was high. It has been estimated that an individual working at the hypocenter 10–20 hours a day for a week beginning at day one would have accumulated a dose of about 0.10 Gy, whereas an individual at 500 m from the hypocenter would have accumulated only 0.01 Gy in the same interval of time, and at a 1,000 m the dose would have been only 0.0002 Gy. Surveys of the body burdens of the early entrants conducted in Nagasaki in 1969 failed to reveal burdens higher than those seen in nonexposed individuals.[27]

Dose Estimation

Once the various contributors to exposure have been evaluated as a function of distance from the hypocenter, this information must still be converted into doses to specific individuals, taking into account the particulars that surrounded that individual's exposure—their distance from the hypocenter, shielding, posture, orientation, and age. An individual's posture, whether lying down, seated, or standing, and facing toward or away from the epicenter, for example, could influence the dose to specific organs of the body even within structures made of the same materials. Similarly, a survivor's age at the time of the bombing is important, for this influences organ size and the shielding around the organ. Although most of our organs grow larger as one passes through infancy and childhood to adulthood, this is not invariably true. The thymus, for example, actually becomes smaller and virtually disappears in later life, and the neurons of the brain grow progressively fewer in number from birth to death.

Literally hundreds of different combinations of these particulars of distance, shielding, posture, and age are possible. Without modern computers, collating the information and examining the alternatives would have been impossible, but even with such computers it is daunting. To calculate for each individual the appropriate kerma in free air, the kerma with allowance for shielding, and the doses absorbed by the 15 organs of the body deemed necessary requires almost 2 minutes on a large computer, or approximately 50,000 minutes to compute the doses on 25,000 individuals. This represents over 20 40-hour work weeks for calculating the dose estimates for the more than 25,000 survivors on whom the information needed to apply the full system exists. At greater distances, where detailed shielding information is not available and the kerma in free air would be low (about 0.33 Gy in Hiroshima at 1,600 m, and 0.13 Gy in Nagasaki at 2,000 m; see Figure 15), an indirect method has been used to assign doses. This is done on the basis of the kerma in free air at the survivor's distance, and the average shielding transmis-

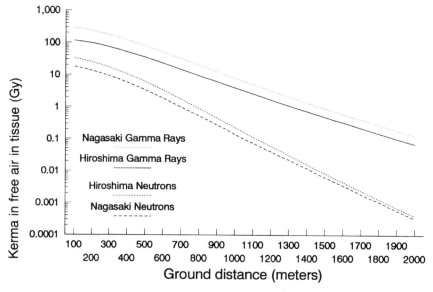

Fig. 15. The distribution of total kerma in tissue in gray (Gy) by ground distance (in meters) for gamma rays and neutrons in Hiroshima and Nagasaki. (Adapted from Kerr et al., 1987, Table 40)

sion factors for those survivors who were closer to the hypocenter (exposed beyond 1,000 m and having a shielding history) if an individual asserted being shielded by a Japanese-style building. Beyond 2,600 m in Hiroshima (2,800 in Nagasaki), doses are presumed to be zero.

Incidentally, it should be noted that the absorbed organ dose computed under the DS86 system is actually an age-specific population average rather than individual specific estimates since the mean dose to an organ depends not only upon the energy spectrum of the gamma rays and neutrons involved but also on its size. This will vary with the size of an individual survivor at the time of the bombing, which is not known. The organ doses are, therefore, based upon a reference man or woman—a hypothetical individual whose size approximates that of the average Japanese man or woman in 1945. And even here, the doses are derived from a mathematical phantom that simulates the human body by a series of simple geometric shapes—ellipsoids, elliptical cylinders, and cones, or parts of these. The size of the phantom can be adjusted to represent individuals of different ages or genders. Indeed, several phantoms corresponding to male and female adults, children, and infants were developed. When one of these phantoms is used in concert with a computer program that models the transport of neutrons or photons through the body by Monte Carlo methods, an estimate of the average absorbed organ dose can be calculated.

The element of uncertainty this method of estimating doses introduces is presumably not serious, save possibly in the case of a pregnant woman. Here the difficulty arises because the reference woman used in the DS86 calculations was assumed not to be pregnant, as was true of the vast majority of women survivors. However, in the pregnant woman the size of the uterus changes dramatically as pregnancy advances, and as the uterus enlarges, the other organs of the abdomen are shifted from their normal position and compressed. Thus the estimated dose to the uterus based on the reference woman describes the actual dose to the uterus more poorly in the later stages of pregnancy than in the early ones before the uterus has undergone much change in size, and may, therefore, be a poorer surrogate for the actual dose to the developing fetus after mid-gestation. It is, of course, possible to develop phantoms that represent the pregnant woman at various stages in pregnancy more faithfully, but this has not as yet been done.

Once all of the individual DS86 doses were computed, it was necessary to examine them singly and collectively to determine, first, that no computational errors had been made, and second, to identify individuals with improbable values. This search did reveal errors, most stemming from the inherent complexity of the program used to derive the individual doses. These errors were quickly corrected, and the computations were repeated. However, there still remained a number of individuals with improbably high doses.

These latter cases were carefully restudied to make certain that no mistakes had been made in the entry of the basic data and to look for inconsistencies in the observations that might account for the unlikely values. Despite all of this study and restudy, 100 or so individuals among the 75,991 on whom DS86 doses were available in 1987 had exposures estimated to be above 6 Gy, which, on the basis of current knowledge of acute radiobiological effects, are questionable. Either the individual had inadvertently erred in recalling the events attending their exposure, or there was some additional shielding that has not been recognized and the true dose was less than the computed one, or the physical parameters that are used in the computations are still not reliable enough to cover all possible contingencies. The occurrence of these cases, small though the number is (less than 0.2% of all individuals with DS86 doses), poses troublesome issues in the analysis of the data on the late effects. Should these doses be accepted as correct although they seem implausibly high? Should they be arbitrarily set to some value that is consonant with other radiobiological evidence, as has frequently been done? Should they be excluded from any analysis? Or should some middle ground be sought, such as analyses that include and exclude them to see what effect there is on the risk estimates. It seems wrong to me, at least, to reject the doses as erroneous automatically. Too little is known about individual variability in response to ionizing radiation to presume that no individual would survive such a dose, and those who survive might very well be more resistant to other forms of radiation-related damage. However, it would be appropriate to reject these cases on the supposition that concern centers primarily on the risk to the "average" person, not the most insensitive. Commonly now, analyses of risk focus on those survivors with doses of less than 4 Gy. But whatever the decision, it is important to bear in mind that the doses of regulatory concern are only one one-hundredth as large as these described, and often less. It is known that doses of 6 Gy are injurious, but what the effects are at the much lower levels of interest in radiation protection is uncertain.

When initially installed, doses could be computed using the new system for only about 80% of the survivors in the major samples under scrutiny. Given the variety of exposure conditions, it was not practical to explore the physical issues in detail in every instance and to develop techniques to cope with all of them. Judgments had to be made as to how to allocate the resources available to provide new dose estimates for the largest number of survivors as quickly as possible. The most common exposure situations involved individuals either in the open or in conventional Japanese wooden structures. Consequently, the new system focused first on those survivors exposed in the open with burns (and therefore presumably in a line of sight with the burst point and no shielding), and those in the open but shielded by a Japanese house, or in a wooden building within 1,600 m of the hypocenter in Hiroshima (2,000 m in Nagasaki) on whom detailed shielding histories were available. As previously stated, doses were directly computed for 15 different organs by modeling the circumstances applying to an individual's exposure. Provision was made in the modeling for structural and tissue attenuation of the energy flux and the survivor's posture, orientation, and age.

Studies continue to develop suitable methods for estimating individual doses in those cases not initially covered when the doses were calculated in 1987. Recently, it has become possible to compute directly the doses on somewhat more than 1,000 additional survivors who were either exposed in factories or were shielded by the terrain in Nagasaki. Rules have also been formulated for indirectly estimating doses for a larger number of survivors exposed beyond 1,600 m in Hiroshima and 2,000 m in Nagasaki, chiefly those individuals exposed in the open or those who were partially shielded. With these additions, it is now possible to calculate the doses on 86,632 of the approximately 91,000 exposed individuals in the major mortality study.

In a few selected instances, such as for survivors from the Chinzei and Shiroyama schools—reinforced concrete buildings near the hypocenter in Nagasaki—for which detailed measurements were obtained in 1945 by the Joint Commission, elaborate attempts have been made to estimate the doses through a modeling of the actual structures themselves. These efforts have involved literally hours of computation on the most powerful computers presently available, and are so formidable as to preclude their broad application.[28] However, they hold promise for specific situations such as the survivors exposed in the Central Telephone Office building in Hiroshima, another reinforced concrete building, where individual doses cannot be computed by the latest dosimetry system as it now exists, or as it is likely to exist in the future. Doses for some of these small clusters of survivors could be especially informative; in the Central Telephone Office building, for instance, of about 28 adolescent girls who were working there at the time of the bombing, no less than six developed breast cancer prior to 1989.[29]

The earlier, T65 system of dosimetry was essentially empirical, based almost exclusively on experimental findings, but the DS86 combines current theories of radiation physics with experimental observation. It does so through elaborate data bases with many different data sets that can be combined in thousands of different ways to model the circumstances of a given survivor's exposure. Out of this modeling emerges a series of different estimates—kerma in free air at the survivor's distance, shielded kerma (kerma in a house) and organ dose (the kerma and fluences in 15 organs). These are computed for each neutron and gamma ray radiation component previously described.

This new system of computing individual doses, comprehensive as it is, is not above fault, nor are future revisions likely to be without error. There lingers an uncertainty about the yield of the Hiroshima bomb, for instance—some investigators contend that 17 kilotons would be a better estimate than the 15 presently used—and about the neutron flux. Although the thermoluminescence measurements previously mentioned support the DS86 dosimetry more strongly than the T65 system, these measurements do not agree perfectly with the theoretical calculations at the distances where samples could be obtained. To obtain an exact agreement, the bomb's yield would have to be increased by about 10% or the output spectrum altered to produce more neutrons of greater energy that will be active at longer distances. In addition, other measurements indicate that the thermal neutron flux was larger than calculated.[30] Further measurements needed to resolve this issue are under way. However, the discrepancy to which we have alluded does not necessarily imply that the new doses are in substantial error, or in error at all. Neutron activation studies measure the number of low-energy neutrons that interact with the detector, whereas the quantity of interest is the energy (the kerma) imparted to the detector. Less-energetic neutrons, the slow ones, activate materials much more than high-energy neutrons, but have little influence on a survivor's dose. If the neutron dose does increase in the future, it would mean that the current risk estimates are too high, but by how much will depend upon the increase in the neutron dose that occurs. Foresightedly, the DS86 dosimetry system has been designed to be more flexible than previous ones, and can, therefore, accommodate changes in information as the remaining uncertainties are resolved.

To summarize briefly, the principal differences between the T65 and the DS86 dosimetries can be stated as follows:

1. The total yield of the Hiroshima weapon is now presumed to have been approximately 20% greater than earlier thought, that is, 15 rather than 12.5 kilotons. There is essentially no change in the estimate of the yield of the Nagasaki bomb with the reassessment.
2. Although the in free air gamma kerma is greater at all distances in Hiroshima, but less in Nagasaki at distances of more than 700 m, the neutron kerma is less in both cities, and substantially so in Hiroshima. Indeed, because it is only 10% of the previously estimated value for Hiroshima, the neutron kerma is now so small that direct estimates of the biological effectiveness of neutrons relative to gamma rays are extremely difficult, if not impossible.
3. Under the DS86 system, transmission of the gamma rays passing through wooden Japanese structures is approximately half that under the T65 dosimetry; however, transmission of the neutrons passing through such structures differs little between the two dosimetries.
4. Transmission of gamma rays through tissue is significantly higher, at least for the deeply situated organs than heretofore supposed.[31]
5. Finally, doses cannot be computed for some 5% or so of the members of the Life Span Study cohort, often individuals exposed in concrete buildings, and the new dosimetry does not clarify all of the implausibly high doses estimated with the T65 system. A number of individuals remain whose estimated whole-body shielded kerma doses exceed 4 Gy, some greater than 6 Gy. These are doses at or above the recently estimated $LD_{95/60}$ in these cities,[32] and, given the virtual obliteration of the immune system at doses to the bone marrow in excess of 7 or 8 Gy, survival under the conditions that existed in these cities is unlikely. Moreover, for some of these individuals the estimated dose is sharply discrepant with biological findings. These incongruities must be reconciled, since the inclusion of these individuals in analyses could obscure the "true" dose–response relationship.[33]

Diagnostic and Therapeutic Doses

Still another source of ionizing radiation should be incorporated into the overall exposure estimates for the atomic bomb survivors; this stems from medically indicated diagnostic or therapeutic irradiation. While this source is probably not important insofar as those survivors who received moderate-to-high doses of atomic radiation are concerned, it could be a significant source of confusion with regard to the risks at atomic bomb doses of a few hundredths of a gray, where medical exposures accumulated since 1945 might equal or even exceed the atomic bomb dose.

Reconstruction of the medical doses in the immediate postwar years is difficult, since many factors contributed to potentially higher exposures to medical X-rays in Japan than would have occurred at the same time in the United States or Europe. The sensitivity of the film, the intensifying screens used, the fluoroscopic screen employed, and the electrical power of most machines were lower in Japan than in the United States or Europe, which led to longer exposures and higher doses on average. Moreover, the cost of X-ray film was such that fluoroscopy, with its higher doses, was commonly used instead of photographs. Finally, with the introduction of obligatory national health insurance in December 1958 and the passage of the Atomic Bomb Sufferers Medical Treatment Law, diagnostic X-ray studies became more routine. Presently, however, the diagnostic dose, for the same radiographic procedure, is no higher in Japan than elsewhere, and may actually be somewhat less because of the smaller size, on average, of the Japanese, resulting in the need for less incident radiation to obtain a suitably exposed film.

The actual impact of these practices, past and current, on the medical exposures the survivors have received is unknown. To provide some assessment, however, a substantial effort, initiated by Walter Russell, has been under way since about 1960 in Hiroshima and Nagasaki, sponsored by ABCC and the Foundation working in collaboration with members of the University of Hiroshima's Research Institute for Nuclear Medicine and Biology (RINMB) and the Department of Experimental Radiology at Kyushu University. Estimating the doses incurred in the course of diagnostic examinations at ABCC or the Foundation has been straightforward, albeit not simple, since the details of the exposure and the estimated dose have been routinely recorded. But it has been an elusive task for exposures that occurred in the numerous small clinics or hospitals elsewhere. Often, especially in the early years, the X-ray machines were poorly calibrated, the records on the duration of exposure and kilovoltage employed skimpy, and these machines were used by a great many physicians who were not radiologists, and less aware of the need for good records and frequent calibration of the equipment.

Therapeutic exposures, on the other hand, have been simpler to document, since the number of medical facilities in these cities at which such exposures occur is small, and the records available substantially better. Moreover, tissue doses are customarily calculated and recorded for each treatment. Of particular interest have been those cases of cancer among the survivors who were treated radiotherapeutically, since a second cancer arising from the therapy could be mistakenly attributed to exposure to the atomic bomb. Collectively, some 1,577 participants in the Life Span Study are known to have received radiotherapy, and 49 of these individuals subsequently developed a secondary malignant neoplasm. Some of these were near the previously irradiated sites and could be causally related to the radiotherapy.[34] For example, instances have been recorded of the death of a survivor due to cancer of the lung arising in the lung beneath a breast treated for cancer. The breast cancer might be atomic-bomb-related, but the cancer of the lung, the cause of death, was more likely to have been secondary to the radiotherapy. Errors of this type could affect the estimates of the frequency of occurrence of cancers at a particular site. To avoid this possibility, recent analyses of cancer incidence have routinely excluded second cancers in the estimation of the risk associated with exposure to the atomic bombing.

Progress has been slow in the evaluation of the diagnostic doses, largely as a consequence of the complexity of the problem and the large number of individuals involved. To estimate the dose associated with the examination of a particular site, one needs to know not only the site involved but the angle of projection of the X-ray beam, the amperage and kilovoltage used, presence and amount of filtration employed, the beam size, source-to-film or source-to-image distance, and the duration of exposure. Given all of these technical particulars, the dose to the various principal organs of the body associated with a specific diagnostic procedure can be reasonably well estimated, either through the use of a human phantom in which thermoluminescent detectors are situated or through using mathematical descriptions of body organs and their elemental densities.[35] However, it is still necessary to know how many examinations have occurred and the sites involved if a cumulative dose to these organs is to be computed.

As noted earlier, exposures occurring in the course of the clinical examinations of the survivors at the Foundation are known, but these examinations are largely confined to the health surveillance sample and do not include other members of the Life Span Study population who are not in this sample. And the exposures of members of the health surveillance sample that occurred at other medical facilities are generally unknown. To obtain some insight into exposures elsewhere, a series of surveys of local hospitals and clinics has been conducted to determine what kind of examinations have occurred, the number of exposures involved in each examination, where they took place, and how frequently.[36] This information must be collated with the exposures at the Foundation if the total cumulative dose to the various organs of the body of an individual survivor from medical exposures is to be computed. Unlike the dose received from exposure to the atomic bomb, the cumulative dose from diagnostic and therapeutic exposures will continue to change over time, increasing more rapidly as the study population ages, and this fact argues for a continuing evaluation of medical exposures.

Despite these difficulties, presumably it will soon be possible to factor these medical exposures into the estimation of the total dose of ionizing radiation each survivor has received. Although the contribution medical exposure will have made is not anticipated to be large—on average only a hundredth of a gray or so over a relatively small area—at the lower atomic bomb doses (under 0.10 Gy) this could be a significant part of the total dose to a particular organ, as previously stated. It must be borne in mind too that there will undoubtedly be some underestimation of the medical dose since it is impossible to compile exposure histories that record every single exposure during an individual's lifetime. To the extent that diagnostic exposures are underestimated, risk will be overestimated, but this overestimation would not be great, given the probable doses involved. Nonetheless, there should be some further improvement in the estimates of radiation-related damage when diagnostic and therapeutic exposures can be added to the atomic-bomb-related doses the survivors in Hiroshima and Nagasaki received.

Finally, this effort to reconstruct medical exposures can have value beyond the immediate needs of the Foundation. It can contribute to the setting of standards and techniques useful in other studies of the health effects of diagnostic and radiotherapeutic irradiation where heretofore the doses have generally been poorly estimated. Commonly these studies have done little more than count the number of presumed exposures and converted these to dose using often arbitrary assumptions regarding the parameters of the exposure.

6

The Postnatally Exposed Survivors

Customarily, a *survivor* is one who experienced an event directly and lived through its immediate consequences. But in a situation as complex as the atomic bombings of Hiroshima and Nagasaki the "relevant event" is difficult to define. Should the expression "survivor" be restricted to just those persons who received doses of 0.01 Gy or more? If interest centers primarily on the immediate effects of exposure to ionizing radiation, this would be a sensible limitation. Or should it include individuals who may not have been exposed to these doses but nonetheless experienced the psychological trauma associated with these disasters? Certainly there were far-reaching economic, social, psychological, and behavioral consequences that cannot and should not be minimized. Many of these, however, are not unique to a nuclear detonation. They accompany all catastrophes—a severe earthquake, a major hurricane or typhoon, or a catastrophic flood.

Since the concerns here are with the effects of ionizing radiation, the term "survivor" has been used thus far, albeit without definition, to describe those persons who were physically within the corporate limits of Hiroshima or Nagasaki at the time of the bombing, irrespective of the dose received, and we will continue to do so. Initially, the Japanese government adopted a similarly restrictive definition when the Atomic Bomb Sufferers Medical Treatment Law was first enacted on March 31, 1957. To qualify for the benefits the legislation provided, one had to have been exposed within 3,000 m of the hypocenter, since beyond this distance there was essentially no exposure to ionizing radiation save possibly in the fallout zones. Over time, partly through political pressure and partly through humanitarian concerns, the expression "survivor" has taken on a wider meaning. It now includes not only those individuals present in the corporate limits of Hiroshima or Nagasaki when the bombing occurred, but also those who entered these limits immediately afterward and could have been exposed to some residual radiation. The current version of the Atomic Bomb Sufferers Medical Treatment Law, for example, accepts as an atomic bomb sufferer the following groups of individuals:

1. A person who at the time of the atomic bomb detonation was directly exposed within Hiroshima City or Nagasaki City or within a specified adjacent area of these cities or a person who at the time was in utero of a mother under the foregoing situation.
2. A person who entered within two weeks after the atomic bomb detonation a zone about two kilometers from the hypocenter of the atomic bomb or a person who at the time was in utero of a mother under the foregoing situation.
3. A person who engaged in relief of atomic bomb survivors or other activities to be physically affected by radiation or a person who at the time was in utero of a mother in the foregoing situation.

Under this law, an individual qualifying under any of these provisions is issued a health handbook by the appropriate community or prefectural authorities. Registration is believed to be relatively complete, since there are numerous economic and health care incentives to register if one qualifies. There is, for example, a health management allowance amounting to some 28,400 yen a month (this allowance, which is about $260 per month at the rate of exchange in the autumn of 1993, was paid to no less than 229,967 persons in March 1989).[1] As of March 31, 1989, in all of Japan, 356,488 handbooks had been issued, a number, it will be noted, that is much larger than the 284,000 survivors revealed by the 1950 census. This difference undoubtedly reflects the broader definition of a survivor, but must also include a number, possibly a substantial one, of individuals whose qualifications cannot be evaluated reliably under the current rules established by the Ministry of Health and Welfare.

But what questions trouble the survivors? Understandably, there are many, differing to some extent with the age of the individual at the time of exposure. Some are health-oriented; others are not. Three loom especially large. First, will they age more quickly as a result of their exposure; more poignantly, will they die prematurely for nonspecific reasons? Second, since cancer is a widely recognized consequence of exposure to ionizing radiation, will they succumb to this debilitating, often lingering, dehumanizing cause of death? And finally, what will be the consequence to their children and their children's children of their exposure?

Unequivocal answers to all of these concerns are not at hand, but there is much to allay, if not to quell their apprehension. What is known follows, but since the anticipated biological effects are dependent upon the age in life when exposure occurred, present knowledge is described in the context of those survivors who were exposed after birth, that is, postnatally, and those who were exposed before birth, or prenatally. However, first a digression to define terms, and to explain, briefly, how risk is estimated.

The Estimation of Risk

As commonly used, the word "risk" merely suggests the possible occurrence of an undesirable event as a result of the exposure to some potential danger. But as used here, it has a more specific meaning. It has been tacitly implied, if not explicitly stated, that risk is the probability of the occurrence of an undesirable consequence of exposure to ionizing radiation, since if differences in radiation-related health effects between individual groups are to be compared and understood, variation in risk with dose must be expressible quantitatively. It must be possible to describe the dose–response relationship (or dose–response function). This implies knowing what will occur at each and every dose, but it is impractical to observe what happens at every dose to which a human being might be subject. The assessment of risk, therefore, necessarily involves the interplay of observations at specific doses and a construct(s) of the events that presumably underlie these observations. This construct or model is used to predict what will happen at those doses where direct observations are not available or are inadequate to provide reasonable assurance that the observed value is the appropriate one.

Most of the biological effects of ionizing radiation that have been observed, either among the atomic bomb survivors or elsewhere, occur only at intermediate to high doses, that is, at exposures generally of 0.2 Gy or more accumulated over a short period of time.[2] However, public and regulatory concerns focus on exposures that are much less, generally under 0.10 Gy and often only 0.01 or 0.02 Gy or even lower. The estimate of what might occur at these doses is an interpolation from what is known to occur at background doses, that is those "naturally" occurring, and at higher ones where effects can be clearly demonstrated.

Many different interpolations can be made. It might be assumed, for example, that the effect of exposure is directly and simply proportional to dose; this is called the *linear dose–response model*. It asserts that if no events occur at 0 dose and five occur at 0.5 Gy, then one event will occur at 0.1, two at 0.2, and so on. Or, it could be assumed that the effect is proportional not to the dose but to the square of the dose; this is termed the *quadratic model*. It implies that if five events occurred at 0.50 Gy and none at 0, then 0.2 would occur at 0.10 Gy, 0.8 at 0.20 Gy, 1.8 at 0.30 Gy, and 3.2 at 0.40 Gy. The following simple table illustrates the linear and dose-squared effects at doses of 0.5 Gy or less:

Dose	Linear effect	Dose squared	Dose-squared effect
0	0	0	0
0.1	1	0.01	0.2
0.2	2	0.04	0.8
0.3	3	0.09	1.8
0.4	4	0.16	3.2
0.5	5	0.25	5.0

Still another common representation is the *linear–quadratic model*. It assumes that the effect is related not only to the dose received by an individual but to the square of the dose as well. This implies that at low doses the effect is essentially linear, since the square of a small dose will add little to the frequency of the effect being observed. As the dose increases, however, the contribution of the dose-squared term, the quadratic one, becomes more important. The point at which the contributions to the total effect of the linear and quadratic terms in the dose–response model are equal is called the *crossover value*. Where this value has been estimated, it has usually been found to lie between 1 and 2 Gy. However, the linear–quadratic model is applicable only if the doses are acutely received and embrace moderate to high values. If the dose is low or protracted, there is little or no opportunity for multiple events of a given kind to occur, since the probability per unit dose of most radiation-related events, e.g., point mutations or chromosome breaks, is relatively small, and it is the interaction of multiple, simultaneously occurring events that gives rise to the quadratic term in the linear–quadratic model.

Although these three simple models are the ones most frequently used, they are not exhaustive; others can be conceived. It could be assumed that a threshold exists—some minimal dose required to produce an effect beneath which nothing happens. If a threshold of 0.25 Gy were assumed, nothing would happen at 0.01, 0.02, or even 0.10 Gy. Above the threshold, effects might increase linearly or in some other manner. What is predicted, then, clearly depends upon the model used. Unfortunately neither observational results nor present biological theory indicates which of these models, or others that can be constructed, is correct. Justification for their use rests largely either on parsimony, that is, they are the simplest conceivable model and presumably, therefore, entail the fewest assumptions which may or may not be true (as in the case of the linear model), or they accord with some observational experimental evidence (as in the case of the linear–quadratic model). But in either event, they are not based on deep insight into the molecular and cellular events that subtend the observations. Customarily, the dose–response model that is used is the one that seems most plausible on current biologic grounds or best describes the observed data empirically. Best in this sense implies that model in which the differences (deviations) between the observed and the predicted values are minimal. This does not mean that it is the correct one. It is merely a convenient descriptor—a means to summarize information succinctly.

Risk can, however, be expressed in a variety of ways. It can be stated in relative terms, such as the ratio of the risk in one population (or exposure group) to that in another, or as the *excess* relative risk (i.e., the difference between the observed relative risk and 1, the value expected in the absence of an effect). As an illustration, suppose two equally sized groups of individuals, all of whom were of the same age and gender at the beginning of the investigation, were studied for five years. Suppose further that one group was exposed to no ionizing radiation, whereas every member of the other group had received a dose of 1 Gy. If one death occurred in the 5 years in each of the two groups, the ratio of the two death rates, the relative risk, would be 1/1 or 1, and the excess relative risk would be (1/1) − 1 or 0, implying the same frequency of death in the two groups, and no dose-related increase in mortality. If, however, five deaths had occurred in the exposed group to only one in the nonexposed group, the relative risk would be 5, and the excess relative risk would be 4. The relative risk (or the excess) need not be based on the entire population (or sample) under study. It can be and frequently is calculated as a function of specific attributes of the exposed individuals, such as their age at exposure or their gender.

Risk can also be measured in absolute terms, as the excess number of occurrences of an effect (for example, cancer deaths or cases) above those "normally expected" in the population of interest. "Normally expected" means in the absence of exposure to other unnatural causes. Thus, it excludes ionizing radiation from the bomb, but includes ionizing radiation emanating from the earth's crust, originating in outer space, or medically incurred. Excess occurrences (deaths or cases) are commonly stated in terms of the number of years the individual (or group of individuals) has lived following exposure to some dose (or the initiation of the study of which one is a part). In radiation studies, it is usually given in units known as person-year-gray, abbreviated PYGy, or person-year-sievert, abbreviated PYSv, if the radiation is mixed. Thus, an individual who lived for 10 years following exposure to 1 Gy would contribute 10 PYGy of experience (years of life lived multiplied by dose), or one who survived 5 years after exposure to 0.5 Gy would contribute 2.5 PYGy. Expression of mortality experience in this manner takes into account the fact that the many different possible causes of death "compete" with one another, since an individual can only die once, whatever the cause. A person who died in an automobile accident 5 years after exposure, for instance, could only contribute 5 years of experience to the estimation of the radiation-related risk. Finally, risk can be defined as the percentage of occurrences of an effect ostensibly assignable to radiation. This is ordinarily termed the attributable risk. Among the atomic bomb survivors, the attributable risk is usually computed by relating the excess cases or deaths among those individuals exposed to 0.01 Gy or more to the total number of cases or deaths seen in this group of individuals.

Each of these estimates of risk has its merits and shortcomings, and it is doubtful whether any single measure captures all of the desired information, especially when risk may depend upon several factors aside from dose, such as age, gender, and time after exposure. Relative risk, since it is proportional to the spontaneous occurrence of the effect of interest, cancer for instance, has often been thought to be more broadly applicable to other populations, and hence to regulatory concern. However, in the absence of knowledge about the spontaneous or baseline rate, it has limited utility, since with it alone the number of events that might occur cannot be predicted. Moreover, as breast cancer illustrates, simple relative risk models can describe the situation within different populations satisfactorily, even though the relative risk can differ substantially with different population backgrounds.

Risk expressed as excess occurrences of a health effect (deaths or cases) has an immediacy readily grasped, but it is dependent upon the spontaneous rate of such events, and the accuracy and completeness of their recognition. The rate of spontaneous cancers and the accuracy of diagnosis differ from country to country, and within a country from region to region, and even from city to city. Consequently, two equally sensitive populations exposed to the same amount of ionizing radiation could vary in the apparent harm that results although, in fact, the number of radiation-related cancers could be the same. However, again, the experience with breast cancer is enlightening; here the excess absolute risk is unexpectedly similar across populations.

Different measures of risk, because of the way they are defined and estimated, respond dissimilarly to differences in the rates at which various health effects occur. When all of the individuals within a study population have not been followed to death, relative risk, since it emphasizes the occurrence of cases, tends to give greater importance to the older members of the population, who have contributed the larger number of cases. Excess deaths on the other hand, emphasizes the number of years at risk of a detrimental health effect following some event, such as exposure to atomic radiation, and since the young will live longer, accumulating more years of risk, this measure gives greater importance to the younger members of the population. Thus these two quantities could be different between two populations merely because of their age compositions and not because of an inherent difference in sensitivity to radiation.

This variance has led some investigators to advocate the use of a standardized mortality or morbidity rate. The latter presumes some standard or reference population (for example, that of Japan or the United States), and computes the expected number of deaths or cases in the population of interest—the atomic bomb survivors—using the age-specific annual observed rates in the referent population and the observed age distribution in the population of interest: the survivors.[3] However, age-specific rates, in the absence of exposure, are imperfectly known in small populations such as a single city, and the use of national values might be misleading because of regional differences in rates. Moreover, standardization, although a convenient means for summarizing mortality, does not always ensure that the effects of extraneous variables are fully removed. For example, standardization will not identify differences in specific rates if these vary in different ways across the groups of interest, as would be true if males had higher mortality than females in some age groups but lower ones in others. One of the great strengths of the Life Span Study is that it contains its own internal comparison group: the survivors who were exposed to low or very low doses and have shared those other vicissitudes of life the other survivors have experienced. Use of an internal comparison also ensures that the data are comparable, since the methods of collecting the data were the same.

As the foregoing remarks indicate, there is no universally accepted "best" way to express risk; the most appropriate one will depend upon the nature of the inquiry itself, the data that are available, and the purpose for which the estimate is to be used. But whatever method is adopted, it is important to express risk in terms that address the concerns not only of regulatory or professional persons but of the general community as well. Risks must be stated in language that the public can comprehend, or misapprehensions and distrust will proliferate to the detriment of understanding and acceptance.

Premature Death

One of the persistent concerns of the survivors has been the likelihood of dying prematurely as a result of their exposure to atomic radiation, or alternatively and less accurately stated, "aging" more rapidly than would have occurred had they not been in Hiroshima or Nagasaki on those fateful August days. As generally perceived, this accelerated aging is not specific in nature. It is not due to some well-recognized effect of ionizing radiation associated with a particular cause of death. The accelerated "aging" feared is more nebulous, and therefore more difficult to define.

There are many signs attributable to aging—the graying and loss of hair, coarsening of features, skin that no longer fits, eyes that accommodate less well, ears that no longer perceive the higher tones, a failing sense of smell, a memory more faithful to the past than the present, a brain that processes visual and auditory stimuli differently and less well, the occurrence of pigmented spots in the skin, etc. These signs singly or in unison do not adequately measure the process of aging; ultimately, premature aging is still recognized through premature death. The latter is not a disease that one can diagnose, however; it merely implies that death occurred prior to the average number of years of life to be expected at the moment of birth, or at a subsequent age in life, such as at the time of exposure to ionizing radiation. This number will vary, of course, from country to country, and within a given country over time. Indeed, the present century has seen a dramatic increase in average life expectancy in all industrialized countries, and many newly industrializing ones as well. The Japanese, as a people, have shared in this increased longevity, currently having the longest life expectancy in the world.

The mortality surveillance central to the Life Span Study provides the age at death of survivors, and is a means, therefore, to address the question of whether the survivors die at an earlier age, on average, than their age- and gender-matched peers who were not exposed, or received doses of less than 0.01 Gy. Until recently, studies of the life experience of the survivors have not revealed consistent evidence of life-shortening aside from that ascribable to radiation-related malignancies. Cumulative mortality from causes of death other than cancer has not heretofore appeared to increase systematically with radiation dose in either city, in either gender, or in any of six different age at the time of the bombing groups (i.e., individuals who were less than 10 years old at the time, 10–19, 20–29, 30–39, 40–49, and 50 years and over).[4] However, as early as 1970 there was some indication that deaths were occurring more frequently than expected at the higher doses, particularly deaths attributed to diseases of the circulatory system. Now the data are more extensive, especially among younger survivors exposed to doses of 1.5 Gy or more, and suggest that deaths attributable to causes other than cancer might be greater than expected.[5] This increase can be illustrated in a variety of ways, but one simple method involves plotting the logarithm of the death rate at specific ages in relationship to dose.

As Figure 16 illustrates, at all ages subsequent to 30 or so, the death rate from causes of death not related to cancer is higher among survivors heavily exposed before the age of 40 than it is among survivors exposed

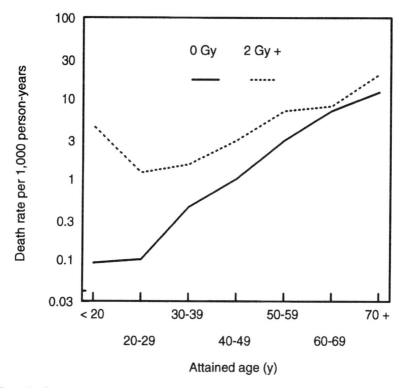

Fig. 16. Gompertz curves for cancer (above) and noncancer mortality (facing page) among survivors in the Life Span Study sample exposed to 0 Gy and 2 Gy by the age of the survivor at the time of death. (After Shimizu et al., 1992)

to doses under 0.01 Gy who were also less than 40 at the time of the bombing. Note that a similarly striking increase is not seen among the older survivors (those more than 40 years old when exposed). Interpretation of the evidence exhibited in this graph is not straightforward, and the data might be misleading. Of special concern is the possibility that the increase is due to cancer, either directly, but not recorded on the death certificate, or indirectly in a preclinical form of neoplastic disease or as a result of successful cancer therapy.

It is often alleged that physicians in Japan, fearful that a patient will lose the will to live, are reluctant to inform them that they have a fatal disease. Cancer might, therefore, be underreported. Whether this is true and the practice is widespread, or merely a matter of medical folklore is not clear. Nor does it follow from this allegation that a physician will consciously falsify a death certificate even if he or she is reluctant to inform a patient of a fatal illness. And if some falsification occurred, it is

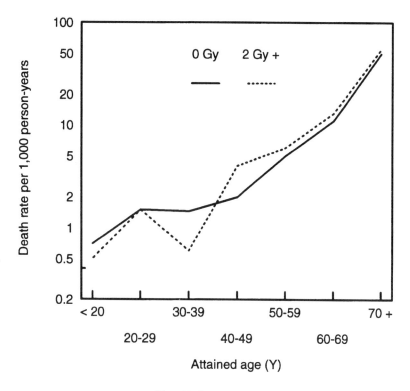

Fig. 16 (continued)

doubtful whether this would have a significant, systematic bearing on the estimated risks, since there is little or no correlation between the survivors' and physicians' "perceived" doses and the actual DS86 estimates. Nonetheless, some deaths due to cancer are likely to be attributed to another cause, either purposefully or through error, but adjusting for possible misdiagnosis, using the accumulated autopsy data pragmatically, or through statistical considerations, does not account for all of the effects that are seen. This suggests that there might be causes of death other than cancer that are radiation-related. Of particular interest is the seeming increase in deaths due to "heart disease," loosely defined as deaths ascribed to diseases of the circulatory system, except stroke, specifically deaths from rheumatic heart disease, hypertensive heart disease, coronary heart disease, and cardiac insufficiency.

Coronary heart disease and strokes are caused by atherosclerosis, a disorder that develops slowly over many years and involves the medium and large arteries of the body. Within the latter, there occurs an infiltration of fat into the smooth muscle cells of the arterial walls, giving rise to discrete plaques, known as atheromas. It has been generally thought that these plaques result from the repair of repeated injury to the arterial wall. A less common view holds that atheromas are actually benign tumors derived from changes in a single cell. The evidence supporting this belief rests on the characteristics of the cells within the plaque. All appear to have the same genetic properties, suggesting they are clonal in origin—that is, they all arose from a single cell—whereas this is not true of normal arterial tissue, where the cells are phenotypically dissimilar. This finding has led to speculation that mutations arising in the genetic material in smooth muscle cells could play a role in plaque formation. Heretofore there has been little direct evidence to support the occurrence of these presumptive mutations. However, experimental studies have recently been published that suggest the changes in the arterial walls associated with atherosclerosis might have a cellular basis not unlike that seen in cancer, including the occurrence of somatic mutation.[6,7]

These studies have shown that DNA extracted from plaques in the coronary arteries of humans (and cockerels) when introduced into the cells of mice can produce tumors. These tumors are histologically indistinguishable from those produced in genetically similar rodents by DNA from human bladder cancer. But unlike the aggressive behavior of most malignant solid tumors, those produced by plaque DNA grow more slowly and result in fewer tumors in the recipient animals. DNA extracted from the normal arterial wall of a human (or a cockerel) does not produce these results. This prompts the belief that a mutation has occurred leading to the activation of one or more of a class of genes (to be described more fully later), known as proto-oncogenes. Many of these latter genes have been recognized, but it has not been established which one, if any, of those presently known may be involved in plaque formation. It has been suggested, however, that platelet-derived growth factor is a worthy candidate.[7]

Since the risk of cancer is increased with increasing radiation dose, these facts tempt one to believe that the increase in heart disease might also reflect exposure. If this is so, since the prevailing belief assumes that there is no threshold in the induction of cancer, can it be assumed that there will be no threshold in the induction of radiation-related atheromatous plaques and, ultimately, coronary artery disease? As already mentioned, the cardiovascular effect is seen only above 1.5 Gy, but this suggestion of a threshold could be misleading. The effect has only recently appeared, and the number of deaths at the various doses are still too few to describe with certainty their relationship to dose. At present, although significantly elevated, the attributable and relative risks are much lower than those seen for cancer, and the number of excess deaths is also much lower in spite of the high background rates.

Only further surveillance will indicate whether or not the effect occurs at lower doses, and other studies of a molecular biological nature will be needed to establish whether these speculations are valid, and whether somatic mutation plays a significant role in the sequence of events giving rise to an atheromatous plaque and ultimately to cardiovascular disease.

In retrospect, these findings make more interesting an earlier study of the incidence of stroke and coronary heart disease in the years 1958 through 1974 among the survivors participating in the Adult Health Study. This investigation reported the incidence of these two circulatory diseases to be significantly higher than anticipated among women in Hiroshima who were heavily exposed—who received a T65 dose of 2 Gy or more. Since an exposure effect was not seen among women in Nagasaki or men in either city, and there was evidence of a higher autopsy rate among heavily exposed women in Hiroshima, which could have led to a higher rate of recognition of stroke or heart disease, there was a reluctance to accept this seeming association as real. Nonetheless, the effect could not be explained by an inadvertent confounding of such known risk factors as smoking, an elevated level of serum cholesterol, or the occurrence of hypertension (high blood pressure), that contribute to the occurrence of cardiovascular disease.[8]

This earlier study has been extended to include the years from 1974 through 1985 using the newer doses. The results confirm those found earlier among heavily exposed women in Hiroshima, but now there is also a statistically significant increase in the incidence of "heart disease" among heavily exposed men in Nagasaki. Within the other two gender–city groups (Hiroshima men and Nagasaki women) the association of exposure with risk of stroke or coronary artery disease remains equivocal and is not statistically significant.[9] While this recent study makes more plausible an association, it does not remove all of the uncertainties. There has been a statistically demonstrable, temporal lowering in the frequency of cerebrovascular disease in postwar Japan, which has commonly been attributed to dietary changes. This has precipitated a concern that the same westernization of the diet (which has presumably contributed to the diminution in cerebrovascular disease) might increase the frequency of cardiovascular disease, although there is little direct evidence to support this apprehension. As judged by participants in the Adult Health Study, serum cholesterol levels have been steadily rising with time in Japan (some 25 mg-percent, on average, in the past 30 years), and elevated levels of serum cholesterol have been associated with a higher frequency of cardiovascular disease, but there has been no consistent upward trend in the occurrence of myocardial infarction. However, the possibility of a temporal trend makes it more difficult to demonstrate a true radiation effect, since the trend itself, if one exists, is so poorly understood. There are other observations that suggest some biological rather than chance basis for these findings. First, X-ray examinations of the Adult Health Study participants have revealed the frequency of calcification of the aortic arch and the abdominal aorta to increase with dose, and second, ophthalmic studies have shown retinal arteriosclerosis to also increase in frequency with dose. As yet, however, it has not been possible to integrate these findings into a coherent biological explanation of the apparent increase with dose of deaths ascribable to causes other than cancer, and it may be some time before this is possible.

Cancer

Cancer is an illness that strikes fear into the most stout-hearted. The name itself is inextricably linked with intractable disease and a slow, wasting, often dehumanizing, death. Substantial strides have been made in the treatment of cancer, and survival following diagnosis is increasing, but the older image prevails, and not without cause. A tumor is said to be cancerous, or malignant, if its growth is potentially unlimited, if it expands locally by the invasion of other tissues, if it is associated with the formation of secondary tumors (is metastatic), and if its natural course is, or has usually been, fatal. Cancerous changes can occur in virtually any organ or tissue where dividing cells occur. They may, for instance, involve the cells of the bone marrow, the breast, or the large intestine.

Cancers arise in all populations but with different frequency, commonly appearing without recognizable cause and more often in the elderly than the young or middle-aged. One's gender can also influence their occurrence as can occupation, lifestyle, and the genes inherited. Among the survivors of the atomic bombings, the relative risk of cancer death for those exposed in the first decade of life appears appreciably higher than the risk for those exposed at later ages. Sufficient data are not now available to determine, without qualification, whether this increased risk obtains with respect to all radiation-related malignant tumors or only selected ones.

The Biological Bases of Cancer

Our defenses against disease—bacterial, viral, and even neoplastic—are many, and complexly integrated, varying from the simple physical barriers to inhaled and ingested noxious agents, including radionuclides, afforded by the linings of the lung and intestine, to a bewildering array of biochemical and immunological responses. Defenses against potentially noxious chemicals are largely metabolic, including the detoxification and excretion of harmful compounds or the repair of damage to DNA from these compounds. Immunological responses range from the development of circulating molecules, known as antibodies, capable of binding specifically to foreign molecules, to the capacity of the various white cells to interact with each other in combating disease. Collectively, the cells and cellular processes involved in these responses are known as the immune system. An individual whose immunological responses are "normal" is said to be immune competent or immunocompetent.

Clues exist suggesting that ionizing radiation impairs immune competence, but no coherent picture has emerged. For example, it is known that under experimental conditions white cells grown in culture and exposed to ionizing radiation, when chemically stimulated will not migrate as far, on average, as similarly grown and stimulated unexposed cells.[10] White cells are normally migratory, and movement is intimately related to their function; under these circumstances, impairment of mobility would seem disadvantageous. However, it is not known whether this same inhibition of movement occurs in the body, nor what happens to the cell to impair its migratory ability, nor what a diminution in mobility of 10 or 20%, or even more, would actually mean in the frequency of disease.

As previously indicated, white cells can be classified by their morphological appearance, but some, the lymphocytes, can also be classified by the tissue in which they differentiate. Somewhat simply put, the cells of our bodies are bathed with a clear, transparent fluid, known as lymph. This fluid is derived from arterial capillary blood, fills the intercellular spaces, exits through a system of lymphatic vessels and circumscribed masses of tissue—the lymph nodes—and eventually returns to the venous blood circulation. Collectively, these vessels and nodes make up the lymphatic system and the tissues involved are called lymphoid. Lymph itself serves as a means of transporting white cells and to circulate or recycle water, proteins, fats, and electrolytes. The undifferentiated precursors of the lymphocytes, known as stem cells, are found in the bone marrow. Early in life many lymphocytes differentiate and mature in the thymus and are called T-cells, whereas those that differentiate and develop in the bone marrow are known as B-cells.

The thymus itself is apparently essential to the establishment of a normal immune system largely through its role in the differentiation and development of T-cells. B- and T-cells have different functions, and can be recognized by their different cellular properties. B-cells, for instance, react with molecules expressed on the outer surfaces of cells (these molecules are referred to as antigens) and are the forerunners of antibody secreting cells; whereas T-cells do not produce antibodies, but are involved in important regulatory functions in helping or suppressing the activity of both B-cells and other T-cells. On the surface of both types of cells are specific receptor molecules involved in the recognition and binding of particular molecules, or in the production of a truly staggering variety of immune responses.

The deposition of radiant energy in a cell can alter the properties of the cell's surface membrane and this alteration could affect the capacity of the lymphocyte to respond to challenges. Although radiation-related leukemias are more commonly B-cell than T-cell in origin, it has not been established that different human lymphocytes have different sensitivities to radiation. In the mouse, however, the functions of B- and T-cells show large differences in radiosensitivity (B-cells are much more sensitive). The deposition of energy within a cell can also induce chromosomal abnormalities and gene mutations, and if these changes involve the genes that are responsible for the immunoglobulins (the proteins with antibody activity), or the receptor molecules on the T-cell, they might also affect the immune response. However this would occur only among those stem cells that can continue to reproduce, that is, can expand clonally.

An incompetent or compromised immune system could, in theory, increase the probability of death not merely from cancer, assuming there is an immune surveillance system for all tumors (which does not appear to be the case), but from those diseases whose suppression rests on immunological responses. Whether exposure to the atomic bombing of Hiroshima and Nagasaki has or has not impaired immune competence in the long term is debatable. The experience of the survivors provides no persuasive evidence that an increase in diseases where immunological competence is important has occurred. However, this assertion rests on the failure to demonstrate an increased probability of death from infectious disease from a year or so after the bombing until the present. If a compromised immune system played a substantial role only in the mortality occurring in the first few weeks or months following the bombing, as is possible, and if a return to immune competence occurred relatively rapidly, the data would be less compelling, since too little is known about the immediate cause of most early deaths to make an informed judgment. From animal studies and the experience with humans subjected to irradiation under other circumstances, mortality due to infectious disease would be expected to occur just when the data in Hiroshima and Nagasaki are weakest: the first few weeks or months after exposure. These are speculations, since the precise nature of the effect of radiation is unknown, or at best very dimly perceived, even for the leukemias, about which possibly the most is known.

It has been recognized for some time that through the human genetic material are distributed specific genes, commonly described as *proto-*

oncogenes (a term derived from the Greek words *protos*, the first or the beginning, and *onkos*, a mass or a tumor), some of which seem to be of viral origin, but many of which have never been found in viruses. Dozens of these genes have been identified. Although individually different, they have in common the fact that they have been genetically conserved, that is, they have changed very little over time and some are shared by species as diverse as man and the fruit fly. It is generally believed that in normal cells they play crucial roles in the control of cell proliferation and differentiation. They are usually regulated or controlled, but it has been observed that in some cancers they become unregulated or "activated." These unregulated or activated proto-oncogenes are termed *oncogenes*. Frequently, particularly in the leukemias, where the phenomenon has been especially well studied, this activation occurs when a rearrangement of the chromosomal material takes place, often a rearrangement that also involves one of the genes responsible for the immunoglobulins or the T-cell receptors, or both. Something within this process leads either to the faulty regulation of the proto-oncogene or a failure to regulate it at all, and the cell begins to reproduce wildly. A variety of different mechanisms could lead to this result. In Burkitt's lymphoma, for example, it is known that a series of specific chromosomal rearrangements place a proto-oncogene under the control of the immunoglobulin promoter, and this apparently turns on the oncogene. Nonetheless, it is widely believed that mutation in a proto-oncogene leads either to a different structure in the protein associated with the gene, or a modification of the mechanisms that regulate the expression of such genes. It might, however, lead also to the production of a normal gene product, but in an abnormal place, or the production of improper amounts of such a product.

The activation of a proto-oncogene does not appear to be enough to produce cancer. Changes must also occur in another class or family of genes known as *tumor suppressors*.[11] One such gene, designated p53, is the most commonly altered gene yet identified in human tumors; mutations at this genetic locus are present in a large fraction, perhaps half, of the total cancers in the United States and Britain.[12] These genes appear to work normally to suppress or inhibit the multiplication of a cell; they are the brakes on the engine of cellular reproduction. Inactivation of these genes through mutation seemingly releases the normal constraints placed on a cell's capacity to divide, and, when accompanied by the activation of a proto-oncogene results in increased cell multiplication, resistance to the final differentiation of cells, and other abnormalities characteristic of a tumor cell. A dynamic equilibrium appears to exist between these two families of genes—one stimulating and the other suppressing growth. It is unlikely, therefore, that a single mutation would be sufficient to result in cancer; there must be an accumulation of more than one mutation, either arising in the presumptive host through exposure to mutagenic agents, or transmitted from the parent(s), or some combination of the two. Among the oncogenes studied to date the mutation has invariably occurred in a somatic cell and is not transmitted from parent to offspring. This has not been true for mutations involving tumor suppressors, such as p53, since these have been seen to be transmissible.[13]

Some DNA lesions are reparable, some apparently are not. The existence of mechanisms in human beings to repair abnormalities in DNA was first demonstrated in connection with a rare inherited disease, xeroderma pigmentosum. Individuals with two copies of the gene associated with this disease, that is, who have inherited a copy of this gene from both their mother and father, are exceptionally sensitive to ultraviolet radiation and beginning in childhood develop multiple cancers of the skin. This hypersensitivity is due to a defective gene product that is unable to remove the DNA damaged by ultraviolet light. If the ultraviolet damage is not repaired before the next cell division, the result is either a cell that will die or have further mutations in its DNA. This type of damage also occurs periodically in healthy individuals, but among the latter there exists a mechanism for identifying and replacing the abnormal pairs with proper ones, using the sister strand of DNA as a model. This repair does not occur in individuals with xeroderma.

Stimulated by this finding, it was soon shown that other repair mechanisms exist, and some of these can correct the mistakes introduced into DNA through exposure to ionizing radiation. Thus, the exposure of a cell to ionizing radiation does not inevitably lead to a persisting mutation. While reassuring, this fact has made more complicated the prediction of what will ensue following exposure, since it is not presently known which DNA damage is repaired or under what circumstances. However, it is generally believed, based on experimental studies, that normal repair is more likely to occur at low doses and low dose rates rather than at high ones where it is presumed that the repair process is overwhelmed or extensive cell killing occurs. If true, this implies that the risk of cancer in an occupational setting, where the dose is normally accumulated in small increments over time, is less than occurs when the same dose is accumulated rapidly, as in the case of an accident. Whether this latter supposition is correct or not (animal studies suggest that it is), it can be said that without these repair processes, if the damaged cells were to survive, cancer would undoubtedly be more common in all human populations than in fact it is.

Although it is now generally believed that gene mutation is important, indeed necessary, in the origin of cancer, other events must also take place if a malignancy is to arise. Currently it is presumed that several separable processes or stages are involved. Four are commonly recognized. First, a normal cell must undergo a change that alters its basic nature, so that it no longer undergoes definitive or terminal differentiation. This process is called transformation or initiation. Second, the transformed cell (or cells) must be stimulated to multiply if a tumor is to appear. This is termed promotion. Third, one or more of these premalignant cells must be converted to one that is malignant. This latter step is known as conversion. Fourth, and finally, among nonleukemic malignancies the converted cell must progress to a recognizable tumor—it must accumulate those other attributes associated with a malignant cell.

Agents that produce transformation are said to be initiators, and those that stimulate multiplication of the premalignant cell are called promoters. It has been commonly thought that ionizing radiation is an initiator and that initiation itself involves a transmissible alteration in the genetic material of the cell. But not all investigators hold to this belief. Some argue that ionizing radiation, by killing cells, impinges, presumably chemically, upon the communication between other cells involved in the suppression of the initiated cell's growth, and should, therefore, be seen as a promoter rather than an initiator (or, alternatively, as a promoter as well as an initiator). If ionizing radiation is actually a promoter, then it is conceivable that a threshold dose might exist, and current notions of what transpires at low or very low doses would have to be revised. Ionizing radiation could, however, also function at the conversion stage by inactivating or deleting certain genes in premalignant initiated cells that are responsible for suppressing cell growth.

There are still other observations that appear to bear on the occurrence of cancer. For example, as stated earlier, specific chromosomal abnormalities are associated with specific leukemias, often involving the T-cells; but whether these chromosomal changes are causal or secondary to the transformation of a normal or even a premalignant cell into a malignant one is not clear. Evidence has accumulated too that various cancer-producing agents, including radiation, tend to break human chromosomes at particular locations rather than randomly. The role these "fragile sites" play in the origin of cancer is not well understood, although some correspond closely to known cancer-related chromosomal break sites. Finally, there is the matter of how cells in a multicellular organism communicate with one another. Each cell within the body is not an independent entity; its activities are integrated with those of other cells in the tissue of which it is a part as well as cells in other tissues. To achieve this integration, cells must be able to influence each other's activities. This is accomplished through a variety of processes collectively known as intercellular communication. Adjacent cells, for example, are known to be able to transport small molecules back and forth through their cell membranes by means of a specific transmembranal channel, known as the gap junction. The capacity to do this, however, is influenced or modulated by a number of factors, such as the calcium levels within the cells. Regulation of intercellular communication through gap junctions has been postulated to play a role in the occurrence of cancer by either increasing or decreasing the exchange of molecules that occurs.[14] The details of this regulatory process are obscure, but a number of chemical and physical promoters have been implicated or postulated.

As yet, it has not been possible with human tumors, malignant or benign, to identify those that are ascribable to radiation and those that are not. When, and if, this occurs, it would become practical to determine whether in a particular individual a cancer arose as a consequence of exposure to ionizing radiation or from some other concatenation of events. This would have profound implications for regulatory purposes, litigation, and more importantly, the study of cancer itself. Today, only probabilistic statements can be made, and while these statements can be improved, for several reasons there is a limit to the added precision that is useful. A risk one individual might see as intolerable can be trivial to another; moreover, statements of risk are based on assumptions that may not be true, and whether they are or are not, they apply to an "average" individual with certain age and gender characteristics, and exposure. However, all individuals with these attributes might not be equally sensitive to radiation-related damage; indeed, it is known that some are more sensitive than others, as, for instance, in the case of xeroderma pigmentosum, to which we have already referred. An average risk would not, therefore, adequately represent the hazard that either the less sensitive or the more sensitive individual faces. It would be too high in the first instance, and too low in the second. Individual specific risks would be more useful but to assess these much more needs to be known. A given individual's risk will depend on the ability to metabolize carcinogens and repair DNA damage, the rate of somatic mutation, the inheritance of mutationally altered genes involved in the suppression of tumor formation, and the multiplication rate of the mutated cells. These are measurable phenomena, but substantial variability exists among individuals in their developmental state and gender. Individual risks cannot be generally estimated yet, but the future promises much in this respect.

Leukemia

As previously noted, in the several years that followed the initial observation of an increased frequency of leukemia, much effort was devoted to the study of the early hematological and preclinical phases of this disease and the effects of age at the time of bombing and calendar time on the incidence and types of cases that were occurring.[15] These studies were hampered by the failure to define a manageable cohort for scrutiny, the absence of a systematic method of case detection, and the inability to assign estimates of dose to individual survivors. The occurrence of leukemia could only be related to the distance of the exposed individuals from the hypocenter, and the presence or absence of symptoms associated with acute radiation sickness. Although these surrogates were helpful in confirming and extending the original observations, they were an inadequate basis on which to derive dose–response relationships. It was difficult, therefore, to support or refute the notion that an increase in the frequency of leukemia was associated with any exposure, however small. It was argued that there was no threshold in leukemia induction. Advocates of this belief acknowledged that an increase might be difficult to demonstrate at very low doses because of the large sample sizes that would be needed and the comparatively greater importance of other factors that could contribute to the occurrence of leukemia, such as the exposure to leukemogenic chemicals in the workplace.

With the establishment of the Unified Study Program, the development of a leukemia registry, and the evolution of means to assign individual doses of gamma and neutron radiation, understanding of the leukemogenic effect of atomic-bomb exposure improved materially. Today, the incidence of leukemia is known to be related to dose—the higher the dose the greater the risk—but this increase is not simply proportional to the dose an individual receives. Risk rises slowly to about 0.5 Gy, and then accelerates; it follows the linear–quadratic model to which we have previously referred, as can be seen in Figure 17.

The frequency of new cases of leukemia among the survivors reached a peak about 1952 and has declined steadily since. It had not, however, completely disappeared as recently as 1985, suggesting a period of risk following exposure of at least 40 rather than the 25 years that has been generally accepted. Moreover, when incidence by dose was examined in relation to age at the time of the bombing and the calendar time of disease onset, it seems that the higher the dose, the greater was the radiation effect in the early period, before October 1955, and the more rapid the decline in risk in subsequent years. The leukemogenic effect occurred later among individuals who were older at the time of the bombing.[16]

As the number of cases of leukemia continued to grow, it was possible to confirm and extend the earlier findings on the relationship of the different types of this disorder to dose. The radiation-related risk of acute lymphocytic leukemia as well as "the other types" of acute leukemia, such as acute myelogenous leukemia, seemed higher among survivors exposed at younger ages, whereas the frequency of chronic granulocytic

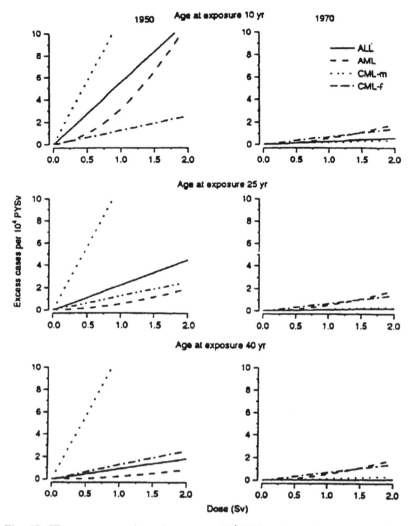

Fig. 17. The excess number of cases per 10^4 PYSv of acute lymphocytic (ALL), acute myelocytic (AML), and chronic myelocytic leukemia (CML) by age of the survivor at the time of exposure. (Adapted from Preston et al., 1993)

leukemia was greater among individuals who were middle-aged or older when exposed. However, it was not clear whether the different types had different dose–response relationships, and as earlier investigators noted, at least one form of leukemia, chronic lymphocytic, did not appear to be radiation-related in the Japanese. However, few cases have actually been seen and no truly reliable statement of the risk of this form of leukemia among the survivors is possible, although it should be noted parenthetically that other groups of individuals exposed to above-background levels of ionizing radiation, such as radiation workers, have also not shown a radiation-related increase in chronic lymphocytic leukemia.

Customarily the risks of malignancy among the survivors are stated in terms of exposure to 1 Gy (or 1 Sv in the case of mixed irradiation); this is largely a matter of convenience and the unit could be different. However, if the risk increases in direct proportion to the dose, that is, linearly, as appears true for all cancers except leukemia as a group, conversion to any other dose is simple. If one was exposed to a 0.10 Gy, for instance, the risk would merely be one tenth that at 1 Gy. Based on the older system of classification of leukemia but the new dosimetry and doses to the bone marrow, the excess relative risk of dying from leukemia is 5.21 per gray, the excess deaths are 2.95 per 10,000 person-year-gray, and the attributable risk is 58.5% when all ages are combined.[17] The meaning of these successive quantities is as follows: An excess relative risk of 5.21 implies that whatever the spontaneously occurring risk (the background), exposure to 1 Gy will increase that risk 5.2 times. Thus, in a population with a "natural" rate of leukemia deaths of 7 per 100,000 persons per year, that number of deaths would increase to 36 if every member of the population were exposed to 1 Gy. Since the average survivor of the bombings received a dose of approximately 0.25 Gy, on a linear scale the risk to the average survivor would not be 7 but 9 (7 × 0.25 × 5.21). A risk of 2.95 excess deaths per 10,000 person-year-gray means that in a population with the same spontaneous age- and gender-specific cancer rates as these cities, and the same age and gender distribution, upon exposure to 1 Gy there would be 2.95 additional leukemia deaths each year among every 10,000 individuals (or 2.95 deaths among every 20,000 exposed to 0.5 Gy). Again, among average survivors, the excess deaths would be one-fourth of this amount, or somewhat less than one extra death per 10,000 survivors per year. Finally, an attributable risk of about 59% indicates that among the 141 deaths from leukemia that occurred in these cities in the years from 1950 to 1985 among the survivors in the Life Span Study sample exposed to 0.01 Gy or more, 83 (or 59%) would not have occurred in the absence of exposure. It should be noted that for illustrative simplicity these assertions assume that the risk of leukemia increases linearly with dose, which is not strictly true, and they ignore temporal variation in risk as well as the fact the actual number will vary with the follow-up period. Thus the values given for doses of less than 1 Gy overestimate to some degree the actual risk.

Recently, it has been possible to reclassify most, although not all, of the cases of leukemia occurring among members of the Life Span Study using the French–American–British system and to reanalyze the accumulated information.[18] This reanalysis has clarified some previously puzzling aspects of the data, but has also raised some new questions regarding radiation-related leukemogenesis. For example, it has been recognized for some time that cases of chronic lymphocytic leukemia have occurred only in Nagasaki. This was puzzling. Reclassification, however, reveals that most of these cases are, in fact, instances of adult T-cell leukemia and it has been demonstrated that infection with the HTLV-1 virus associated with this form of leukemia is common in areas of Japan's westernmost major island, Kyushu, including Nagasaki, but is relatively rare in the western part of Honshu island where Hiroshima is located.

This newer analysis and an even more recent one, [18] based on 231 cases of leukemia occurring between 1950 and 1987 among survivors receiving doses of 4 Gy or less, suggest that the effects of irradiation differ depending upon the type of leukemia involved. Preston and his colleagues find the effect of exposure, as measured by the excess absolute risk, to be somewhat more pronounced on the occurrence of acute myelogenous leukemia (1.1 cases per 10^4 PYSv) and chronic myelogenous leukemia (0.9 cases) than upon acute lymphocytic leukemia (0.6 cases); whereas the excess relative risk is greater for acute lymphocytic (9.1) and chronic myelogenous leukemia (6.2) than for acute myelogenous leukemia (3.3). This difference in the absolute and relative risks is not unexpected, since acute lymphocytic leukemia is less common than the other two subtypes and a smaller absolute risk can give rise to a higher relative risk under these circumstances. Moreover, they find that only acute myelogenous leukemia exhibits a distinct nonlinear dose–response function; there was no evidence of nonlinearity for the other subtypes. If this apparent difference in the dose–response function is real, it would have interesting implications for radiation-related leukemogenesis, but it must be borne in mind that the number of cases of acute lymphocytic leukemia and chronic myelogenous leukemia is small relative to the number of cases of acute myelogenous leukemia and the capacity to discriminate among different dose–response functions correspondingly poorer. These analyses further confirm earlier observations that survivors exposed before the age of 20 are more likely to develop acute leukemia (notably acute lymphatic leukemia) than older survivors, but the latter are more prone to develop chronic myelogenous leukemia. Finally, it should be noted that these studies focused on primary tumors and did not include secondary cases of leukemia where chemotherapy, especially alkylating agents which readily combine with many different molecules, including DNA, may have been used. Leukemia secondary to chemotherapy is markedly different from leukemia arising spontaneously, that is, de novo.[19] No cases of chronic myelogenous leukemia have been reported among chemotherapy-related leukemias, and the types of leukemia that have been seen are different in other respects as well. Some, for example, have shown a simultaneous malignant proliferation of nucleated red blood cells as well as white cells, a condition described as erythroleukemia, and others exhibit exceptionally large white cells, identified as young megakaryocytes, that are normally present in the bone marrow but not seen in circulating blood. The origin of these differences between chemotherapy-related and spontaneously occurring leukemias is not clear.

Death from Cancers Other Than Leukemia

About 1960, it was noted that in addition to leukemia some malignant solid tumors were also more frequent among the survivors. Conventionally, cancerous solid tumors are classified on the basis of the tissues involved into either carcinomas or sarcomas. The former are malignant growths involving the cells that cover the inner and outer surfaces of the body, the epithelia; whereas sarcomas involve tissues that bind together and support the various structures of the body—bone, muscles, and the like. Each of these broad classes is further subdivisible on the basis of the location of the malignant growth and the specific cells involved.

Lung Cancer

Judged from death certificates, the risk of dying of any cancer other than leukemia, based on the DS86 doses, is about 8 excess cancer deaths per 10,000 person-year-gray. This increased risk was first observed in 1960, as earlier noted, when Tomin Harada and Morihiro Ishida, in a study of the Hiroshima Tumor Registry data, reported that the incidence of lung cancer was significantly higher among those survivors who were exposed within 1,500 m of ground zero.[20] Subsequent studies, particularly of mortality ascribed to lung cancer among members of the Life Span Study sample, have extended this evidence. The carcinogenic effect seen thus far has been most pronounced among individuals 35 or older at the time of exposure. This is not unexpected, however, since the evidence that has accumulated suggests that radiation-related malignancies increase in frequency only at those ages when the specific cancer normally occurs. Leukemia is the only striking exception. Most cancers are diseases of middle and later life, and the failure to see an increase in lung cancer among the young may merely reflect the fact that they have not yet reached the age at which their increased risk will be manifested. An alternative explanation of the age at time of bombing effect, which cannot as yet be rejected, is that exposure to ionizing radiation accelerates the appearance of some cancers that would have appeared later in life in the absence of exposure. If this is true, these cancers would not be additional ones in the usual sense, but rather earlier manifestations of predestined tumors.

Breast Cancer

Another malignancy whose relationship to radiation became apparent at about this time is cancer of the breast. Clifford Wanebo and his colleagues were the first to suggest this.[21] However, their study was restricted to women who were members of the Adult Health Study, a group too small to determine whether the apparent effect was real, or whether the different tissues in the breast were equally sensitive to radiation damage. Subsequent investigations have focused on all of the women in the Life Span Study sample, some 63,000. The most recent of these studies finds that the distribution of histological types of mammary cancers does not vary significantly with radiation dose, and that among all women who received at least 0.10 Gy, those irradiated before age 20 will experience the highest rates of radiation-related breast cancer.[22] This significantly elevated risk is seen even among those females who were exposed before the age of 5, suggesting that breast cancer can be induced by irradiation of stem cells, well before breast budding actually begins.

Retrospectively, it is clear that both breast and lung cancer began to increase about 10 years after the bombing, in the period 1955–1960, but the additional cases were too few then to make this fact demonstrable. However, it is now recognized that there is a delay, a "lag" period, following exposure before radiation-related cancers become evident. This interval of time between exposure and the clinical manifestation of the tumor is called the *latent period*. Why this period should differ for leukemia and other cancers is not known. But, it is generally thought to arise as a result of the difference in the rate of cell division, differentiation, and loss of hematopoietic stem cells, on the one hand, and the cells of other tissues, on the other.

Most organs, such as the breast, are made up of a variety of cells, each with a different function and a recognizably different location or morphological structure. Cancer could develop in any one of these with somewhat disparate consequences, and radiation might have a greater effect on one cell type than another. Heretofore, no difference has been seen in histological type nor tumor size by age at exposure. Nor, have atypical changes or residual proliferative lesions been found in the breasts of women exposed to radiation but free of cancer.[23] Recently, however, continuing studies of the histological changes in the breast following exposure to ionizing radiation suggest that an "atypical" metaplasia might occur before the actual appearance of a radiation-related malignancy.[24] This metaplasia seems to increase in frequency with the dose of ionizing radiation to the breast, suggesting that it may be preliminary to the occurrence of a tumor. If subsequent studies confirm this finding and if "atypical" metaplasia can be easily established, it could provide a means to identify those women who are more liable to develop breast cancer as a result of their exposure, and could afford a means for earlier intervention and better prognosis. At the moment, it seems unlikely that the presence of this atypical metaplasia could be established through mammography, the most commonly used technique for identifying potentially malignant changes in the breast.

Stomach Cancer

Stomach cancer is the most common cancer among the Japanese, accounting for over 30% of all cancer deaths. But data linking its frequency to exposure to the atomic bombing were slow to accumulate. Studies conducted in the 1960s failed to reveal consistent evidence of an association, and it was not until 1977, with Kuniomi Nakamura's publication on the occurrence of cancer among the survivors in the years from 1950–1973, that a relationship could be documented.[25] He was able to show that in Hiroshima the standardized mortality ratio for stomach cancer increased steadily and significantly with dose up to 5 Gy (T65) and then declined precipitously; but in Nagasaki the only significant increase was seen at doses in excess of 5 Gy. Over the succeeding years, the mortality data on stomach cancer have become much more persuasive in both cities, and can now be supplemented by the incidence data derivable from the tumor registries.

An analysis of the histological types of stomach cancer among the survivors did not find evidence of a radiation-related histological type. It is known, however, that the stomach is often populated by cells usually found only in the small intestine. This phenomenon, which occurs in the gallbladder too, is termed intestinal metaplasia. It involves the transformation of the mucous membrane, particularly in the stomach, into a glandular tissue resembling that seen in the small intestine. There was evidence in this study that the populating of the stomach with intestinal-like cells increased in frequency with increasing dose.[26] This is important, since cancer of the stomach often begins in these metaplastic areas. More recently, Suminori Akiba and his colleagues, studying members of the Adult Health Study sample, have described two biological markers useful in the identification of individuals at an increased risk of stomach cancer.[27] The markers are achlorhydria—the absence of hydrochloric acid from the stomach's secretions—and low levels of an iron–protein complex, known as ferritin, which is measurable in serum. Ferritin itself is involved in the storage of iron in the body and has been implicated in the metabolism of iron in the gastrointestinal tract. In combination, low levels of serum ferritin and achlorhydria are associated with a tenfold increase in the risk of stomach cancer. Early recognition of individuals at increased risk of cancer can not only influence judgment with regard to treatment but can enhance the prospect of cure or at least longer survival.

Thyroid Cancer

In a study conducted in 1959 in Hiroshima of diseases of the thyroid, Dorothy Hollingsworth and her associates noted that carcinoma constituted some 7% of the total number of cases of thyroid disorders seen, and that these malignancies were found more commonly among the heavily exposed individuals.[28] Anatomically, the thyroid is a ductless gland situated astride the anterior base of the throat. Physiologically, it is the source of a number of hormones in which iodine is an important constituent. Indeed, most, if not all, of the metabolism of iodine within the body occurs in this gland. As a consequence, radioactive isotopes of iodine released in a nuclear accident or produced through the detonation of a nuclear weapon are largely sequestered by the thyroid. While the most common of these isotopes are relatively short-lived, their absorption can result in doses to the thyroid that are disproportionate to the total amount of radioactivity released. This is particularly true in children, where the gland is very active and small, so that the dose per cubic millimeter of tissue can be much higher than would be expected in adults exposed to an equivalent absolute amount of radioisotopes, since the radioactive material would be distributed over more tissue.

Although thyroid cancer is not commonly fatal (less than 10% of individuals with this malignancy die from it), a succession of studies has removed any doubt of the association of this tumor with radiation exposure. The most recent one, utilizing the DS86 doses and the data from the tissue and tumor registries accumulated in the years from 1958 through 1987, found 225 cases of thyroid cancer (162 in Hiroshima and 63 in Nagasaki) among 79,972 study subjects.[29] The estimated excess relative risk using a linear dose–response model was 1.15; the overall excess number of incident cases per 10^4 PYSv was 1.61; and the attributable risk when age at exposure is ignored was 26%. However, these latter two values do not summarize the data well, since there is a very strong dependence of the risk on the age of the individual at the time of exposure. Among survivors under the age of ten, the excess number of incident cases per 10^4 PYSv was 4.37, but among survivors who were 10–20 years old or older than 20 years, the excess cases were 2.67 and 0.21, respectively. The striking radiosensitivity of the young is further shown by the fact that over half of the cases occurring among individuals who were less than 20 when exposed were attributable to their exposure; whereas among individuals over 20 years of age only 3% of the cases were ascribable to radiation. Finally, although the background incidence rate for thyroid cancer was more than three times higher in females than males, this study found no effect of sex on the excess relative risk.

Other Cancers

Recently, an increase in mortality from cancer of the colon, esophagus, ovary, and urinary bladder has appeared in the survivors. Multiple myeloma, a malignant tumor of the plasma cells of the body that commonly originates in the bone marrow but involves the bony portion of the skeleton too, has also been reported to increase with radiation dose. Multiple myeloma is a malignancy largely confined to older individuals—persons in their 60s or so. The suggestion that there might be an increase in multiple myeloma first appeared in 1964,[30] but several years elapsed before the number of cases was sufficient to examine the dose–response relationship more closely.[31] However, it should be noted that with the recent reclassification of the leukemia data, some of the earlier cases of multiple myeloma have been rejected, and the evidence now supporting a dose–response relationship is not compelling. There is an increase too in salivary gland tumors and, as we have seen, an increase in thyroid tumors.

An increase in mortality from the cancerous tumor of the lymph nodes, known as lymphoma, remains uncertain. Customarily, lymphomas are divided into two groups—Hodgkin's lymphoma, first described in 1832 by Thomas Hodgkin, an English physician, and non-Hodgkin's lymphoma, actually a grouping of ten or more other forms of cancer involving the lymphatic system. Although both types of lymphoma generally begin with a swollen, possibly painful, lymph node in the neck, armpit, or groin, they differ in the aberrant cells seen in the affected lymph node, in the age groups they affect, in the course and severity of the disease, and in prognosis. Among the survivors, the evidence that malignant lymphomas might be increased has been equivocal at best, and even the most recent analysis of the lymphoma incidence data fails to reveal a significant elevation in the excess relative risk of non-Hodgkin's lymphoma with increasing dose, although the excess absolute risk is marginally significant statistically for males but not females.[18]

No increase has been seen in cancers of the bone, gallbladder, nose and larynx, pancreas, pharynx, prostate, rectum, skin (except melanoma), small intestine, and the uterus. In some of these instances, such as cancer of the small intestine, this apparent absence of a radiation-related effect might merely reflect the rarity of the malignancy under normal circumstances. Less than one individual per 100,000 will die of a cancer of the small bowel or duodenum in a year in Japan; the rarity of this malignancy is presumably attributable in part, at least, to the rapid turnover of the cells comprising the epithelial lining of the small intestine, known as the brush border. When such tumors do develop they usually involve cells embedded within the intestine, either the endocrine cells (those cells that generally secrete their products directly into the circulatory system) or the supporting muscle. At other sites, however, such as cancer of the gallbladder—which is rare in most populations—when the Japanese data are combined with the data from other studies, specifically those from Britain of individuals with the deforming rheumatic disease known as ankylosing spondylitis, an apparent effect of exposure to radiation has been reported.[32] This is a provocative finding, but it should be noted that recent incidence studies in Hiroshima and Nagasaki have not confirmed this tentative effect. However, the populating of the stomach with intestinal-like cells previously described also occurs in the gallbladder. It is not known whether in this organ the metaplastic areas are related to the site of cancer, but if this is true, then, again, there could be a means to identify the tissue at risk. Similarly, gallbladder cancer shares some similarities with cancer of the colon, in that in both instances there is an elevation in serum levels of the carcinoembryonic antigen (CEA), which although not restricted to cancer is a useful marker. Polyps or polyp-like structures also arise in both organs. In the colon these polyps can become cancerous; however, in the gallbladder the "polyps" are either adenomyomatoses (a benign tumor usually of the smooth muscle) or more commonly sites of deposition of cholesterol associated with gallstones. This combination of stones and cholesterol polyps, especially if accompanied by pain and fever, is frequently seen as indicating a need for removal of the gallbladder (cholecystectomy), which makes it difficult to determine whether a malignancy would or would not have occurred if the bladder had not been removed. Overall, the prevalence of gallbladder polyps in Japan is about 3.2%.

Present evidence fails to suggest an increase in brain tumors and is equivocal with regard to tumors of the central nervous system other than the brain. Whether an increase in deaths due to cancer of the liver occurs has been unclear. When the analysis is restricted to only those cancers known to be primary, liver cancers do not increase significantly with dose; however, if the cancers termed "unspecified" are included, there is a dose-related increase. The liver is, however, a common site of metastasis for cancers arising elsewhere, in the breast or lung, for example, and the unspecified tumors might be metastatic ones that should be assigned to other organs where an effect of radiation is known to occur. But a recent analysis of the tissue and tumor registry data suggests that primary cancers of the liver increase in a dose-related manner among the survivors. To avoid the inaccuracy of diagnoses of liver cancer on death certificates, this analysis focused on histologically confirmed cases where the excess relative risk at 1 Sv was found to be 0.66 [confidence interval (CI): 0.11; 1.14].[29] This finding is of special interest since liver cirrhosis and hepatic cancer are commonly related to the presence of a hepatitis virus, in particular the hepatitis B virus. The surface molecule associated with this virus has been shown to increase in prevalence in those survivors exposed to 1 Gy or more.[33] This would appear to suggest that either radiation-related damage of the liver leads independently to both liver cancer and an increased likelihood of infection with the hepatitis B virus or possibly that the damaged liver is more prone to infection with the virus, which leads, in turn, to a higher risk of liver cancer.

Factors Affecting Cancer Risk

As previously stated, the increase in mortality from cancers other than leukemia becomes significant, generally, when individuals reach the usual age of onset for a given cancer. In addition, the distribution of time from radiation exposure to death does not differ significantly by radiation dose for solid tumors, but it does vary depending on the age of the individual at the time of the bombing. Both the relative and absolute risks for cancers other than leukemia are higher for younger age at the time of bombing cohorts at the same attained age. Among individuals over the age of 20 when exposed, and certainly over the age of 30, the relative risk has changed little with time although the absolute risk has continued to rise. However, in the two youngest groups of survivors— those individuals who were 0–9 or 10–19 years old at the time of the bombing—the relative risk has been declining (significantly among those 0–9), whereas the absolute risk has steadily increased. These findings are not inconsistent statistically, since if the relative risk is declining with age while the baseline rate is increasing (as it does with age), even a smaller relative risk, applied to a larger baseline, will produce a larger absolute risk. But this fact does not tell us what the biological basis for the decline in relative risk is.

Earlier studies have suggested significant differences in the frequency of cancer in the two cities following exposure to the same amount of radiation. These differences were attributed to what was thought to be the far greater exposure to neutrons in Hiroshima than in Nagasaki, especially since experimental work suggests that neutrons are appreciably more carcinogenic than gamma rays at the same absorbed dose. Analyses using the newer DS86 doses reveal these city differences to be no longer statistically significant; however, at the same dose, mortality remains generally higher in Hiroshima than in Nagasaki. The origin of these differences is unclear.

The Impact of Cancer on the Population of Survivors

What do these measures of risk and statements about cancer mortality mean to members of the Life Span Study sample and to the survivors generally? Perhaps more immediate substance can be given to the assertions we have made through a consideration of the additional cancers that have presumably occurred among these groups as a consequence of their exposure. In the years from 1950–1985, 141 individuals in the Life Span Study who have been assigned DS86 doses of 0.01 Gy or more died of leukemia. As noted earlier, 83, or slightly less than 59% of these deaths, were attributable to radiation exposure. These same years saw 5,734 deaths from cancers other than leukemia among the members of the Life Span Study sample, and 3,172 of these deaths involved survivors exposed to 0.01 Gy or more. Approximately 8%, or 254, of these 3,172 deaths were presumably due to radiation exposure. These are estimates, of course, and must be interpreted in this light since, as stated, it is presently impossible to separate a radiation-related cancer from one due to some other cause. Moreover, these values can be expected to change as the period of follow-up increases, and more of the survivors succumb.

While it can be argued that even one radiation-related death from malignancy is too many, there is obviously no epidemic of cancer deaths among the exposed. Most survivors who will die of cancer will do so as a result of exposure to other factors—smoking, alcohol consumption— and not from exposure to atomic radiation. No more than about 8% will have cancers directly attributable to exposure to ionizing radiation.[17] And even among these unfortunate individuals, exposure to ionizing radiation was but one of the contributing factors in the multistage process that culminates in a cancer.

The Incidence of Solid Malignant Tumors

Estimates of cancer risk based on the tumor and tissue registry data—and hence on incidence rather than death—have not been generally available in the past, although these data have been used in the study of specific tumor sites, such as the breast and thyroid. This situation is changing. Recently, for example, there has appeared a comprehensive assessment of solid tumor incidence in Hiroshima and Nagasaki in the years from the inception of the registries, 1958, through 1987.[29] Among members of the Life Span Study sample, after the removal of those individuals with unknown dose, or who were not in one or the other of these cities at the time of the bombing, or are not currently residents, some 8,613 cancers were identified and enrolled in the registries in these years. This is 3,080 more cancers than were identified through death certificates in the same years and sample. Much of this difference is due to cancers of the stomach (795 more cases), the breast (386), the thyroid (182), the skin (152), and the uterus (347), but other sites contribute as well. Some of the discrepancy is undoubtedly due to under reporting of cancer on the death certificates reflecting the limitations of death certificate data, which have been mentioned, but some is also due to the registry identification of a malignancy that, while not yet fatal, will be in time.

Broadly speaking, the incidence data support most of the findings based upon the death certificates. For example, neither set of data reveals a significant difference between the cities in the estimates of radiation-related cancer risk. Both suggest that the best fitting dose–response model is a linear one, and both reveal the risk to be higher among those survivors exposed early in life (before the age of 20) than at later ages. There are differences, however—some expected and others not. Thus, the incidence data generally lead to risk estimates with smaller errors, as would be expected, since the number of incident tumors seen is greater than the number of deaths at the same site. These data also show a significant increase of thyroid tumors and nonmelanomatous tumors of the skin with increasing dose, and this increase has not been clearly seen in the mortality data, presumably because these tumors are not commonly the cause of death. But as previously described, the incidence data do demonstrate a significant increase in primary cancers of the liver, a finding that has proven equivocal when analyzed on the basis of the mortality data. Although the estimates of the excess relative risk at 1 Sv do not differ greatly, the attributable risk based on the tumor registry data is often higher than that seen on the basis of the death certificates, and particularly for some organs such as the breast, where the incidence data suggest an attributable risk of 32% as contrasted with about 22% for the mortality data. Overall, the attributable risk is about 12% using the registry data, and about 8% with the mortality findings. That the former should be higher than the latter is not unexpected, since the mortality findings include deaths from cancer in the years 1950 to 1958, before the beginning of the registries, when the contribution of radiation-related cancers to all cancers seen was small.

Uncertainties in the Estimates of Cancer Risk

Even after many decades of study, the uncertainties that surround the estimates of the carcinogenic effects of radiation described are many and fundamental. They are both general, in that they are common to all studies of the carcinogenic effects of ionizing radiation—as well as to studies of chemical carcinogenesis—and specific, in that they apply only to the investigations of the survivors. Among the general uncertainties is the absence of a compelling biological model of the underlying process involved not only in radiation carcinogenesis, but carcinogenesis more broadly. But why are models of carcinogenesis necessary? A variety of reasons can be adduced. First, a model that is consistent with all of the data could stimulate new observations that might provide still deeper insight into the biology of cancer. Second, a biologically realistic mathematical model is essential to a meaningful, quantitative assessment of cancer risk. Finally, insofar as ionizing radiation is concerned, the need for a good theoretical model is greatest where the data are weakest—at low doses and low dose rates of irradiation. Here the estimates of risk are more tenuous, and will undoubtedly continue to be, since the excess risk is apparently small and the sample size needed to demonstrate an effect is prohibitively large.

The majority of all cancers appear to be environmental in origin in the sense that exposure to environmental factors is involved, but such exposures often embrace a multiplicity of different agents and the nature of their interaction is unknown. In addition, many host factors, such as a person's genes, age or developmental stage, hormonal status, and gender, affect risk. It is to be expected that in any population exposed to ionizing radiation there will be variation in exposure to these other risk factors as well. At low levels of radiation exposure, the effect of this variation might be greater, perhaps much greater, than the risk produced by the radiation itself. It is not surprising, therefore, that it is difficult to estimate risk at low doses or that differing results often occur, since the methods used to estimate risk tacitly assume that all exposed individuals in a given category (e.g., age, gender, dose) have equal risk, which seems unlikely to be true.

The importance of these extraneous modifiers of radiation risk has been particularly well demonstrated in the case of breast cancer, where Charles Land and his associates have detailed the role of reproductive factors in altering the risk of this malignancy in Hiroshima and Nagasaki. They have shown, for example, that a woman exposed to 2 Gy of atomic radiation in the first decade of life, who has her first child at the age of 18, has a risk of breast cancer that is only one-sixteenth that of a similarly exposed woman whose first child is born when she is 32.[34] Presumably the observed differences in risk are ascribable to hormonal changes subsequent to the woman's exposure, but precisely how these changes and radiation interact is unknown.

Another important risk factor is tobacco smoking, but its effect in the presence of ionizing radiation is not clear. Based on the experience of the survivors, and specifically with respect to cancer of the lung, smoking does not multiply the effect of exposure to irradiation, but merely adds to it.[35] That is to say, an exposed survivor who smoked would have a risk equal to the sum of the risk from the dose received plus that from smoking. Whether this will be true for other sites of cancer related to smoking, for instance, cancer of the urinary bladder, is still not known. The relationship between smoking, dose, and lung cancer seen among the survivors may not apply to other exposures to irradiation, in particular those from uranium mining, where radioactive particles are actually deposited in the lung and serve as foci of irritation. Indeed, some investigators have alleged that smoking and radiation exposure interact multiplicatively rather than additively among miners. However, neither the Foundation's data nor that available on the uranium miners are extensive enough to establish the relationship of smoking and radiation to lung cancer incontrovertibly.

Most of the other uncertainties that attend the estimates of risk resulting from the mortality surveillance in Hiroshima and Nagasaki are not unique to the atomic-bomb data. They are common to all epidemiological studies of the effects of ionizing radiation in humans. Some relate to the doses assigned to particular survivors, but others are more general. We will focus on the latter group, since the survivor-specific uncertainties, dependent as they are upon an individual's ability to recollect precisely the details of their exposure—where he or she was at the time of the bombing, the presence of shielding, and the like—will never be fully resolved.

First, as has been previously noted, unreconciled differences between the two cities remain. These differences include not only the increased mortality seen in Hiroshima as contrasted with Nagasaki for the same presumed dose, but also the frequency of epilation, chromosomal aberrations, and lenticular opacities. Whether these differences imply residual inaccuracies in the dosimetric system itself or still unrecognized extraneous sources of variation has yet to be determined.

Nonetheless, their existence has prompted some controversy about the estimated yield of the Hiroshima weapon, as well as the neutron flux itself. Indeed, as discussed in Chapter 5, at distances beyond about 800 m, the observed data on thermal neutrons does not agree well with the theoretical values, suggesting that either the source term is still not correct insofar as neutrons are concerned, or the modeling of the transport of neutrons in the atmosphere at greater distances is flawed. Hopefully, with time, this issue will be resolved, but at the moment it remains contentious.

Second, although a linear relative risk model is a simple, suitable descriptor of the actual observations to date on cancers other than leukemia, as judged by a comparison of the number of cases predicted by such a model with the observed data, it is unclear whether an alternative dose–response model would be better. Generally, it is difficult to discriminate between other plausible alternatives, such as a linear–quadratic model. All of these models are, however, merely convenient descriptions of what is observed, and might have no deep causal meaning. Radiobiological considerations could suggest a dose–response relationship based upon cellular or molecular events, but it does not follow that the same dose–response relationship will be seen when measured in terms of case occurrences or relative (absolute) risk of death. The prospect of early clarification of the "true" dose–response relationship is not encouraging. Presumably, as a larger and larger proportion of the lifetime experience of the atomic bomb survivors accrues, some clarification will occur, but it is unlikely that the appropriate model can be defined solely on the basis of epidemiological data (only 38% of the 75,991 members of the Life Span Study sample included in the mortality analysis spanning the years 1950–1985 were dead; the number of deceased continues to increase and had reached 42% in 1990).

Third, there is the matter of the extent to which the Life Span Study sample represents a selected one, conditioned by the changing probability of survival with changing dose. At high doses, few individuals survived; whereas at low doses most did. These facts have a number of implications. It has been argued,[36] largely on statistical grounds, that one consequence of these differences in survival is overestimation of the true doses at the higher levels and underestimation at the lower. This could lead to an underestimation of the risk, obscuring the true dose–response relationship through introducing a curvature where none exists.

Other investigators have repeatedly contended that exposure resulted in a compromising of the immune system, and consequently individuals who might have succumbed from cancer had they survived died instead of infectious or other diseases where immune competence is important. It is difficult to test this thesis rigorously, since little is known about the causes of death in the first 9 months following the bombings, as noted earlier. Resistance to bacterial infection is known to be impaired for weeks, possibly months, following exposure to substantial amounts of ionizing radiation.[37,38] It is not the plausibility of impaired immune competence in the short run that is questioned, but its tenuousness as a late (or long lasting) effect in the light of the failure to find direct evidence among the survivors of a dose-related change in the frequency of specific, clearly infectious diseases, e.g., tuberculosis, or in a variety of laboratory studies seeking to establish a significant diminution in the bactericidal activities of the white cells of the survivors.[39] But there are other studies of the survivors that are difficult to reconcile with this hypothesis as well. For example, when lymphocyte survival time is contrasted in persons exposed to DS86 doses of 1.5 Gy or more with those receiving doses of less than 0.01 Gy, there is no evidence of the selective elimination of a radiosensitive group of individuals among the more heavily exposed.[40] However, studies of the survival of lymphocytes irradiated in vitro, as a test of possible population bias, have not been conclusive. No consistent change in survival times was observed among the 201 survivors studied. This could be interpreted as implying no preferential elimination of radiosensitive individuals, but the failure to find a change could also be ascribable to the fact that these cells may not fully express variations in radiosensitivity. Similarly, in a simple but ingenious study, Peter Gregory and his colleagues divided the surface of the body of exposed members in the Adult Health Study into four equal quadrants. Using scarring of the skin from flash burns in these quadrants as indicative of the direction of the radiation, they asked whether there was a diminished frequency of individuals whose spleens had been irradiated directly. Since the spleen is the reservoir for many of the lymphocytes involved in immune response, if a compromising of this response increased the likelihood of premature death, one would expect to find fewer individuals with direct splenic irradiation. This was not found to be true.[41]

This potential selection of unusually healthy individuals is often called the "healthy worker effect." It derives its name from the fact that, in occupational cohorts, workers are often healthier than the general population of which they are members, and thus their baseline cancer rates might differ from population rates, complicating the estimation of the expected number of cases in the exposed cohort. While the possibility of such selection in the Life Span Study cannot be peremptorily rejected, it is important to note that the "healthy worker effect" has been shown to be most pronounced in the early years of an investigation and to wane with time. Thus, if there were an effect on the data from the Life Span Study it would involve leukemia primarily, since the onset of this malignancy occurred early, but would have little impact on the risk of solid tumors, where a significant increase did not occur until ten or more years after the bombings.

Fourth, although tumor registries were established in Hiroshima and Nagasaki more than 30 years ago and tissue registries have existed for over two decades, in the main, the risk estimates that have been available are based largely upon mortality, and might underestimate the risk at specific sites, as is usual for those cancers that are not commonly fatal. However, this situation is changing. As has been observed, data on risks based upon incidence rather than mortality are becoming available, and moreover, the results from the Autopsy Program provide a basis for assessing the extent to which cancer may be inaccurately reported on the death certificate, since this program included a substantial number of the members of the Life Span Study.

About 4,900 of the 16,000 or so Life Span Study subjects dying between 1961 and 1975 were autopsied. These postmortem examinations have served as the bases for a succession of evaluations of the reliability of death certificate diagnoses of cancer and other causes of death. Invariably, they have shown the confirmation at autopsy of cancers reported on death certificates to be high.[42] Confirmation rates vary, however, with cancer site, gender, the age of the individual at death, and whether death occurred at home or in a hospital. Although relatively more of the heavily exposed individuals came to autopsy than the lightly exposed ones, the confirmation and detection rates appear essentially independent of radiation dose. The confirmation rate is the frequency with which an autopsy verifies the cause of death stated on the death certificate; whereas the detection rate is the actual frequency of a specific cause of death as revealed by autopsy. Most studies of the reliability of death certificates, whether in Japan or elsewhere, have shown that when the death certificate states that cancer was the cause of death, this remark is usually correct;[43] however, these studies have also shown that possibly one out of every four cancers (all sites combined) goes unreported on the death certificate.[42] Among those frequently not recorded are malignancies of the prostate, the thyroid, and those other organs which are rarely the cause of death. Some commonly fatal cancers go unrecognized too—only about 1 out of 6 malignancies of the gallbladder or bile ducts were detected in the years of the autopsy program, and no more than 1 in 5 cancers of the cervix of the uterus.

The detection rate can be as low as 15% for cancers of the liver, gallbladder, and bile ducts to as high as 78% for cancer of the breast. Detection rates do not vary significantly with dose, but may vary with age at occurrence of the cancer, declining as age increases. It should be noted, however, that under reporting will not affect the estimation of the relative risk importantly if under reporting itself is merely a reflection of the general standards of medical diagnosis and is not related to dose. But it would affect the estimation of the number of excess deaths. Moreover, as the treatment of cancer improves and the years of life following diagnosis increases, as it has, estimates of risk based on time to death could be changed, suggesting a diminished risk with time where no diminution actually exists, or a flawed distribution of time to expression following exposure. To some extent, this possibility can be explored in the population of survivors. If survival following diagnosis is increasing, then a real reduction in risk of cancer death at fixed times after diagnosis should be occurring. If this were found to be true, it could have an important bearing on the modeling of the time from exposure to death in the projection of risk to a lifetime in other populations with different medical standards. But how one would apply this knowledge to other populations is problematic.

Even the estimates that are now becoming available based on the tissue and tumor registries have their limitations, since the registries are necessarily geographically based and cannot, therefore, provide incidence cases on the full Life Span Study sample. As yet there exists no general national registry that could be employed to supplement the local data and provide information on those survivors who have migrated to other areas of Japan since the establishment of the Life Span Study. To some extent, the limitations imposed on the registry data as a result of this migration can be offset, even in the absence of a national registry, if sufficient information is available on the demographic particulars of the migrants—the age at migration, gender, city, year of migration, and the like. Given this information, it is possible to adjust the population at risk statistically to compensate for the migration that has occurred over time. Intuitively, such adjustment is less satisfying than complete coverage, since it necessarily involves assumptions—such as health status at the time of migration—the validity of which cannot be rigorously assessed. Moreover, the depth of detail on the migrants that is needed is at present available only on the members of the Adult Health Study, a stratified rather than random sample of the Life Span Study population. Nonetheless, if it can be assumed that the pattern of migration seen in the Adult Health Study is similar to that of all of the members of the Life Span Study, adjustments can be made to take into account the effect of migration on risk estimation as, indeed, has been done in the recent analysis of the incidence of cancer among the survivors. Unquestionably, judicious use of the registry and mortality information can provide a more balanced perspective than the one in the past, and case-control studies based on incident cases (especially those making use of the stored tissue samples) might provide important clues into the mechanisms of radiation damage and the role of concomitant sources of variation in the risk estimates themselves.

Benign Tumors and Related Disease

A benign tumor, unlike a malignant one, does not by definition have the property of uncontrollable growth and dissemination or recurrence after removal and is generally viewed as nonfatal. However, this distinction is sometimes nebulous. First, some ostensibly benign tumors can be life-threatening if occurring in a rigidly confined area of the body, as for example in the skull, where their expansion impinges upon vital function. Second, it can be difficult to determine histologically if a tumor is expanding "uncontrollably" if its growth is not particularly aggressive. Although many benign tumors are recorded in the tumor registries, their relationship to radiation has not been systematically studied. There are exceptions, however, and the disease known as hyperparathyroidism is one.

The parathyroid is a set of four small glands, situated in pairs, near the outside borders of the thyroid gland. The secretions of these glands are involved in the body's metabolism of two essential elements or minerals—calcium and phosphorus. In hyperparathyroidism these secretions increase, resulting in a loss of calcium. The affected individual commonly experiences spontaneous bone fractures (due to the loss of calcium), muscular weakness, abdominal cramps, and an inflammation of the bone characterized by degeneration of the bone fibers and formation of fibrous nodules on the affected bones.

Among the members of the Adult Health Study in Hiroshima the observed prevalence of hyperparathyroidism after doses of 1 Gy or more is 7 times as great as the prevalence among those exposed to less than 0.01 Gy, and the increasing prevalence with dose is well described by a linear dose–response model.[44] The commonest tumor of the parathyroid was an adenoma, a tumor of epithelial origin, and usually benign. Again, as is so frequently seen, the risk of hyperparathyroidism following exposure is higher among those survivors exposed as children or adolescents than later in life; specifically, for individuals in the first decade of life at the time of exposure, the relative risk at 1 Gy is more than 4 times greater than that among persons who were 20 years of age or older when exposed (11.1 versus 2.75).

These findings are intriguing, and suggest an alternative explanation to the one previously advanced to explain the radiation-related increase in cardiovascular disease. This alternative argument proceeds as follows: First, the parathyroid glands are endocrine glands, that is, they produce one or possibly more hormones that find their way into the circulatory system through osmosis rather than a direct release into a duct. The parathyroid hormone (PTH) is known to play a significant role in the regulation of blood pressure. As PTH increases, blood pressure increases. One of the diagnostic features of hyperparathyroidism is an increase in PTH. Thus, if the increase in hyperparathyroidism is dose-related, one should also expect a dose-related increase in the incidence of hypertension. Second, since the increase in hyperparathyroidism with dose is more pronounced among females than males, it is conceivable that the earlier report alluding to an increase in CVD among Hiroshima females not seen in other gender–city groups, which was rejected as a chance phenomenon, could, in fact, have reflected the greater number of Hiroshima females with hyperparathyroidism.

Psychosocial Concerns

Robert Jay Lifton, the Yale psychiatrist, has written movingly, though dramatically, in his book "Death in Life—Survivors of Hiroshima" of the fears of some of the survivors. While the sample of individuals interviewed was not representative, the worries expressed are presumably shared by many survivors. From the epidemiological data, we can say that they do not seem balefully preoccupied with their own health, since there is no increase in suicide. Their age-specific death rates for suicide are in fact lower than those of individuals not exposed.[45] And limited use of health assessment questionnaires patterned after the Cornell Medical Index Health Questionnaire has not identified a clear relationship between anxiety and the estimated dose of the individual. However, these studies have shown that as a group the survivors have anxieties not shared with Japanese who were not exposed.[46] They often complain of what has been termed *hibakusha bura-bura*—the occurrence of lingering fatigue and medically ill-defined symptoms for which no biological basis can be found. Nonetheless, it is clear that they report symptoms and complaints more frequently than nonexposed individuals, but this could reflect their suffering immediately after the bombing and the long lack of adequate social support rather than actual ill-health. But whatever its origin, a perception of ill-health is important, since it can influence decisions to seek medical attention and can conceivably affect attitudes toward the hopefulness of ameliorating disease when it occurs.

But the survivors have other concerns not related to their immediate state of health. For example, when younger, some of the survivors wondered whether, if unmarried, their opportunities to marry would be compromised by the knowledge of others that they were survivors or the children of survivors. And if married, whether they could or should have children. These fears can be allayed to some extent, since these issues have been examined.

Most Japanese marry as young adults, in their twenties. Traditionally, these marriages have been, and still are, often arranged, in spite of the sentiment expressed in Article 24 of the postwar Constitution, which states "Marriage shall be based only on the mutual consent of both sexes and it shall be maintained through mutual cooperation with the equal rights of husband and wife as a basis." Marriage has been viewed as a contract between families, an arrangement too important to be determined solely by the prospective bride and groom, whose assessments of one another are apt to be clouded by their romantic involvement. Arrangements are normally made through *a nakodo*, an agent who may be a relative, a family friend, a boss, or someone else. Ordinarily, as a part of this process, the biologic, economic, and social background of each family is made known to the other. It has been alleged that in this process atomic-bomb survivors, and even their children, fare poorly. Their exposure to atomic radiation is thought to make them undesirable mates.

It would certainly not be unexpected if some survivors felt discrimination and were fearful of stigmatization. Surely, one who was physically scarred or maimed in some visible manner might do poorly in this marital marketplace, but this would be a consequence of their physical status and not because they are a survivor per se; others similarly limited, for whatever reason, would undoubtedly fare no better. Evidence to support the belief of discrimination, common in the Japanese media in the 1960s, is largely anecdotal; there are few systematic studies that meet serious scientific scrutiny. However, Scott Matsumoto, a social scientist, found that a greater percentage of male and female survivors exposed within 2,000 m of the hypocenter in Hiroshima married than was true either of individuals exposed at greater distances, or who were not present in the city at the time of the bombing.[47] Possibly, the main deterrent to greater self-esteem is their own psychological adjustment, and the credibility placed in the anecdotes and rumors heard.

Fertility

There are significant differences in gametogenesis in the two sexes, and furthermore, there is evidence that the female sex cells, the oocytes, of some animals, possibly including human beings, are uncommonly sensitive to ionizing radiation. A single traversal of the membrane of an oocyte by an ionizing electron in these sensitive animals might be sufficient to kill the cell.[48] Cell sensitivity varies among species, and with the age of the exposed animal and the maturational stage of the germ cell. In the mouse and rat, for example, it is the small, immature oocytes that are the most sensitive; whereas in the guinea pig, it is the maturing oocytes that are more vulnerable. The data that are available do not suggest that the oocytes of the human female are especially sensitive when compared with the stem cells of the bone marrow. But the picture is incomplete—quantitative measurements on the number of sex cells remaining in a woman's ovary following irradiation are limited. However, since ionizing radiation can certainly kill germ cells, there is a basis for believing that in the short-term the fertility of some of the survivors might have been affected by their exposure to atomic radiation and their general health status. Parenthetically, irradiation of the ovaries was used in the years immediately before and after World War II to treat infertility that was intractable to hormonal therapy, and with some apparent success. It was generally assumed that exposure stimulated the formation of small blood vessels, and that enrichment of the blood supply to the ovaries resulted, in turn, in more normal ovarian function. This procedure is rarely, if ever, used now, since the absorbed dose to the ovaries was about 0.75 Gy and concern was expressed about the possible mutagenic effects of the therapy as well as the subsequent risk of ovarian cancer.

The first steps toward establishing a program of research on the effects of atomic radiation upon the reproductive capacity of the survivors were taken in March 1949 under the supervision of Masamichi Suzuki. These early studies were primarily clinical rather than demographic in nature, focusing upon diseases or disorders that could impair fertility or possibly lead to sterility itself. Among the endpoints considered in females were tubal patency, genital tuberculosis, the occurrence of *Trichomonas vaginalis* (a protozoan parasite that can inhabit the vagina), and a hypoplastic (aberrantly small) uterus. In males, semen specimens were obtained to count the number of normal and abnormal sperm, and snippets of testicular tissue provided an opportunity to study the histology of the testes and material for the chromosome studies of Kodani, previously described. Information of this type is socially sensitive, and the requisite specimens not easily obtained. It seemed likely, therefore, even at the beginning of these studies, that the difficulties inherent in making the necessary observations would be so large as to overwhelm the program. This did, indeed, prove true, and in 1953 it was terminated. But the question of possible radiation-related infertility remained an important one, and emphasis was shifted from clinical examinations of a small number of individual survivors to studies of a statistical nature, embracing long-term fertility and far larger exposed groups.

None of these latter studies has found evidence of permanent impairment.[49] Because there is no single faultless measure of the capacity to procreate, fertility was measured in a variety of ways—as the chance of fertilization, the chance of successful reproduction (the birth of a viable child), the time between the beginning of cohabitation and the first pregnancy, and the time to the first liveborn delivery. However, none of these parameters appears systematically and significantly altered with exposure.

Radiation-Related Cellular and Molecular Events

Human beings are composites of hundreds of billions of cells. Ill-health begins with an impingement on the normal functioning of one or more of these cells. And while it may be easier to enumerate events, such as the occurrence of cancer, at the organ level, understanding what went awry rests upon knowing the nature of this impairment of function. As previously indicated, it is now generally believed that cancer begins with a change in a single cell, most probably a mutation, although a malignant tumor is the culmination of many other events as well. Similarly, it is known that malignant cells often exhibit peculiarities in the structure or number of chromosomes they contain. These and related observations have fostered a great many studies of the survivors aside from those described thus far. Some of these have focused on chromosomal aberrations—either changes in number or structure—and others on the occurrence of transmissible changes in somatic cells, that is, on somatic mutations. It is to the findings with respect to these cellular and molecular events that we now turn.

Chromosomal Abnormalities

Chromosomes are the vehicles through which genetic information is transmitted from one cell to another, and ultimately from one generation to another. Each human being is normally endowed with 23 pairs of chromosomes, or 46 in all, half derived from the mother and half from the father. One of these pairs, commonly referred to as the sex chromosomes, plays a major role in determining the gender of an individual; whereas the other 22, the autosomes, are primarily involved in transmitting other inherited characteristics. Through much of the life of the cell, these chromosomes appear as very fine, long filamentous threads in the cell nucleus with little clear structure. However, with the onset of cell division, they begin to condense through coiling and each chromosome takes on a well-defined appearance. It is at this time that individual pairs of chromosomes can be identified and their structure studied.

Under normal circumstances, an orderly sequence of events leads to the reduplication of these chromosomes in the course of cell division and the separation of the reduplicated sets equally to the two daughter cells that are formed. As the chromosomes condense, the nuclear membrane previously surrounding them disappears, and the chromosomes align themselves near the center of the cell along what is frequently called the equatorial plate. This is the metaphase stage in division. Soon thereafter the chromosomes begin to separate, one member of each reduplicated pair moving to opposite ends of the cell. This stage is termed the anaphase. Regularity in the process of separation and distribution is achieved through the aid of a specialized region of the chromosome, the centromere, and the location of this centromere is an important landmark in identifying a particular chromosome. Once all of the daughter chromosomes have gathered at opposite ends, they begin to uncoil, a new nuclear membrane is formed about each set, and in time the remainder of the cell, the cytoplasm, will divide to form two new cells indistinguishable from the original. This process continues throughout life, more frequently in some tissues of the body than in others, but more slowly in all tissues with age.

To the extent that transmissible changes in these aggregations of genes occur either in somatic or germinal cells following exposure to ionizing radiation, the daughter cells could inherit an abnormal complement of genes. This abnormal complement might manifest itself in numerous ways detrimental to the cell and the individual involved. Accordingly, as has been previously noted, studies of the chromosomal complements of the survivors began as early as 1948, but foundered amid technical difficulties. Once these were solved, new, larger, more systematic investigations were initiated, based on short-term cultures of the white cells in the peripheral blood of the survivors, specifically the T lymphocytes described earlier.

While the precise nature of the events that give rise to radiation-related chromosomal damage remains speculative to a degree, it is usually presumed that as ionizing particles (electrons or, in the case of neutron irradiation, protons) traverse the cell—or more particularly a chromosome or its still unseparated daughters (the chromatids)—breaks are produced in the chromosome's normally continuous structure. These breaks might or might not be spontaneously repaired, and if repaired, the result might or might not be a faithful copy of the original. Portions of the chromosome (or chromatid) could be inadvertently omitted, or if two or more chromosomes (or chromatids) were broken simultaneously, the wrong ends might be rejoined. In the first instance, a deletion of genetic material would occur, and in the second, a translocation of material from the normal chromosome to another. Either of these events is generally recognizable through a study of the structure of the chromosomes in the daughter cells, provided the amount of material involved is sufficiently large to be seen. To aid the investigator in the search for these translocated segments, special stains that bind only to specific chromosomes can be used, and when these stains are coupled with a dye, fluorescein, the abnormally located segments will glow when examined in a suitably darkened background. This technique, which has been introduced quite recently, is sometimes referred to as "chromosome painting," or more technically as fluorescent in situ hybridization (FISH). As yet, it has not contributed much to the chromosome studies but its promise is great, as Figure 18 illustrates. Even the untutored eye can discern the misplaced chromosomal segments in this cell from an exposed survivor.

The nature and number of these aberrations produced within a cell will depend upon the density of the ionizations along the track of the particle as it traverses the cell. Gamma rays, for example, penetrate tissues deeply but generate few ionizations per unit path length (or alternatively per cell encountered); whereas alpha particles are less penetrating but produce a higher density of ionizations along their track than do gamma rays and therefore, on average, do more damage to the cells encountered. At low doses, when the number of particles traversing a single cell is small, it is assumed on probabilistic grounds to be un-

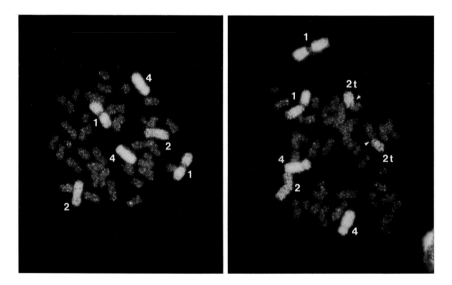

Fig. 18. An illustration of the use of fluorescent-in-situ hybridization, popularly known as chromosome painting, to detect radiation-related chromosomal damage. The left panel illustrates the appearance of the chromosomes in a normal cell derived from a woman exposed to an estimated DS86 kerma dose of 1.4 Gy. The three chromosomes, namely 1, 2, and 4, that appear yellow are the target chromosomes, which have been stained with FITC-avidin. The nontarget chromosomes, which appear red, are stained with propidium iodide. The right panel illustrates an abnormal cell derived from the same woman in which a translocation of a portion of chromosome 2 to another chromosome, indicated with the white arrow and labeled 2t, has occurred. (Courtesy of Akio Awa)

likely that more than one break would occur in a particular chromosome, or that more than one chromosome would be broken. Therefore deletions of the terminal segments of chromosomes would predominate through the failure of the broken ends to rejoin properly. However, as the dose increases and the number of particles or photons striking a cell grows, it is more likely that two or more different chromosomes will be broken or that two or more breaks in the same chromosome may occur that might not rejoin.

These notions have led to the supposition that the frequency of translocations will increase as the dose increases and so will the frequency of certain other abnormalities that hinge on the occurrence of two breaks within the same chromosome or chromatid. The segment between the two breaks might be accidentally omitted in the repair process, giving rise to an interstitial deletion, not a terminal one, and the ends of the omitted segment might rejoin of their own accord, resulting either in a ring structure if the segment is sufficiently large, or in a pair of minute "dots" of chromosomal material if the segment is very small. Since the ring has no centromere to guide it to a daughter cell, it is left behind as cell division proceeds. It is also possible that the broken ends might rejoin in a manner that leads to two rather than one centromere being on a single chromosome, and that when cell division occurs this dicentric chromosome, as it is called, will attempt to move to both daughter cells simultaneously, producing a bridge between them. Or finally, the segment between the breaks might be reincorporated in the chromosome but in such a fashion that the normal order of the genes within the segment has been altered by 180° (degrees), resulting in what cytogeneticists term an inversion. Figure 19 diagrammatically illustrates how these events occur. The heavy figures are those that the cytologist actually sees, whereas the stick figures represent the events that are thought to occur that give rise to the abnormality observed. The centromere appears either as the constricted region in the condensed chromosomes or as a heavy dot in the stick figures.

These considerations suggest that over the range of doses from low to high the dose–response curve should rise slowly at first, and then rapidly; it should be linear–quadratic in shape. This is what has been observed in studies of cells recently exposed to ionizing radiation. However, the study of the shape of the dose–response curve among the survivors is complicated—first, by the fact that when the newer studies were initiated almost 20 years had elapsed since the survivors were exposed, and second, because some radiation-related chromosomal abnormalities, such as those culminating in a ring or a dicentric chromosome, will prevent the affected cell from undergoing further division into cells that will survive. Therefore, any chromosome abnormalities seen among the survivors at this late date would likely either be those that

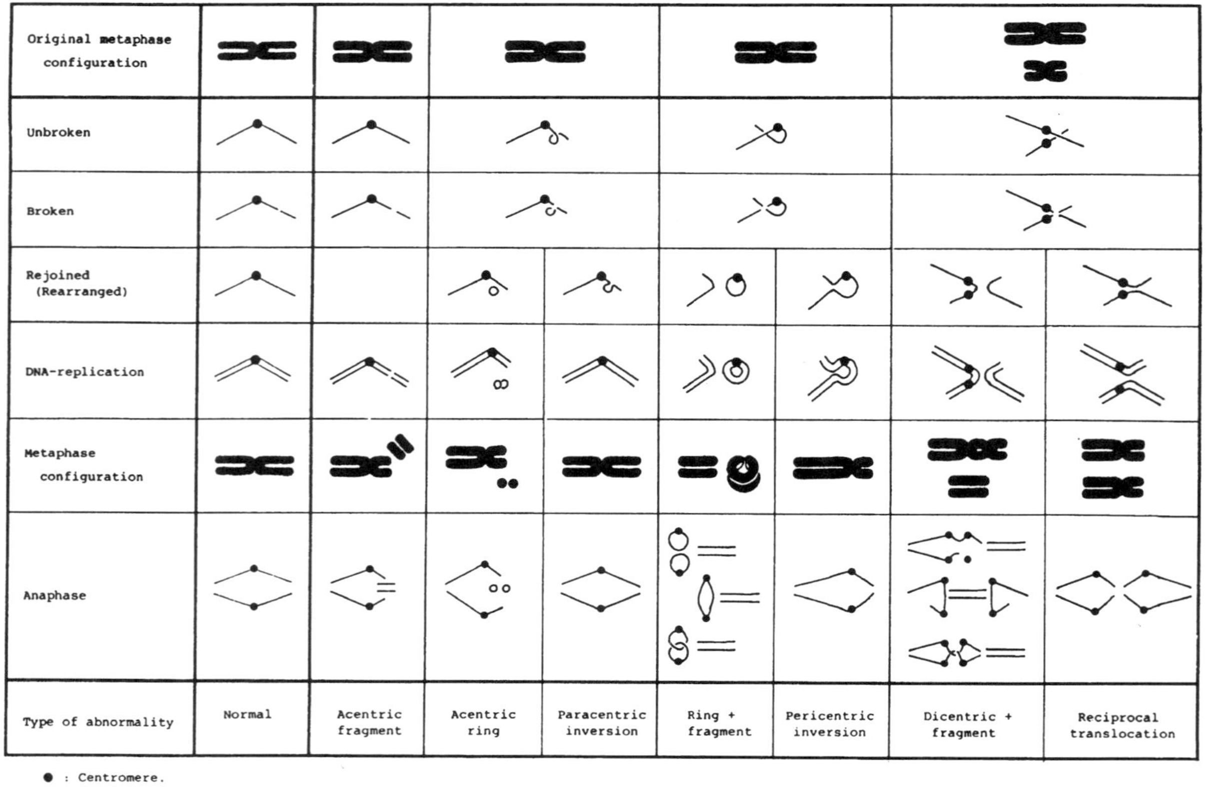

Fig. 19. A diagrammatic representation of different types of chromosomal aberrations and their presumed origin following exposure to ionizing radiation. The heavy images represent what the cytogeneticist sees, and the stick figures the presumed events that give rise to the heavy images. (Courtesy of Akio Awa)

● : Centromere.

have led to no loss or gain in genetic material—so-called stable aberrations, which do not impinge upon a cell's capacity to produce viable daughters—or have arisen in a cell that, for whatever reason, did not undergo its first postexposure division until quite recently. These events comprise only a fraction of those that would have been seen if chromosome preparations could have been made shortly after the bombing. Thus the frequency of unstable aberrations, those that compromise cell division, will be underestimated by an uncertain, possibly large amount.

The cytogenetic findings among the survivors have been summarized on a number of occasions,[50,51,52] but since many of these summarizations involved the earlier dosimetry, our remarks will be restricted to those that have employed the DS86 doses.[53] The data described involve 1,245 individuals exposed to an estimated kerma of 4 Gy or less who were studied between 1968 and 1980. Collectively, 119,497 cells derived from lymphocyte cultures of their blood have been analyzed (roughly 100 cells per individual), and some 5,646 of these cells exhibited an aberration. The distribution of these aberrations as it relates to dose is indicated in Figure 20.

This figure reveals the following: The frequency of aberrations, given as the percent of cells studied, rises from about 1% at 0 dose to over

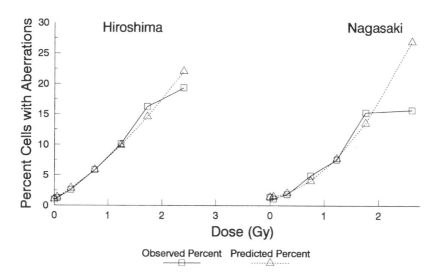

Fig. 20. Percentage of cells with chromosomal aberrations by marrow dose (Gy) among exposed survivors in Hiroshima and Nagasaki. The observed percentage at each dose is indicated by an open square, and the expected percentage under a linear–quadratic dose–response curve by an open triangle. (Adapted from Preston et al., 1988)

20% at 3 to 4 Gy in Hiroshima. The rise in frequency with dose, expressed either in kerma or absorbed by the marrow, is significantly nonlinear, as theory predicts. At the same dose, the frequency of aberrations is generally higher the younger the individual at the time of exposure, even when the individual's age at the time of study is taken into account. And lastly, at virtually all doses above 0.01 Gy the aberration frequency is greater in Hiroshima than in Nagasaki, and the difference in the rate of increase with dose between the cities is statistically significant. It is obviously tempting to attribute this difference to the greater neutron dose in Hiroshima than in Nagasaki, since for the same energy deposition per gram of cell neutrons are more densely ionizing (albeit indirectly) and, therefore, more likely to produce a chromosomal abnormality. Attempts to estimate the effectiveness of the neutrons emitted by the Hiroshima bomb, as contrasted with gamma rays, in producing chromosomal aberrations have not, however, been especially rewarding. Numerous reasons can be advanced for this failure to estimate effectiveness reliably, not the least of which is the small fraction of the total DS86 dose represented by neutrons.

These cytological studies of the survivors have been or are contributing to two other issues of moment to biologists. First, chromosomal abnormalities are frequently found in association with malignant tumors, as stated earlier, but it is unclear whether this relationship is a causal one or secondary to the malignant transformation of the cell that gives rise to the cancer. The cytogenetic information on the survivors might, with time, contribute importantly to this issue, since the chromosomal studies are prospective in that the abnormalities were found before cancer had occurred, and in the case of nonleukemic malignancies involve a tissue not immediately related to the one in which the cancerous changes presumably took place. As yet, despite the enormous amount of cytogenetic information that exists, a resolution of this problem is not at hand, since too few of the individuals studied cytogenetically have developed a fatal malignancy.

Second, there is interest in the extent to which chromosomal studies can be used to estimate the dose of an individual exposed to ionizing radiation when the dose is not known or cannot be estimated from the physical considerations associated with exposure. Such estimates can be and have been made, and are reasonably reliable for recent exposures, but the estimates are less satisfactory when the exposure occurred some years before the cytological studies.[54] Here there appears to be so much variability between individuals exposed to the same dose that cytological data alone are less informative than would be desired. Although the frequency of aberrations increases significantly with dose, the range of doses compatible with the frequency of aberrations observed in a given person is too great to provide a precise dose. This can be seen from Figure 21, where the frequency of chromosomal abnormalities within a sample of individuals is plotted against their DS86 doses. If the response to a given dose were the same for all individuals, and if all doses were exactly known, there would be only one point at each dose; however, as can be seen, this is not true. Some of the variation is due to random errors in the dose estimates but the variability observed is too great to be accounted for through these errors alone. Undoubtedly, a greater contributor is the error inherent in the estimation of the percent of aberrations. The results are based on the examination of a limited number of cells and are affected by differences in the baseline proportions of aberrations in different subjects. Other possibilities can also be envisaged. Individuals could vary in their response to a given dose, or in their capacity to repair damage, since what is observed is a relic of damage induced in 1945 at the time of the bombing. However, the evidence in support of either of these latter two alternatives is weak, at best.

Somatic Mutations

Geneticists distinguish between those mutational events, spontaneous or induced, that occur in cells that give rise to gametes—eggs and sperm—and those in cells that will not be transmitted but comprise the

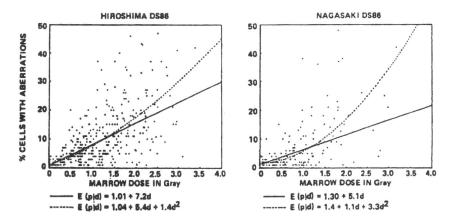

Fig. 21. The variation observed in the percentage of chromosomal aberrations among postnatally exposed survivors exposed to the same dose (Gy) for differing reasons. The expected percentage at each dose is indicated under a linear (solid line) and a linear–quadratic (dotted line) dose–response model. (Adapted from Preston et al., 1988)

rest of the body. The former are termed germinal mutations and the latter, somatic. Since somatic mutations appear to play a significant role in the occurrence of malignant tumors, their frequency among the survivors of the bombing of Hiroshima and Nagasaki is, and has been, a matter of interest. The frequency of somatic mutation is generally measured in the context of a specific gene whose function is known. Four types of genes (or gene families) have been studied—the HPRT gene, the glycophorin A gene (GPA), the alpha and beta genes responsible for the T-cell receptor (TCR), and more recently, the major histocompatibility complex (MHC). The first of these is located on the X chromosome and defines the structure of an enzyme known as hypoxanthine-guanine phosphoribosyl transferase, which appears to play a controlling function in the synthesis of precursors of nucleic acids, the purines (two of these, adenine and guanine are common constituents of DNA). The second, located on chromosome 4, is involved in the specification of a protein whose structure determines two common red blood cell surface molecules, known as M and N. The third determines the nature of a specific site of immunological activity on the membrane of the T-lymphocytes previously described. Finally, the major histocompatibility complex, which is located on human chromosome 6, specifies a series of cell-surface molecules important in determining tissue compatibility, as for example, in the transplantation of an organ from one individual to another.

While an explanation of the details of the assay techniques lies outside our purpose, it should be noted that different cells are used—lymphocytes in the case of the HPRT gene, the T-cell receptor site, and MHC, and red cells (erythrocytes) in the case of glycophorin.[55] What is measured is, in the main, the frequency of cells deficient in the activity of one of the genes associated with these specific genetic loci. For example, in the case of the glycophorin gene, if one selects individuals who have inherited from their parents one copy of each of the genes associated with the M and N cell surface molecules—that is, who are heterozygous for the genes responsible for these molecules—then normally their red blood cells will exhibit both the M and N molecule. These two molecules are proteins and have been shown to differ from each other by only two amino acids. However, if a mutation has occurred in one of the progenitors of the red blood cells (mature red cells do not have a nucleus, and hence the mutation could not have occurred within the red cell itself), and if this mutation has led to a loss in the expression of one of the glycophorin alleles, the cell will exhibit only the M or N molecule, but not both. It will appear to be either NO or MO instead of the expected MN. MM or NN cells could also be found in these heterozygotes, and would represent changes either in the N allele in the first instance, or the M in the second. These latter changes could arise through several different genetic mechanisms, leading to an alteration in the expression of the gene without its inactivation. It should be noted, however, that since the approach described, which was developed at the Lawrence Livermore National Laboratory, requires that the individual tested be MN in blood type, and since these individuals normally represent only about half of the persons in a population, this assay cannot be applied to all of the survivors.[56]

Through these methods it is possible to estimate the frequency of cells deficient in specific genetically controlled activities and to derive a mutation rate that is unique to an individual, or if averaged over individuals receiving a particular dose, to the population of individuals exposed to that dose, assuming, of course, that the deficiency arose through mutation in the relevant gene and not through some other mechanism. In a strict sense the direct demonstration of a mutational basis requires cloning of the putative mutant cell, its propagation in vitro, and evidence that the molecular structure of the gene has been altered. This is not always possible, and as a result the estimated mutation rate could be too high, contaminated by the occurrence of aberrant cells arising through nonmutational events. As a further caveat, these mutation rates are accurate representations of what occurred in the stem or precursor cells that were exposed (and gave rise to the cells that are studied) only if the altered precursor cells survive and reproduce as well as normal cells do. If they are less viable or reproduce less well, then the mutation rate will be underestimated.

To date, the number of individuals who have been studied is small, but numerous enough, nonetheless, to demonstrate a significant increase in the frequency of somatic mutation with increasing dose (either bone marrow or kerma). For example, in the glycophorin system where the frequency of aberrant cells of a particular type (NO, MO, or MM) varies from 20 (MO) to 27 (NO) cells per 1,000,000 cells in normal individuals or individuals exposed to less than 0.005 Gy, the estimated frequencies of NO and MO cells are about two times higher in individuals with a bone marrow dose of 1 Gy—50 (MO) and 60 (NO) cells per 1,000,000 cells.[57] An increase is also seen in MM variants, but it is much smaller—less than twofold. This is not unexpected on genetic grounds, since a more highly specific change in DNA is required to produce a particular variant gene than a gene inactivation.

Studies that have compared the frequency of variant glycophorin cells with the frequency of chromosomal abnormalities suggest a reasonably high correlation between these frequencies; the correlation coefficients vary from 0.42 to 0.63. However, individuals have been found with high frequencies of one but not the other of these measures of radiation-related damage, suggesting that there is a factor or factors confounding the measurement. There is, for example, some variability between individuals in their sensitivity to particular kinds of genetic damage, but the frequency also changes with time, possibly as a reflection of different stem-cell pools engaged in differentiation in the bone marrow.

The findings with respect to the HPRT locus are much less dramatic; indeed, it is questionable whether a significant dose–response can be demonstrated among the survivors. The spontaneous frequency of aberrant cells is about 3–4 cells per 1,000,000 and the increase at 1 Gy is approximately one additional cell, which is marginally significant statistically, at best. The difference between the two sets of findings is presumed to reflect differences in the events that normally intervene between the stem cell and the formation of functional erythrocytes and lymphocytes in vivo. In addition, replenishment of HPRT-deficient T-lymphocytes from mutant stem cells might be more limited than the replenishment of aberrant red cells, since some further processing in the thymus is required for T-cell differentiation (but not for the differentiation of red cells), and such processing is known to diminish with age. This is an important consideration, since the aberrant cells that are counted are the descendants of stem cells exposed in 1945, and suggests that the mutant frequency based on the HPRT-deficient cells becomes less reliable the longer the time after exposure to ionizing radiation.

Mutations at the T-cell receptor (TCR) site are among the most interesting and challenging, but perplexing, of the somatic mutations that have been studied.[58] This site, which is found on the surface of the vast majority of T-lymphocytes, is instrumental in the recognition of cell surface molecules and signal transduction, that is, the transfer of information among different biologic units, such as the nucleus of a cell and its cytoplasm. The receptor itself is made up of two different polypeptide chains (alpha and beta) associated with a still larger molecular complex, the cellular differentiation molecule commonly referred to as the CD3 complex. Unlike the case with respect to the glycophorin locus, normally only one of the two TCR alpha genes (located on chromosome 14) and one of the two TCR beta genes (chromosome 7) is functional. As a result, since there is no "normal" TCR alpha or TCR beta allele to mask the expression of a mutation, if one occurs in the functional gene, it will lead to a cell that is deficient in receptor activity. It is a situation analogous to that in X-linked inheritance, such as red–green color blindness in males. This naturally raises the question of how X-linkage affects interpretation of the data for HPRT also.

Over the past half dozen years, numerous mutant cell lines have been established from human cells exposed to ionizing radiation, but only recently have mutations at the T-cell receptor site been studied with a view toward assessing radiation damage. Mitoshi Akiyama and his colleagues have shown that among unexposed individuals about 2–3 cells in every 10,000 T-cells lack expression of the receptor, and that in cells exposed in vitro to 2 Gy this number rises to 20–40 in 10,000. These values, although clearly demonstrating that receptor site mutations can be identified and increase in frequency following exposure to ionizing radiation, cannot be used directly to estimate the dose of an individual, since the number of mutations that can be recovered diminishes as the interval between exposure and measurement increases. Specifically, when T-cell receptor site mutants are measured repeatedly over time in individuals who have received radiotherapy, the number of mutations recovered is highest soon after exposure and then declines with time, with a half life of about 2 years, but eventually returns to the normal level.

How uniform this approach to stability might be is uncertain, since there are age-related changes in the number of T-cell mutants as well.[58] Given these complexities, this system may not be useful for quantitative estimates of radiation exposure on an individual basis, but it could still be of use in ascertaining whether or not a group of individuals received significant exposure, provided the examinations are performed sufficiently soon after exposure (within two years or so), and an appropriate control group can be selected. As a system it has a number of advantages over those previously described. One of the most important of these is the fact that the frequency of mutations can be measured in surprisingly small quantities of blood, as little as one milliliter (about a fourth of a teaspoon or about a dozen drops). This is obviously appealing to study participants, especially in countries such as Japan where blood is not enthusiastically donated for any purpose.

At present only two surface molecules associated with the MHC locus have been measured; these are the ones designated as HLA-A2 and HLA-A24. However, this system has seen only limited application, and its ultimate utility in measuring the rate of somatic mutation following exposure to ionizing radiation has still to be determined. Preliminary results, based on the atomic bomb survivors, suggest that the age-related increase in mutations might be so pronounced as to obscure radiation effects in the survivors. Whether this will also be true in recently exposed individuals is under study.[59]

Techniques to measure mutations in somatic cells are in their infancy. It is reasonable to expect that more genetic systems will be available for study in the future, and that some of the troubling aspects posed by possible differences in the survival and reproduction of aberrant cells in vivo will be clarified. However, there will likely remain two questions, analogous to those that concern cytogeneticists, namely, "Can these techniques be used to estimate an individual's dose when this is unknown?" and "What are the long-term health implications for an individual with an elevated mutation rate?" As to the first of these questions, the present consensus among the users of these so-called biomarkers is that individual responses are too variable to provide acceptable dose estimates, particularly at low doses, but this situation may change in the future. As to the second, it is tempting to believe that if the mutation rate is higher for those genes that can be studied, it might also be higher for those genes associated with the occurrence of malignant tumors. If this correspondence does exist, somatic mutation could prove to be a means to identify, albeit imperfectly, those individuals who are at greater risk of developing a malignant tumor. This would obviously be a boon not only to the individual concerned but to science as well. Only further prospective studies will be able to support or refute this speculation.

The Outcome of Pregnancy among Exposed Pregnant Women

Given the physical trauma associated with the bombings as well as the exposure to ionizing radiation, it is reasonable to believe that women who were pregnant when exposed would have had an increased likelihood of an abnormal pregnancy outcome. Indeed, observations made in Nagasaki shortly after the bombing by members of the Faculty of Medicine of Kyushu University suggested that miscarriages were common. Among a total of 182 pregnant women attended by these physicians during the first 3 months following the bombing, 33 had miscarriages and another 12 delivered prematurely. James Yamazaki and Stanley and Phyllis Wright also explored this matter among a different sample of women known to have been pregnant at the time of the bombing of Nagasaki.[60] Their principal findings were as follows: Among 30 mothers with one or more "major" signs of radiation exposure—epilation, oropharyngeal lesions, purpura, or petechiae—who were within 2,000 m of the hypocenter, there were 7 fetal deaths (23.4%), 6 neonatal and infant deaths (26%), and 4 instances of mental retardation among 16 surviving children (25%).[61] The overall morbidity and mortality is approximately 60%. This is in sharp contrast to the group of 68 mothers without "major" signs but also within 2,000 m, in which the overall mortality was only 10%, and in the control group of 113 women exposed between 4,000 and 5,000 m, in which it was about 6%.

A similar study has not been done in Hiroshima; however, it is apparent from other sources of information that in this city too many of the embryos and fetuses exposed during approximately the first 4 months following fertilization failed to survive. Unfortunately, the data that are available in Hiroshima and Nagasaki are too sparse to permit the quantitative estimation of the probability of the loss of a pregnancy either as it relates to dose or to the stage in gestation at which exposure took place. Nonetheless, it appears that the risk of pregnancy loss was greater in the first 4 months, and particularly in the first 8 weeks, than later, and that therefore pregnancy wastage must be seen as one of the hazards of exposure to moderate to high doses of ionizing radiation.

The Future

The fabric of effects of exposure to the bombing of these cities is not fully woven. Some heretofore unsuspected consequences will surely emerge as the studies continue, and others will be better defined as the survivors age further. Over half of the survivors are still alive, and what their future holds can only be judged in terms of what the past has revealed. The past may, however, be an untrustworthy guide, since it is dominated by events involving individuals who were, on average, older when exposed than those now surviving, and whose lives were completed at a time when the practice of medicine in Japan had not achieved its current standards of diagnosis and treatment. It is obviously of the utmost importance that the morbidity and mortality surveillances continue, since only through these will answers be found to such issues as the effect of age at exposure on subsequent risk and the duration of expression of that risk.

Some effects seen among the survivors have not been observed in other exposed populations, and effects have been reported in these other groups that have not been seen in the atomic bomb survivors. The origin of these differences, if real, must be resolved if our understanding of the biological effects of ionizing radiation is to be complete. For example, an increase in brain tumors has been reported in some groups of individuals exposed to diagnostic X-rays involving the head but these tumors have not been shown to increase in the survivors. Similarly, when the results of the studies of the survivors are contrasted with those of individuals with ankylosing spondylitis treated with X-irradiation, both sets of data suggest that the risk of leukemia has largely disappeared with time, but differ with regard to the risk of other malignancies. The survivors show no abatement of this risk, whereas among the spondylitics the risk no longer appears elevated. Do these differences imply that one study is correct and the other wrong? Do they suggest a racial difference, since the other studies do not involve Japanese? Is the difference, attributable to the fact that few of the spondylitics were under the age of 25 when exposed, whereas many of the survivors were, and the risk has been shown to be higher among the young? Is there, in short, a plausible explanation that could reconcile the disparate findings?

Still other unsolved problems involve a better characterization than has yet been achieved of the contribution of host and environmental factors to the occurrence of radiation-related malignancy. A variety of studies from many areas in the world indicate that some cancers aggregate in families, suggesting that genetic or familial factors play a part in their etiology. But it is not known whether those individuals in Hiroshima and Nagasaki who have developed malignancies, presumably related to their exposure to ionizing radiation, come from families who, even in the absence of such exposure, are cancer-prone. If this should be so, how could this information be used to identify those persons with the greater risk? And how much greater is that risk? For example, breast cancer, especially the bilateral occurrence with early onset, appears to occur more frequently among women who are related, commonly as sisters, suggesting a genetic basis, but Ban and his colleagues have studied the radiosensitivity of skin fibroblasts from women who were atomic bomb survivors with and without breast cancer and have found no difference in the means or variances in radiosensitivity of their fibroblasts.[62] Does this imply that individuals genetically prone to breast cancer are not more sensitive to radiation, or does it mean that the assays of radiosensitivity used were not adequate to test the hypothesis of a genetically related sensitivity, or was the failure to find an effect attributable to chance? The BRCA1 gene, the one most clearly associated with breast cancer, has an allele frequency of only 0.003 or so, and as a result might not be represented at all in a small sample of individuals.

Recent advances in molecular biology hold promise of eventually providing answers to these and other questions pertinent to radiation-related carcinogenesis. As a possible illustration of what might be anticipated, consider the role of p53 mutations in the origin of undifferentiated thyroid gland carcinomas. It has been shown that among a sample of 10 differentiated papillary adenocarcinomas in Hiroshima not a single one exhibited a mutation in the segment of the p53 gene identified as exons 5-8 whereas such mutations were seen in 6 out of 7 undifferentiated thyroid carcinomas. This finding prompted the investigators to conclude that their results "strongly suggest that, in human thyroid glands, p53 mutations play a crucial role in the progression of differentiated carcinomas to undifferentiated ones."[63] The importance of this finding relates to the prognosis associated with differentiated and undifferentiated thyroid malignancies—differentiated tumors have a high curability whereas undifferentiated ones usually kill the host soon after diagnosis.

Improving the estimates of cancer risk as well as other radiation-related damage will necessarily remain a central activity of the Radiation Effects Research Foundation through the years immediately ahead, but better estimates without an understanding of the underlying molecular and cellular processes that are involved is an empty victory. This is not to denigrate the importance of risk analysis since it can be helpful in more ways than just through the quantitative expression of risk. It can identify issues that should be of concern to experimentalists, and this argues for a more dynamic interaction than commonly occurs between epidemiologists and statisticians, on the one hand, and experimental biologists, on the other. Nonetheless, in the final analysis, intelligent intervention and the amelioration of risk must be based on biological understanding. Developments in biology have lifted a corner of the curtain that has obscured this understanding. To further these advances, and in particular their application to radiation-related damage, it is not only important that current tissue repositories, which focus primarily on malignant tumors, be supported, but that means be found to collect and store tissues, such as peripheral lymphocytes, on a wider sampling of exposed individuals, most of whom will not die from a malignancy. These tissues and cells can serve as the bases for future molecular and cellular studies as newer, more insightful techniques become available. It is known even now that the cells of some malignant tumors exhibit specific chromosomal changes, and that these chromosomal aberrations can be seen in other tissues as well, including lymphocytes. But as previously stated, it is not known whether these changes can be seen before a malignancy is clinically apparent. The Adult Health Study has been, and will undoubtedly continue to be, the primary source of much of this biological material. This argues not only for the continuation of these examinations but means, in turn, that the Foundation must maintain an active laboratory program, one with a staff and facilities capable of using the newer techniques as they evolve.

There are still other ways that the studies in Hiroshima and Nagasaki can contribute to the betterment of human health and the quality of life. Over the thirty-odd years the Adult Health Study has continued, an enormous body of data has accumulated pertinent to the process of aging among the Japanese and presumably other ethnic groups. Analysis of these data holds promise of insights into childhood precursors of subsequent cardiovascular disease, for example, or into events premonitory of the occurrence of senile dementia later in life. As stated earlier, until recently the analysis of longitudinal data was difficult, since neither the statistical techniques nor the means of managing such data were available. This is no longer true. Suitable statistical techniques exist as does the computing capability to implement them. It can be presumed, therefore, that in the years ahead the Foundation's data will see important uses in the study of variation among individuals in the process of aging.

7

The Prenatally Exposed Survivors

The early clinical studies, save those of leukemia and lenticular opacities, revealed only one other conspicuous radiation effect. This exception involves a special group of survivors—those exposed prenatally, that is, in utero. It was evident as early as 1949 that among these individuals the prevalence of severe mental retardation was sharply elevated, particularly among those exposed in the earlier months of gestation. These retarded individuals are not numerous, but the problems they and their parents confront are especially affecting, as we have said. Their experiences are, however, pertinent not only to an understanding of the effects of ionizing radiation but to the consequences of prenatal exposure to methyl-mercury, alcohol, lead, and possibly other agents that can impair normal development of the brain. The instantaneous nature of the exposure of these survivors to ionizing radiation makes identifying the critical, vulnerable, stages in embryonic or fetal development easier than in the other instances cited, where the commonly chronic nature of the exposure makes the stage of their effect difficult to identify, if not impossible. Yet knowledge of these vulnerable periods could contribute significantly to the avoidance of exposure at those times of greatest risk and the development of effective ameliorative or preventive measures.

Insofar as other birth defects are concerned, despite much experimental evidence suggesting a greater frequency of occurrence following exposure to ionizing radiation, no increase has been observed among the prenatally exposed in any of the abnormalities that arise spontaneously in all populations, such as cleft lip with or without cleft palate. It is known, as has already been noted, that many women who were pregnant at the time and exposed to relatively high doses of radiation lost their infants.[1] Presumably this was more likely to have happened if the embryo or fetus was abnormal, particularly if the abnormality involved some vital center. But ionizing radiation can impair only those developmental events that are actually occurring at the time of exposure. Some of these, for instance, the closure of the primitive neural tube or the palate of the mouth, take place rapidly, in a few days, and a failure to find these abnormalities might merely reflect the small number of embryos or fetuses who were exposed at the appropriate developmental juncture and survived.

Occasionally, although the number of cases may be very small, the correspondence between the time of occurrence of exposure and the pertinent embryological events suggest an association of irradiation with an observed developmental anomaly, albeit one that cannot be established unequivocally. For example, among the prenatally exposed, three individuals, all females, have congenital dislocation of one or both hips. The socket of the pelvis into which the femur, the long bone of the upper leg, inserts normally develops in the 10th through the 13th week following ovulation,[2] when all of these cases were exposed. This might be coincidental, but it could be causal. These uncertainties notwithstanding, it is prudent to assume that other malformations arise, but do not present as imminent a risk as does damage to the developing brain. It has always seemed reasonable also to assume that the prenatally exposed would confront the same spectrum of risks seen among the postnatally exposed survivors—cancer, lenticular opacities, and possibly sterility or impaired fertility—but were there other risks, possibly related to the developmental events transpiring at the moment of their exposure?

Mental Retardation

Few population-based studies of the effects of prenatal exposure on the developing human embryo and fetus exist. Among these, however, the size of the population and the variability in dose and prenatal age at exposure make the experiences in Hiroshima and Nagasaki by far the most consequential. The members of these populations were presumably exposed at a variety of developmental stages and sensitivities. Many of these individuals have been followed clinically since shortly after their birth to the present. Two conspicuously abnormal effects of a developmental nature have emerged. First, some 21 of the approximately 500 pregnancies of mothers in Hiroshima and Nagasaki exposed to more than 0.01 Gy in the study sample terminated in a child with severe mental retardation, a significantly greater number than the 4 or 5 normally expected, and second, many of these pregnancies gave rise to an infant with a disproportionately small head. In this context, severe mental retardation implies an individual who is unable to form simple sentences, to solve simple problems in arithmetic, to care for himself, or is unmanageable or institutionalized.

As has been stated, the most important single biological factor in determining the nature and extent of the insult to the developing embryo or fetus following exposure to ionizing radiation is the developmental age at exposure. This age is not, and presently cannot be, known accurately on an individual basis. But there is a high correspondence between developmental age and gestational age. Today, gestational age can be estimated by more precise means than those available when the individuals described here were exposed. Fifty years ago, it could only be inferred from the first day of the last menstrual period. Under normal circumstances, ovulation occurs about two weeks after the onset of menstruation, and fertilization, if it takes places, soon thereafter. Gestation will last for another 38 weeks. Thus, given an infant's date of birth and the date of the bombing, an estimate can be made of the age at exposure expressed as weeks following ovulation.[3] It is important to note that this method of computing age assumes that all pregnancies develop at the "average" rate, but embryological studies tell us that this is not true. In specific instances, the developing embryo may be a week or more ahead or behind the norm. These errors in the estimation of developmental age are presumed to be independent of exposure. This may not be true—survival could depend upon developmental age, and in a small sample, such as exists at the higher doses, chance by itself can play curious tricks.

Individual developmental ages have limited accuracy and are, therefore, frequently grouped to reflect broader stages in normal embryological and fetal development. Insofar as the developing brain is concerned, four discretely different stages are often recognized. Measured from the presumed moment of ovulation, the categories commonly used are: the moment of ovulation through the 7th week (for tabular convenience indicated as 0–7 weeks), the 8th through the 15th week (8–15), the 16th through the 25th week (16–25), and, finally, in the 26th week or later (26+). These groupings correspond to the timing of the following biological events: In the first period, the cells that will give rise to the neurons and neuroglia, the two principal types of cells that will populate the cerebrum, emerge and are actively dividing. In the second, the production of neurons increases rapidly through cell division; they lose their capacity to divide further, and migrate to their final location in the primitive cerebrum. In the third, differentiation of the immature neurons accelerates, synaptogenesis (the process of establishing connections with other neurons) that began about the 8th week increases, and the definitive architecture of the cerebrum unfolds. The fourth period is one of continued architectural and cellular differentiation and synaptogenesis of the cerebrum concurrent with accelerated growth and development of the cerebellum.

When the frequency of mentally retarded individuals is examined in the light of their doses and the age groupings at which they were irradiated, the following emerges.[4]

First, as Figure 22 illustrates, the risk of severe mental retardation is greatest when exposure occurred 8 to 15 weeks after ovulation. This exceptionally vulnerable period coincides with the most rapid production of neurons and when all, or nearly all, of the newly formed immature neurons migrate to the cerebral cortex from the proliferative zone where they originated. There is no apparent increased risk prior to the 8th week or after the 25th, but insofar as the earlier period is concerned, the data could be biased because of the relatively high mortality revealed by the studies of Yamazaki and his colleagues.

Second, within this period of greatest risk, damage expressed as the frequency of cases of severe mental retardation can be described by a model that relates frequency to dose linearly. However, as shall be seen, other models, including those that assume a threshold, will fit the data. In the linear model, about 45% of fetuses exposed to 1 Gy in this period

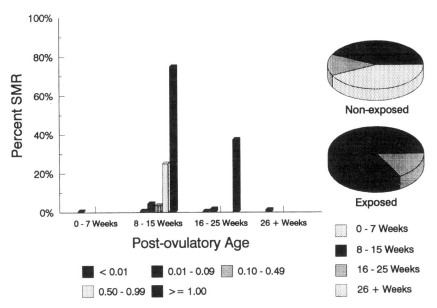

Fig. 22. The frequency of severe mental retardation (SMR) among prenatally exposed survivors by dose (Gy) and postovulatory age at exposure in weeks. The pie diagrams indicate the proportion of cases of severe mental retardation in the nonexposed and exposed samples at various postovulatory ages. (Adapted from Otake, Yoshimaru, and Schull, 1987)

will be mentally retarded. This risk is more than 50 times greater than that seen among individuals who were not exposed or who were exposed to doses of less than 0.01 Gy.

Third, a period of lesser vulnerability appears to exist in the interval 16–25 weeks after ovulation. Here a threshold seems to exist; no apparent radiation-related cases are seen at exposures of less than 0.50 Gy.

The presence of a threshold at this later developmental stage, and its uncertainty in the 8–15 week period should not be construed as contradictory. These differences are consistent with the supposition that the biological events involved in the induction of mental retardation are different in the earlier period of development than in the later. In the first instance, the neuronal cells are immature, undifferentiated; whereas in the second, when neuronal production lags and migration has been largely completed, the cortical cells at risk are differentiating or already differentiated, and differentiated cells are less vulnerable to ionizing radiation than are immature ones.

Finally, it warrants noting that three of the severely retarded children, all in Hiroshima (estimated uterine absorbed doses: 0, 0.29, and 0.56 Gy), are known to have, or have had (one is dead), Down Syndrome; a fourth, also in Hiroshima (estimated uterine absorbed dose 0.03 Gy), had a viral disease of the brain (Japanese encephalitis) in infancy; and a fifth, in Hiroshima, had a retarded nonexposed sibling. In these instances, the mental retardation may merely be a part of the former syndrome or secondary to an infection, or inherited, but in any event not radiation-related. It can be argued, therefore, that these cases should be excluded from the estimation of the effects of radiation. If this is done, virtually the same relationship of mental retardation to dose obtains; the increase at 1 Gy is now 40%, instead of 45%. Thus the main conclusions are not dependent upon the inclusion or exclusion of these individuals.

Small Head Size

Although earlier investigators commented on the occurrence of small head size among the prenatally exposed, the first systematic study was that of Robert Miller.[5] Small head size here implies a head circumference that is two or more standard deviations below the average circumference of all of the individuals in the study sample. Miller's findings have been reaffirmed and extended using the new doses.[6] The relationship of small head size to dose and gestational age revealed by the latest analysis is shown in Figure 23.

In this analysis, among 1,473 individuals on whom the head circumference was measured at least once between the ages of 9 and 19,

62 had a small head by the criterion just stated. Some of these individuals were mentally retarded, but many were not. There was, however, little difference in the average IQ of the mentally retarded with small heads and those without; the averages being 64, and 69, respectively. Although some depression in IQ appears to exist among individuals with small heads but not recognized to be retarded clinically, it is not large. Their average IQ was 97, which is lower than the average of the overall sample, but the difference is not significant statistically.

The rubric "small head size" may, indeed probably does, cover a variety of different developmental "abnormalities." Among the individuals with a small head and severe mental retardation, for example, some clearly invite the clinical diagnosis of microcephaly, since the head is not only unusually small but misshapen—often pointed. Still others, and they are more common, have a head that is proportionate in all dimensions, albeit small. Moreover, since head size varies in all populations, some of the individuals here designated as having a small head may merely represent the lower extreme of normal variability. Indeed, based

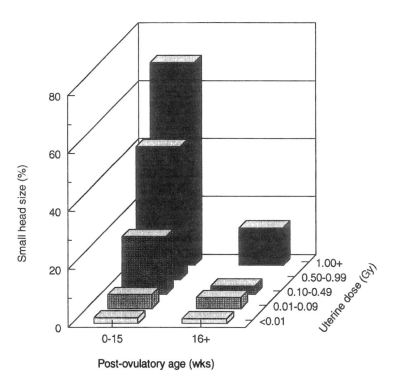

Fig. 23. The frequency of small head size by dose (Gy) among the survivors exposed prenatally in the first 16 weeks postovulation and thereafter. (Adapted from Otake and Schull, 1992)

on the criterion for small head size that has been used, if head sizes are approximately normally distributed, some 2.5% of "normal" individuals would be so classified.

Since the mean IQ and its standard deviation among the 47 individuals having a small head without severe mental retardation approximate the values seen in the entire clinical sample, it is conceivable that a significant fraction of these individuals are the "normals" mentioned above. Accordingly, an attempt has been made to estimate the excess number of individuals with small heads ostensibly attributable to exposure to ionizing radiation. Among the 62 individuals with small head size, 37 would be expected on the basis of the normal variability in head circumference, and the observed and expected numbers agree reasonably well when exposure occurred in the 16th postovulatory week or later. However, there is a striking excess before this time—where 16 individuals with a small head were expected, 46 were observed, an excess of 30 cases. If the small head size among the 12 individuals with

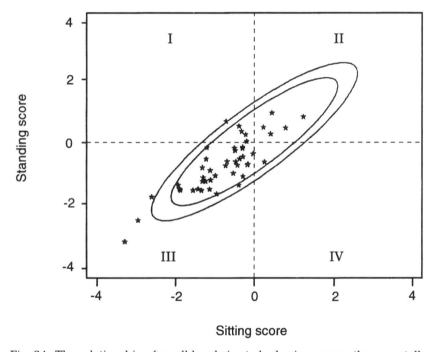

Fig. 24. The relationship of small head size to body size among the prenatally exposed survivors. Note that the preponderance of individuals with small heads lies below the average sitting and standing height for all individuals in the study sample, that is in quadrant III. (Adapted from Otake and Schull, 1992)

severe mental retardation is secondary to brain damage, this leaves 18 cases that might represent radiation-related instances of growth retardation without accompanying mental impairment. Can these latter individuals be distinguished from those to be expected normally?

To explore this possibility the 47 cases of small head size without mental retardation were located in a plot of standing versus sitting height expressed as age- and sex-standardized deviates based upon the full sample of 1,473 individuals (see Figure 24). Note that the individuals with small head size but no apparent mental retardation are disproportionately represented in quadrant III, which defines the values below the mean sitting and standing heights in the ellipses that encompass 95% or 99% of the sample, suggesting that small head size reflects a generalized growth retardation. As will be noted, three individuals lie outside the 99% ellipse; the uterine dose (DS86) in these three cases was 0, 0.04, and 0.49 Gy. These observations support the notion that small head size is not an independent teratogenic effect, but is either secondary to mental retardation or to a generalized growth impairment.

Other Evidence of Brain Damage

The findings that have been briefly described have encouraged a search for other evidence of changes in brain function, possibly less dramatic than the occurrence of mental retardation but that would, in a sense, corroborate the results or shed light on the nature of the damage to the brain associated with ionizing radiation. Several sources of such data were available. One of these involved the performance of the prenatally exposed on one or both of two standard Japanese intelligence tests—the Koga and the Tanaka-B—and another was their academic performance in school. The intelligence tests were administered at the clinical facilities of the Commission by trained psychometrists in 1955–1956, when the children were approximately 11 years old, and the school performance scores were the evaluations of their teachers in the first four years of primary school.

Intelligence

It is presumed that one can recognize intelligence, but its definition is nebulous. It has been described as the ability to manage oneself and one's affairs prudently; to combine the elements of experience; to reason, compare, comprehend, use numerical concepts, and combine objects into meaningful wholes; to have the faculty to organize subject-matter experience into new patterns; or to have the capacity to act purposefully, think rationally, and deal effectively with one's environment. Given such differences in definition, it is natural that the bases of measurement should vary.

Most intelligence tests seek to measure "general aptitude" through the use of a standardized series of questions, problems, or both, but precisely what a given test measures is debatable, since the importance given to verbal ability, psychomotor reactions, and social comprehension, for example, differs. Thus, the score attained by an individual will depend to some degree upon the type of test used; however, generally, individuals scoring high on one type of test tend to obtain high scores on other similar tests. This correspondence in scores is unlikely to be fortuitous, but as said, whether the tests can be construed as broad measures of intelligence is moot. This is not especially important in the present context, since these tests are seen only as comparative measures of cortical function and not as statements about intelligence.[7]

Analysis of the intelligence test scores, using estimates of the DS86 uterine absorbed dose, revealed the following:[8] 1) There is no statistically significant evidence of a radiation-related effect on intelligence among those individuals exposed within 0–7 weeks after ovulation or in the 26th or subsequent weeks; 2) for individuals exposed at 8–15 weeks after ovulation, and to a lesser extent those exposed at 16–25 weeks, the mean test scores but not the variances are significantly dissimilar among dose categories; 3) the cumulative distribution of test scores suggests a progressive shift downward in individual scores with increasing dose; and 4) within the group most sensitive to the occurrence of clinically recognizable severe mental retardation—individuals exposed 8–15 weeks after ovulation—the relationship of intelligence score to estimated DS86 uterine absorbed dose appears linear; the diminution in intelligence score under the linear model is 25–31 points at 1 Gy. If the clinically diagnosed cases of mental retardation are excluded from the sample analyzed, the estimated diminution in intelligence is less—21–25 points. The results are less clear-cut with regard to the period 16–25 weeks; the increase with increasing dose is statistically significant only if the mentally retarded cases occurring in this time interval are included (see Figure 25).

The progressive shift downwards in intelligence scores with increasing dose suggests that the effect on the developing brain is a quantitative rather a than qualitative one. If this is so, the observed effect on intelligence test scores could arise either as a result of the increased radiation-related killing of immature neurons or their precursors, or a quantitatively greater impairment of their capacity to move from their sites of origin to those of function, or possibly both.

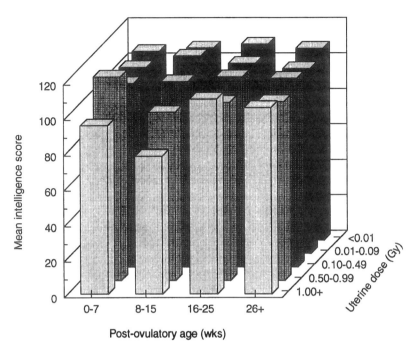

Fig. 25. Bar diagrams illustrating the difference in mean IQ among prenatally exposed survivors by dose (Gy) and postovulatory age at exposure in weeks. (Adapted from Schull, Otake, and Yoshimaru, 1988)

School Performance

Japanese regard for and motivation toward an education are proverbial; both parents and children look upon schooling as "a primary avenue for bettering one's self and one's family, and the urge to achieve betterment is strong."[9] This urge for improvement results in a strong emphasis upon school attendance and performance. The Japanese child rarely misses school without good cause, and while in school places high value upon achievement. If, then, a child's attendance record is highly correlated with illness and performance with innate ability, attendance might be correlated with exposure as a consequence of more (or less) frequent illness, and performance might reflect the nature of the developmental events that obtained when exposure occurred. Accordingly, with the approval and assistance of the Municipal Board of Education in Hiroshima and the written consent of the parents, the school records of the prenatally exposed were microfilmed in August and September 1956.[10] At that time, these survivors were 11 years old, and most had completed their fourth year of schooling. The records themselves included information on school attendance, performance in various subjects, their behavior, and physical status.

Data were sought on all qualifying children in both cities. However, for reasons no longer clear, this information was not collected in Nagasaki. The intent of collecting these data was to correlate school performance with the results of the intelligence tests conducted at the clinical facilities of ABCC from August 1955 through June 1957.[11] Retrospectively, it is uncertain whether these early investigators recognized that school performance itself could be used as a measure of radiation-related brain damage, but presumably they did not, since no analysis of these data occurred until recently.

Attendance. The attendance records indicate the actual number of days of school in each third of the academic year, and within each such segment, the total number of days of school missed by a specific student, the number of days tardy, and the number of days he or she left school prematurely. The days missed are further subdivided into absence because of illness and absence for other reasons, such as death or illness of another member of the student's family. The ratio of the days missed through illness to the total days of school (that is, days at risk of an illness-related absence) affords a crude measure of the health of a given student.

With some 250 school days per academic year (Japanese children attend school on Saturday mornings as well as weekdays), the average child in these years failed to attend school less than 5 days a year—a remarkable performance. On average, school absences for illness tend to increase generally with increasing dose when the clinically recognized cases of mental retardation are within the sample analyzed. This holds true under several different dose–response models, including a simple linear one. However, the evidence, particularly when gestational age groups alone are considered (genders combined), supports a linear–quadratic model more consistently. The latter continues to be true when the mentally retarded, who are more likely to be ill than their normal age peers, are excluded, but the overall effect of radiation on attendance becomes more equivocal.

When one turns to age-specific categories, the following is observed: first, absences diminish in number as the child advances in school for all irradiated postovulatory age groups. However, the trend is significantly different among all four age groups when considered two at a time. Second, with and without respect to gender, the largest and most consistent effect associated with radiation involves the 8–15 week age group, suggesting that these children experience more health problems.

Performance. Elementary schooling in Japan has been coeducational since shortly after the beginning of the reign of the Emperor Meiji in 1868. Presumably, then, the education of boys and girls should not differ. However, the Japanese culture is decidedly male-centered and it may be assumed that, in fact, somewhat more attention focuses upon the education of boys than girls. In school years 1–4 the Japanese student receives training in seven subjects: Japanese language (*Kokugo*), science (*Rika*), arithmetic (*Sansu*), social studies (*Shakaika*), music (*Ongaku*), fine arts (*Zuga kosaku*), and physical education (*Taiiku*).

Every student's performance with respect to these subjects is evaluated, and a score, varying in unit steps from $+2$ to -2, is assigned for each. The highest and lowest five percentiles of the class are assigned scores of $+2$ (very good) and -2 (poor), respectively. The next highest and lowest 20 percentiles are given $+1$ (somewhat above average) and -1 (somewhat below average), and finally, the middle 50% are given zero (average). In the analysis, these values were converted to a 5-point scale, giving the highest and lowest groups the values 1 and 5, respectively, etc. Clearly, this system of scoring does not give rise to a continuous distribution of performance scores for any one subject, and hence an element of approximation, to be described shortly, is introduced into the tests. However, when all of the scores on an individual are averaged, and these averages distributed over all students, the result is a more continuous and symmetric distribution.

If all children within a score group were precisely the same age and if gender did not enter into the teacher's rankings of the children, the percentile to which a child is assigned should not be related to age or gender. However, within a given grade in school, under normal circumstances, children can vary by as much as a year in age; this might be reflected in performance in certain subjects, notably physical education. It is reasonable to assume that the first-grader who is a year older than some of his or her classmates would surpass most of them in running, jumping, and other physical activities, and school performance, particularly in physical education, could show an age dependence. This should not be an important consideration, however, since the analysis relates performance to dose within relatively narrow gestational age groups. Nonetheless, the means within all of these groups might not be precisely 3 as a consequence of this dependence.

Since seven different school performance scores existed, one for each subject, the question naturally arose whether the scores should be analyzed separately or whether some single measure should be employed. The individual scores are known to be correlated and, therefore, an analysis based on the separate scores would be difficult to interpret, since one would be repeatedly analyzing some of the same information. The average of the seven individual scores seemed a better choice as the basis for analysis. This decision was not capricious, and can be justified by a statistical technique known as principal component analysis. The latter is a way of determining the most important weighted combinations or factors among a series of variables; these factors can then be used to analyze the variability that is observed among individuals. In the present instance, the most important factor, the one that captures the greatest amount of the interindividual variability, is related to the mean, or average of the seven school scores.

The findings can be summarized as follows.[12] As Figure 26 indicates and a simple regression analysis shows, damage to the 8–15 week-old fetal brain appears to be linearly related to absorbed dose. This is true with or without inclusion of the 14 cases of retardation in the study sample. As reflected in these scores, damage to the fetus exposed at 16–25 weeks after ovulation appears similar to that seen in the 8–15 week group.

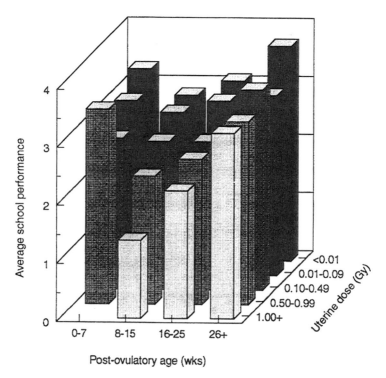

Fig. 26. Bar diagrams illustrating the difference in average school performance among prenatally exposed survivors by dose (Gy) and postovulatory age at exposure in weeks. (Adapted from Otake, Schull, Fujikoshi, and Yoshimaru, 1988)

In addition to the results of a single analysis, other more elegant statistical methods that take into account the nonindependence of a child's performance in various subjects were also used. These too show a highly significant relationship of exposure to achievement in school in the periods 8–15 weeks and 16–25 weeks after ovulation. This trend is stronger, however, in the earliest years of schooling, suggesting that some amelioration of the effect occurs as the child grows older. Whether this is attributable to more teacher attention or to greater opportunities for socialization, both of which can affect performance, is unclear. In the groups exposed within 0–7 weeks or 26 or more weeks after ovulation, there is no evidence of a radiation-related effect on scholastic performance.

Are these findings compatible with those on the occurrence of mental retardation and the diminution in intelligence test score, and if so, is there an argument that makes them coherent? With respect to the second question, if radiation is seen as operating on a continuum of qualities of brain function, and if the latter qualities are reasonably well measured by the analytic models that have been used, the results are coherent. As to their compatibility in terms of magnitude of damage occasioned by exposure to 1 Gy, within the most vulnerable period, 8–15 weeks after ovulation, prenatal exposure to ionizing radiation increases the frequency of mental retardation among those exposed to about 45% at 1 Gy (background frequency: 0.8%); whereas the loss in IQ is approximately 25 points at 1 Gy. Exposure to 1 Gy prenatally appears to imply a decrement in average school performance score of about 1.6, which is tantamount to the shift of an average individual from a score of 3 to about 1.4, that is, from the middle 50th percentile of the class to the lower 5th or 10th percentile.

Seizures, Neuromuscular Performance, and Perceptuo-Motor Maturation

Still other observations exist bearing on the nature and extent of the damage to the developing brain following prenatal exposure to ionizing radiation. Three in particular warrant mention. These are the occurrence of unprovoked seizures (convulsions), performance on two neuromuscular examinations that assess fine motor coordination and strength, and perceptuo-motor maturation.

Seizures. Seizures are frequently associated with impaired brain development, and therefore, could be expected to occur with greater frequency among children with radiation-related brain damage than among children without such damage. Seizures in early childhood are dramatic events rarely forgotten by the mother, but their occurrence could be related to a variety of events. They might result from birth injuries, from congenital malformations involving the central nervous system, from inherited diseases, or from febrile episodes associated with extracranial infections, but may also appear unprovoked—to be attributable to none of the factors just enumerated. It is this latter class that is of particular interest here.

Many studies have demonstrated that there is a high risk of seizures and epilepsy in mentally retarded individuals, and both mental retardation and epilepsy have been associated with areas of misplaced or ectopic gray matter. If a continuum of brain abnormalities stemming from developmental errors underlies the biological origin of seizures and mental retardation, it is legitimate to postulate an increase in the frequency of children with seizures who were prenatally exposed to ionizing radiation during a developmentally vulnerable period.

No cases of seizures were observed among individuals exposed 0–7 weeks after ovulation at doses higher than 0.10 Gy.[13] The highest frequency is seen in the highest dose groups in postovulatory weeks 8–15 (see Figure 27). When the 22 mentally retarded individuals are included

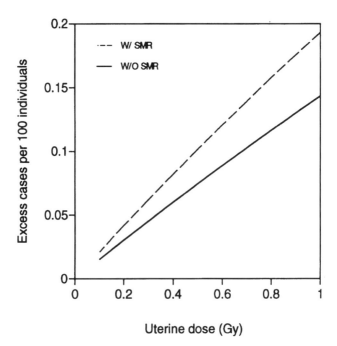

Fig. 27. The frequency of seizures among survivors with and without severe mental retardation (SMR) exposed prenatally in weeks 8–15 after ovulation. (Adapted from Dunn et al., 1988)

in the study sample, a statistically significant increased risk of seizures with increasing dose is consistently observed only between 8 and 15 weeks after ovulation. This is true for all seizures without regard to the presence of precipitating causes and for unprovoked seizures (those without a known cause). Within this critical period, their occurrence is linearly related to the absorbed dose received by the uterus. Since the simultaneous occurrence of seizures and mental retardation is high (20%), and since mental retardation is known to be associated with exposure in this same gestational period, the cases of severe mental retardation were excluded to see if the effect was attributable solely to their inclusion in the analysis. When this was done, a suggestively significant risk is detected only at 8–15 weeks after ovulation and only in the group with unprovoked seizures.

The relative risk of unprovoked seizures following exposure to 0.5 Gy or more at this time in brain development, as contrasted with those individuals who were not exposed, is 24.9 when the mentally retarded are included and 14.5 when they are excluded. At face value, then, there still seems to be an effect not ascribable to the association of mental retardation with dose and seizures.

It is arguable which of these two analyses best describes the data biologically. The answer hinges on the mechanisms underlying the occurrence of seizures and mental retardation following prenatal exposure to ionizing radiation. If, for example, seizures and mental retardation arise through two different mechanisms, both possibly dose-related—one of which causes seizures and the other mental retardation in some individuals who are then predisposed to develop seizures—the mentally retarded must be excluded if one is to explore the dose–response relationship associated with the first mechanism. If, however, mental retardation and seizures arise from a common brain defect manifesting itself in some instances as mental retardation and in others as seizures, the mentally retarded should not be excluded.

At present, the only evidence arguing for a common radiation-related developmental defect is the occurrence of ectopic (abnormally situated) gray areas in some instances of both disorders.[14] This evidence is difficult to put into perspective, however, for while it is known that ectopic gray matter occurs among some of the radiation-related instances of mental retardation, the observation of ectopia in individuals with seizures is based on other studies. As yet, there has been no investigation of the frequency of occurrence of ectopic gray areas among the prenatally exposed with seizures but no mental retardation using contemporary methods of imaging the living brain.

Neuromuscular Performance. In 1961–1962, coincident with the periodic physical examination of the prenatally exposed children, two measures of neuromuscular performance were obtained.[15] One of these, the grip test, measures the grip strength of an individual's hands. The other is a repetitive action or a speed test, and is generally thought to be a measure of neuromuscular coordination.[16] Poor performance on either or both of these tests could signal cerebral or cerebellar damage or both. The child might do poorly because cerebral damage impaired the ability to fully comprehend what he or she was expected to do, or because of the loss of cerebellar neurons and the consequent inability to implement intent.

The data from these tests illustrates many of the problems inherent in the analysis of epidemiological observations. Unlike the experimentalist, the epidemiological investigator must accept the fact that factors other than the variable(s) of primary interest can influence the data being analyzed. These other factors could produce a spurious effect or obscure a true one. Customarily, allowance is made for such potential sources of ambiguity either through matching comparison groups as closely as practical, or through the choice of the statistical procedure employed, or often both. Body size, for example, could affect performance on either or both of these tests, and since exposure to the atomic bomb retarded growth among some of these children, poor performance could reflect this retardation and not represent an independent functional effect. To address this possibility, the body size of the children—their weight, stature, sitting height, and chest circumference—was recorded at the time of the tests. With the appropriate statistical techniques, these measurements can be used to determine the effect on the developing brain of exposure to ionizing radiation mediated indirectly through variation in body size rather than directly through exposure of the brain.

Among the 941 children on whom neuromuscular tests exist, 888, including 15 cases of mental retardation, have all of the other requisite information—a DS86 dose, weight, stature, sitting height, and chest circumference. These 888 children provide the basis for the analysis to be described. Since mentally retarded individuals are known to perform poorly on tests of neuromuscular skill, two analyses were undertaken, one with and one without the retarded individuals.

Several different analytic methods were used. These included conventional regression analysis, using the body size measurements as potential sources of error, as well as regressions using standardized variables, that is, with the prior removal of the effects of gender and city within gestational age groups. Standardization is tantamount to putting the observations on a common scale. Since the findings are substantially the same for the observed and the standardized variables, for convenience only, the findings with respect to the standardized variables will be summarized.

However, it should first be noted that no change in the means of the "raw" (body size ignored) grip test score could be demonstrated between dose categories within a gestational age group, either including or excluding the individuals with mental retardation. However, the means of the "raw" repetitive action test score were significantly different between dose categories in the 8–15 week age group, but only when the mentally retarded were included in the analysis.

When the standardized scores were used the following was observed:[17] If no correlation exists between absorbed dose and the individual hereditary and environmental factors that contribute to body size, as seems likely, the analysis without allowance for differences in body size should expose both the direct (on the test score itself) and indirect (through the effect of body size on the score) effects of ionizing radiation. Interestingly, there is a highly significant depression in the grip test score with dose in the 8–15 week age group when body size is ignored and the individuals with mental retardation are included, but not when allowance is made for radiation-related differences in body size. Similarly, the repetitive action test score declines significantly with dose in the 8–15 week age group, when body size is ignored and the mentally retarded individuals are included. In the latter instance, absorbed dose is highly and significantly negatively related to test score even with the body size measurements added to the regression model. But the latter holds true only if the individuals with mental retardation are included in the analysis. This suggests that prenatal exposure affects the repetitive action test score directly, and that the association between the measurements of body size and the test score is produced via independent radiation-related effects on the score and body size. The change in the grip test score with dose, however, seems to be related to a change in body size, or alternatively stated, the effect of ionizing radiation on the grip test is mediated through a reduction in body size.

Again, when the cases of mental retardation are excluded, no statistically significant effect of ionizing radiation on the results of either test is observed. However, the decline in the repetitive action test score with increasing dose in the group exposed in the 8th through the 15th week following ovulation is large and negative, but only marginally significant statistically. It should be borne in mind, nonetheless, that exclusion of the mentally retarded, who do poorer on both tests than the average child, considerably diminishes the discriminatory power of the analysis since a high proportion of individuals exposed to 1 Gy or more are retarded. Alternatively stated, most of the nonretarded individuals who were exposed at the critical juncture in development were exposed to lower doses, on average, at which the effect is expected to be smaller and more difficult to demonstrate.[18]

Perceptuo-Motor Maturation. It has been recognized for over a century that the drawings of children evolve through a well-defined sequence of increasing complexity as the child ages. Six stages in this evolution have been identified, beginning with unrecognizable scribbling and culminating in readily apprehended drawings, correct in orientation and profile.[19] The existence of these stages has been repeatedly confirmed, but it was not until 1926 that Florence Goodenough devised a scale for evaluating drawings of the human figure as expressions of mental development.[20] Her test found quick acceptance because of its attractiveness to children, its ease of administration, its permanence as a record of response, and its correlation with standard intelligence tests. The test she constructed, now commonly known as the Goodenough–Harris "Draw-a-Man," is presumed to represent a distillation of the cognitive, affective, and perceptual complex from which the image the child reproduces is derived.

Customarily, the test is administered in individual sessions. The examiner places paper and a pencil (or crayon) before the child, and asks him or her "to draw a person, draw it the best you can." As used today, the Goodenough–Harris procedure usually calls for the child to draw two or possibly three figures—a man, a woman, and themselves. The drawings are then evaluated using a point scoring system originally devised by Goodenough and later modified by Dale Harris.[21] This modification consisted mainly of adding new scoring categories to the original 51 categories developed by Goodenough. Her scoring system involved assigning a single point for each of a series of features or parts represented in the drawing (see pages 85–106 in her monograph[20]).[22] These features, including whether a head, trunk, legs, and arms are present, whether the body is proportional, whether eyes, nose, mouth, and lips have been drawn, and whether clothing is present, were selected empirically to reflect the frequency of occurrence of the feature in the drawings of children in each successive yearly age group. A total score was achieved by summing the individual points attained.

Our testing differed from current practice in two respects. First, only one figure was requested and the child was free to draw either a "man" or a "woman." Most of the girls drew a girl and the boys a boy. Second, the tests were scored using Goodenough's original system since they were administered prior to the introduction of the Harris modification.

The Draw-a-Man test is often administered in conjunction with other tests aimed at uncovering perceptuo-motor difficulties, and this was true of the children prenatally exposed to atomic radiation. The ancillary test used is known as the Bender Visual Motor Gestalt, named after the American psychiatrist, Lauretta Bender.[23] This test involves a series of geometric designs made up of dots or lines. These are reproduced on cards that are shown to the child one at a time, who is asked to copy the design on a piece of paper using a pencil or crayon. The figures the child produces can be scored with respect to a variety of attributes, such as rotation, angulation difficulty, fragmentation, cohesion, collision, and the like.

The utility of these simple tests, specifically the Draw-a-Man, is illustrated in Figure 28, where the sketches of four different children, all about 9 years old, are shown. The upper two panels are the handiwork

Fig. 28. Four Goodenough draw-a-man tests illustrating the differences between normal and retarded 9-year-old children. The drawings of the normal children are indicated in the upper two panels and the retarded in the lower two.

of two normal girls; those at the bottom of two boys, one retarded and the other "dull," but not judged to be retarded clinically. As is immediately apparent little girls generally draw better than little boys at this age, but note that the figure on the lower right has no arms or hair, and the child seemed not to know which way was up and which down; whereas the drawing on the left by the "dull" child is somewhat more developed. Nonetheless, this boy obviously had difficulty knowing how to begin, as indicated by the several false starts, and the drawing is not nearly as representational as those of the girls nor of other males his age. Each of these boys had his intelligence tested on several occasions using standardized tests. The "dull" boy's IQ varied from 77 to 88; the retarded child had an IQ at age 15 of 55 and when retested two years later it was 59. The difference in the two successive tests, 4 IQ points, is within the range usually seen with retesting.

Interest in these tests stems from the following observations: First, the cerebral cortex is known to be divisible into a series of functionally and usually structurally distinct areas termed cortical fields. At least two of these are of moment here, namely, the regions identified as the motor and the sensory or somatosensory cortex.[24] It is known that stimulation of the sensory cortex can produce motor effects, and similarly that sensory impulses from the thalamus produce effects not only in the sensory cortex but in the motor one as well. These findings suggest that the two areas are functionally interrelated, and this fact has led in turn to their designation as the sensorimotor cortex. This domain is further subdivisible into a series of cortical fields, but the precise number of these fields is not clear in any species, although the number may be as large as a dozen.[25]

Second, synaptogenesis in this area of the cortical plate is first seen between 19 and 23 weeks after fertilization.[26] These synapses are still immature at birth, and the morphological characteristics associated with maturity are not attained until 6–24 months postparturition. These connections must be sufficiently precise to accommodate the changing needs of the developing individual. This implies that the appropriate connections must be established and retained, but that there must also be sufficient plasticity to accommodate functional interactions of the developing organism with the external environment.

Third, these tests are generally considered more "culture fair" (albeit not culture free) than conventional intelligence tests, and are more widely used than the Koga or Tanaka-B Intelligence Tests on which the previously described results rest. These two tests are essentially nonverbal and provide measures of behavior that do not depend upon linguistic maturation. Both offer insight into perceptual and motor maturation, and it is generally accepted that the effects on these tests of differences among individuals in previous educational, cultural, and social experiences are much smaller than those seen with reading and arithmetic achievement tests on children of the same age.[27] Presumably, therefore, the findings on the Goodenough–Harris and Bender Gestalt tests can be more readily extrapolated to other exposure situations and populations.

Analysis of the results of these two tests revealed the following. (1) The observed means and standard deviation accord well with the "norms" for the tests published by Goodenough and Bender for children of the same age as in our sample, suggesting that perceptuo-motor maturation among Japanese children is similar to that seen in other groups of children.[27] (2) Girls do consistently better on the Goodenough test than boys but there is no difference on the Bender test. This obtains with or without inclusion of the mentally retarded cases. The difference between the sexes on the Goodenough Draw-a-Man was expected; it has frequently been reported by other investigators. (3) There is a significant effect of city (Nagasaki children doing more poorly than children in Hiroshima on both tests). Two explanations for this difference exist. First, the observers were not the same, and while there is a high correlation between observers in scoring the tests, some difference does exist. Second, the children in Nagasaki were slightly younger at the time of examination, and age at examination has not been taken into account in the analysis. (4) Finally, with respect to dose, when the mentally retarded children are included, performance decreases with increasing dose, marginally so in the 0–7 and 8–15 week period, but strikingly in the 16–25 week interval. In the 8–15 week interval, where all of the previously reported effects have been seen, statistical significance disappears when the mentally retarded are excluded. However, the effect in the 16–25 week period remains highly significant.

Since the tasks required in both the Draw-a-Man and the Bender Gestalt involve a coordination of sensory (visual) and motor responses, the radiation-related defect described here could involve either the visual cortex, or the motor cortex, or conceivably both. It is not possible with the information presently at hand to determine which of these alternatives is true. However, it might be possible, using techniques now available but not at the time these specific tests were done, to determine whether the basic "lesion" is in the sensory (visual) or the motor areas.

Uncertainties in the Estimates of Risk

Although the plausibility that ionizing radiation accounts for the observations on the prenatally exposed survivors of the bombing of Hiroshima and Nagasaki must be seen in the mutual coherence and correspondence of the findings with known neuroembryological events, there are uncertainties that should be addressed. These include the limited nature of the observations, particularly on mental retardation and seizures, the appropriateness of the comparison group, errors in the estimation of the absorbed doses and the prenatal ages at exposure, and the presence of other factors in the postbomb period that could be causally involved but are difficult to measure, either now or in the past.

The limited data and the appropriateness of the comparison group. The clinical observations described are based on a sample, not a cohort of births. And the number of retarded individuals is quite small: only 21 of the 30 received absorbed doses of 0.01 Gy or more, and there are only 18 cases without known cause for the retardation other than exposure to ionizing radiation. However, the cases and numbers of individuals at risk are known to be incomplete for at least three reasons. First, the primary source of ascertainment of the sample was through births registered in Hiroshima and Nagasaki. Prenatally exposed survivors whose births were registered elsewhere are not included. Second, presence in the clinical sample entailed residence within contact areas (essentially the limits of the two cities), and thus migrants from the contact area after birth are not included. Third, limited clinical space and personnel were determinants of the size of the study sample. Thus, the sample does not include survivors exposed at distances of 2,000–2,499 m, since at the time of the definition of the sample these individuals were presumed to have received little or no irradiation, and given the space and personnel limitations alluded to above, their study seemed likely to be unproductive. Nonetheless, their omission makes the sample of survivors prenatally exposed to doses of 0.01 to 0.10 Gy limited in number, and the estimate of the frequency of mental retardation in this dose range correspondingly less reliable statistically.

While it is impractical, if not impossible, to estimate the incompleteness of the clinical sample, through a comparison of the Foundation's roster of cases with those of a variety of special survivor societies in Hiroshima, five exposed mentally retarded individuals have been identified who are not within the clinical sample. Four of these five were exposed in the 8–15 week interval, and all were within about 1,000 m of the hypocenter. While these cases cannot be readily factored into the dose–response relationship—the number of individuals at risk from which they are drawn is unknown—they do provide information on the sensitive period, since prenatal age at exposure is not a criterion for enrollment in the groups for which rosters are available.

Insofar as the comparison group is concerned, the bombings admittedly resulted in exceptional circumstances that could have either altered the normal frequency of severe mental retardation or interacted with exposure in complex ways. However, exclusion of the comparison population does not generally alter the dose–response relationship seen in the clinical sample appreciably, suggesting that the effects that are seen are not due to an inappropriate comparison group.

Errors in the estimation of absorbed dose. All estimates of the doses to survivors of the bombing are subject to at least three sources of error—those that stem from the air-dose curves themselves; the attenuation factors for tissues, materials, positions, and the like; and the survivors assertions as to their location. Some of these—the assertions of the survivors—can never be evaluated rigorously for all of the individuals concerned. Errors of this nature can affect inferences on the shape of the dose–response relationship as well as the parameter values defining that shape, but would not affect the identification of vulnerable periods of development. However, randomly occurring errors in the individual dose estimates will lead to an underestimation of radiation effects in dose–response analyses. In the case of the excess risk for cancer mortality, for example, estimates that make allowance for such errors are 5–15% greater than those estimates that do not.[28] Presumably, a similar error obtains with respect to the estimates of radiation-related risk presented here.

Errors in the estimation of prenatal age at exposure. The apparent timing of vulnerable events in development can be affected by errors in prenatal age, and possibly seriously so in specific cases. As earlier stated, postovulatory age is estimated from the onset of the last menstrual period, and adjustment is then made for the difference between that date and the probable date of ovulation. Women with irregular menstrual cycles or who miss a menstrual period for any reason—the failure to menstruate while lactating (lactational amenorrhea), illness, or malnutrition—could erroneously identify the onset of their last cycle. All of these possible sources of error were present immediately before the cessation of hostilities in Japan. Women nursed their infants longer then than now, and lactational amenorrhea could have been more common. Some were undernourished due to the economic stringencies that obtained, and infectious diseases were more frequent in the surviving population, although the Joint Commission noted no epidemics of disease in the months following the bombings.

Similarly, it has been tacitly assumed that all of the pregnancies giving rise to the sample of individuals studied proceeded to term, but if any terminated prematurely, as was surely true for some, the estimated age of the child at exposure would be incorrect. Prematurity is generally determined by an infant's size and weight at birth, but these measurements were not routinely made and recorded in the months immediately after the bombing. At the initial physical examination of these survivors the mother was asked about the child's weight at birth. The reliability of their recollection is uncertain, however, since a subsequent mail survey often revealed large discrepancies between the weight obtained at interview and that given in the survey.

No less important than these sources of error in the estimation of the age after ovulation is the normal variability in developmental age, the critical measure of vulnerability, for fixed intervals of time after ovulation.[29] Although the impact of the factors enumerated on the estimated ages is impossible to assess with confidence, some, possibly much, of the effect on mental retardation seen in the 16–25 week interval could be ascribable to individuals whose developmental age was less than their estimated post–ovulatory age, or their age has been overestimated, as would be true if their birth was premature.

Other factors that could be causally related and produce a spurious effect of radiation or obscure its true importance. Identification of the molecular or cellular processes that culminate in brain damage, or for that matter most other forms of radiation-related damage, in the prenatally exposed is more difficult than in adults or children exposed directly. One is, in a sense, examining a well integrated system involving two individuals—the mother and the developing embryo or fetus. Since neither can be exposed without exposing the other, it is necessary to dissociate direct effects on the developing embryo or fetus from those that arise from the exposure of the mother herself and intrude on her capacity to nourish the developing child properly; or, within the developing child, radiation-related effects on one organ can impinge on the development of still another organ. For example, the brain is exquisitely sensitive to oxygen deprivation and injury to the brain could stem from a diminished blood supply resulting from damage to the mother's or the fetus's blood-forming organs, thus compromising delivery of oxygen to the brain as it develops. Effects of this latter nature are termed *abscopal*, but they can be difficult to establish unambiguously.[30]

Alternative explanations to radiation as the basis for the effects that are seen can be imagined. These include genetic variation, nutritional deprivation, bacterial and viral infections during the course of pregnancy, socioeconomic differences among the dose groups, and lastly, as stated above, an embryonic or fetal deprivation of oxygen secondary to radiation damage to the hematopoietic system of the mother and/or her developing child. Assessment of the contribution these potential causes could or did make in the present instance is daunting. But it is dubious how viable these alternative explanations are, since the confounding sources of variation would have to be highly correlated with dose to produce the responses that are seen. It seems unlikely that this would be so save, possibly, for fetal hypoxemia, where the likelihood of a significant bone marrow depression in the mother would be dose-dependent. As to possible demographic or socioeconomic differences among the comparison groups, an early survey within the in utero mortality sample, which includes the members of the clinical study group, failed to find significant dose-related differences in these variables that could influence not only mortality but nurturing of the handicapped.[31]

Three other unanswered, dose–response-related questions warrant special consideration.

First, within the most vulnerable period, as revealed by the Japanese data, at least three important events are occurring embryologically—the accelerated production of neurons, their migration to their functional sites, and the establishment of their connections with one another. Each of these could have a different dose–response relationship. Since these events occur concurrently in the human, this suggests that the dose–response that has been seen could be an amalgam of different dose–responses and possibly inapplicable to any one of the basic biological events. However, it warrants noting that within the period of maximum vulnerability, virtually without exception, the data can be satisfactorily approximated by more than one dose–response function, generally a linear or a linear–quadratic model. Since there is little or no prior basis for assuming that one or the other of these models better describes the fundamental biological events involved, the "true" model is conjectural, and it is unlikely that epidemiological studies alone will ever determine what the "true" model might be. Of necessity, therefore, the estimation of risk rests on a series of considerations, not all of which are biological, and this is likely to continue to be so for some time.

Second, there is the matter of whether or not a threshold to the biological effects exists. Neither the experimental nor the epidemiological data are compelling in this regard. Nonetheless, this question is central to radiation protection, and warrants more exhaustive study. Although a linear or a linear–quadratic dose–response relationship describes the observed frequency of mental retardation in the 8th through the 15th week adequately, there could be a threshold. However, the estimated value of this presumed threshold depends upon how the data are analyzed. Within the most vulnerable period, 8–15 weeks after ovulation, when all of the cases of mental retardation are included in the analysis, a threshold cannot be shown to exist statistically. But if the two cases of Down's syndrome observed in this period are excluded, as seems appropriate since they are unlikely to have been radiation-related, the estimated 95% lower limit of the threshold ranges from 0.12 to 0.23 Gy. Parenthetically, this range encompasses the lowest doses at which significant effects have been seen in the developing brains of experimental animals.[32] As previously noted, the DS86 doses suggest a threshold in the 16–25 week period of 0.21–0.70 Gy.

A similar search for a threshold in the occurrence of seizures suggests that, if a threshold does exist, it is at a lower dose than that seen for severe mental retardation. The central values of the threshold for all seizures range between 0.11 Gy and 0.17 Gy in the most critical period, that is, 8–15 weeks after ovulation, and the estimates are even lower for unprovoked seizures (0.04 to 0.08 Gy). In all of these instances, the estimated lower 95% limit of the threshold includes zero, and hence the data provide no compelling evidence for a threshold.

Third, there is the matter of the applicability of these findings to exposures at low dose rates. The relative brevity of the especially vulnerable developmental periods precludes the accumulation of a large dose at low dose rates, but this does not imply that effects could not occur. At issue, however, is whether these effects would be commensurate with those associated with an equivalent dose received at a high dose rate. There is as yet little human evidence that speaks to this situation, and the experimental data are not conclusive.[33] It is reasonable, however, to assume that reducing the dose rate or dose fractionation will have some effect. The hippocampus, for example, and the cerebellum continue to have limited neuronal multiplication, and migration does occur in both organs. Changes continue in the hippocampus and cerebellum into the first and second years of life. Continuing events such as these may show effects of exposure differing from those associated with the acute irradiation of the multiplying cells of the ventricular and subventricular areas of the cerebrum, or the migration of neurons to the cerebral cortex, since a larger number of cells will be exposed, albeit to lower doses.

The Biological Nature of the Damage to the Brain

Could this association of mental retardation and the other measures of cortical dysfunction with exposure be happenstance? What is known about the biological bases of the observed effects on the developing brain? Can one distinguish between several alternative explanations for their occurrence? It has been suggested, for example, that the distribution of cases of severe mental retardation among the prenatally exposed survivors in Hiroshima and Nagasaki could be explained either on the basis of a large radiation-related effect on a relatively small number of survivors (presumably more inherently susceptible to radiation damage), or a small effect on virtually every survivor, an effect that merely shifts downward the normal distribution of functional potentials.

Although these are not mutually exclusive alternatives, they suggest different susceptibilities and possibly different mechanisms for radiation-related brain damage. Can one distinguish between a dose-related large effect on a small number of individuals—a malformation—or alternatively, a small effect on a large number of individuals—maldevelopment—or conceivably a mixture of the two? In maldevelopment, mental retardation presumably arises through a shifting downwards of the distribution of capacity for intelligence, and results in more individuals whose capacity falls below that associated with clinical or societal judgments of retardation than normally expected. An analogy involves small-statured individuals, some of whom are small because of the effect of a single inherited gene, such as achondrodysplastic dwarfs, but others merely represent the lower end of the normally occurring variation in stature among individuals.

Answers to these questions, if they are to be found, must be sought through knowledge of the normal development of the brain, arguably the most complex organ in the human body. Structurally and functionally it sets humans apart from other animals. Developmentally it is the culmination of a long and interrelated series of molecular, cellular, and tissue events. These include multiplication of the brain cells, their movement from their site of origin to that of function, their differentiation and aggregation into functional groups, the growth of specific cell connections, neuronal death, and consolidation of the stimulus-conducting projections of the nerve cells, known as axons or neurites. Some of these events happen before birth, others after. Those that take place before birth are divisible on the basis of the time after fertilization when they occur, and are said to be either embryonic or fetal. Conventionally, the embryonic events occur in the period from the appearance of a primitive disk of cells shortly after fertilization to the end of the 8th week, during the phase of development when the characteristic shape of the embryonic body evolves. However, most of the architectural complexity of the brain unfolds later through an exquisitely choreographed set of events, when the developing individual is known as a fetus. To understand the ways in which ionizing radiation could intrude upon the normal processes of brain development, one obviously must know the nature of the processes and their time of occurrence.

The paragraphs that follow set forth briefly, selectively, and in simple terms the normal development of the human brain.

Development of the Brain

Viewed broadly, the brain includes the *cerebrum*, the seat of conscious activity; the *cerebellum*, a sponge-like structure located behind and beneath the cerebrum concerned with the coordination of movements; and the *brain stem*, which connects the spinal cord with the cerebrum and is involved in many vital events such as the regulation of breathing. Much of what is now known about the early development of these portions of the brain and their function has been learned in the last two decades or so, and comes from studies of aborted human fetuses, experimental investigations of other mammals—particularly primates, such as the rhesus and squirrel monkeys—and the development of new techniques, including the use of radioactive isotopes and autoradiography, and instruments, such as the scanning electron microscope, that permit following the fate of individual cells.

Tissue destined to give rise to the central nervous system can be identified in the human embryo as early as 16 days after fertilization. The events that ensue in this tissue are strikingly similar among all higher animals.[34] First a plate forms, followed by a groove, and soon thereafter a tube develops. Initially, this tube has open ends but these soon close. The closed tube, which will give rise to all of the major elements of the central nervous system, rapidly increases in length—the cell population doubles every eight hours or so—and flexes, most pronouncedly in the area that will evolve into the brain. The walls of the tube then quickly thicken, and four distinct layers or zones appear—(1) a ventricular one immediately surrounding the open center, the lumen, of the tube; (2) a subventricular one; (3) an intermediate zone (often called the migration layer); and (4) a marginal zone. The portion of the tube that will become the cerebrum is at this time not much different from the remainder of the tube, but this soon changes as two bulbous vesicles begin to develop. Eventually, these vesicles will become the cerebral hemispheres. These early events are illustrated in Figures 29 and 30.

As these events are transpiring, two functionally different groups of cells emerge—the forerunners of the neuronal and neuroglial cells. When and where the divergence of these occurs is still unknown; however, through special techniques, it has been possible to establish their presence as actively reproducing classes of cells well before they become morphologically distinguishable. At this time, the division of both of these classes of cells is symmetric; each neuronal or neuroglial precursor (or stem cell) produces two similar precursor or stem cells.

About eight weeks after fertilization, a period of rapid cell multiplication begins that lasts about two months, and the division of the ventricular and subventricular neuronal stem cells ceases to be symmetric. Each division now gives rise to one stem cell and one immature neuron. Differentiation of these immature cells into mature neurons puts an end to their multiplication, since differentiated neurons do not divide. This results in the establishment, a little before mid-gestation, of adult numbers of permanent neurons.[35] At about the 16th week following fertilization, there follows a second, numerically greater period of cell division that lasts longer, well into the second year following birth.[36] These are the neuroglial cells, which retain a capacity to divide throughout life and provide the supportive network for the developing cerebrum and cerebellum.

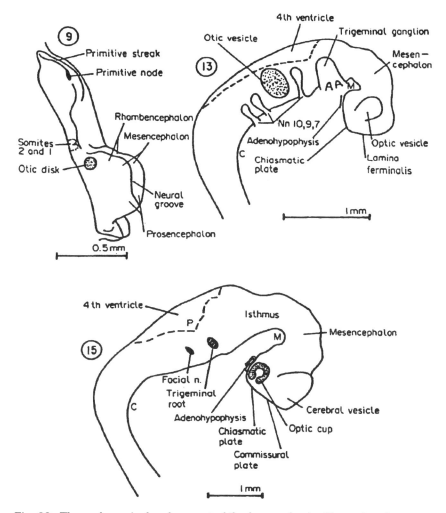

Fig. 29. The embryonic development of the human brain. These drawings correspond to the appearance of the brain at 3, 4, and 5 postovulatory weeks (stages 9, 13, and 15, respectively). The letters C, M, and P indicate the cervical, mesencephalic, and pontine flexures. Beneath each drawing the size of the brain is indicated in millimeters. (Adapted from O'Rahilly and Gardner, 1977)

Simultaneously with these changes, the immature neuronal cells migrate toward what will become the outer, gray layer of the brain—the cortex, their eventual site of function. Two waves of migration take place.[37] The first commences at about the 7th week, and involves cells from the inner area of proliferation, the ventricular zone. The intermediate zone through which the cells move is then sparsely structured,

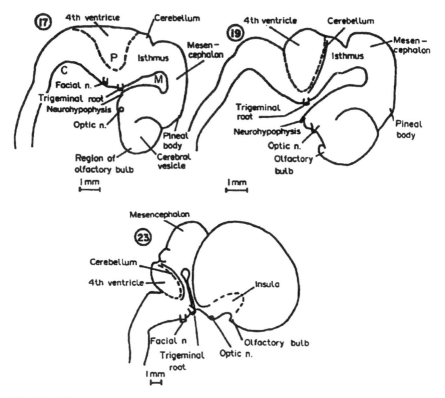

Fig. 30. The embryonic development of the human brain. These drawings correspond to the appearance of the brain at 6, 7, and 8 postovulatory weeks (stages 17, 19, and 23, respectively). The letters C, M, and P indicate the cervical, mesencephalic, and pontine flexures. Beneath each drawing the size of the brain is indicated in millimeters. (Adapted from O'Rahilly and Gardner, 1977)

and contains few impediments to their movement. This wave ceases at about the 10th week, when numerous nerve fibers appear in the intermediate zone, which thickens markedly. The second and larger wave begins about the 10th week, normally terminates about the 16th, and involves cells from the subventricular zone, the one further from the lumen than that from which the earlier migrants emerged. This time the migratory cells traverse a much denser intermediate zone and are assisted en route to the cortical plate by the long processes of specialized cells, the radial glia. At a later time, these cells will release their attachment to the ventricles and move to the cortex and will be transformed into star-shaped cells known as astrocytes. However, at this juncture, they seemingly serve two functions, first, to guide the migrating neurons through the densely packed intermediate zone, and second, to ensure a faithful mapping of the surface from whence they came onto the expanding and increasingly convoluted cerebral cortex by minimizing the intermixing of cells arising in different regions of the proliferative zone (see Figure 31).

While most clonally derived neurons appear to migrate radially following the processes of the radial glia, as just described, there is a recognizable dispersion of some of these cells laterally, that is, non-

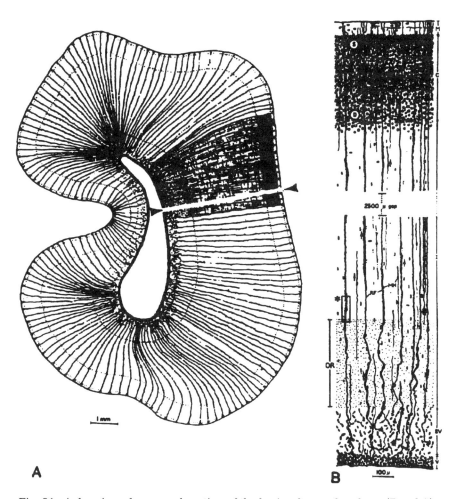

Fig. 31. A drawing of a coronal section of the brain of a monkey fetus (Panel A). The radial fibers are inscribed in slightly darker lines than would actually be true to illustrate their arrangement. Panel B is a drawing of the cerebral wall at the area between the two arrowheads indicated in Panel A. (After Rakic, 1978)

radially. Most of this dispersion appears to occur in the intermediate zone, and during the later stages of neuronal migration.[38] It is believed that this nonradial dispersion of some cells endows the brain with greater plasticity through delaying the commitment of a cell until such time as it has achieved its final functional location; whereas the radially deployed cells have their commitment defined by their starting points in the ventricular proliferative zone. It is not clear whether these two different patterns of migration are equally vulnerable to radiation damage, and if they are not what the functional implications of this difference might be. The fate of these migrating cells, once migration has ceased, will depend upon their ability to connect with other neuronal cells. Those cells that do not form the connections that allow stimuli to be passed from one cell to another will die and disappear leaving no trace of their onetime presence.

Although the migratory process itself was once thought to be passive—cells were merely pushed toward the outer zones of the brain as other cells continued to divide beneath them—the process is actually an active one. Not all of the factors involved in this phenomenon are known; however, there is evidence that a modification of cell surfaces or spaces on or near the glial processes, which guide the neurons during their movement to their final destination, is involved. The cells probably advance by making adhesive contacts at their leading edges, called the growth cones, and then pulling themselves forward. Figure 32 illustrates the appearance of one these cells as it proceeds along a bundle of fibrils. Although this scanning electron micrograph was taken of a neuron migrating in culture, the appearance would presumably be the same or very similar in the living organism.

Gerald Edelman, as well as others, has shown that specific molecules, known as cell adhesion molecules, whose structure and function are under genetic control, play a central role in these movements and do so through alteration of the cell surface.[39] A variety of these molecules have been functionally and biochemically characterized, and the chromosomal locations of some of the genes involved are known. For example, the cell adhesion molecule, identified as N-CAM, which is present on neuronal progenitor cells as well as neurons, is ubiquitous within the human fetal brain. The gene that specifies the structure of this molecule is located on human chromosome 11. Still another such surface molecule, designated L1, appears only on neuronal cells after they have lost the capacity to divide, and the distribution of this molecule changes as the cell matures, so it can be used to study the progress of differentiation.

Precisely how many different molecules are involved in this migratory process and in the final situating of a neuron and the establishment of its connections is not known. Some investigators believe the

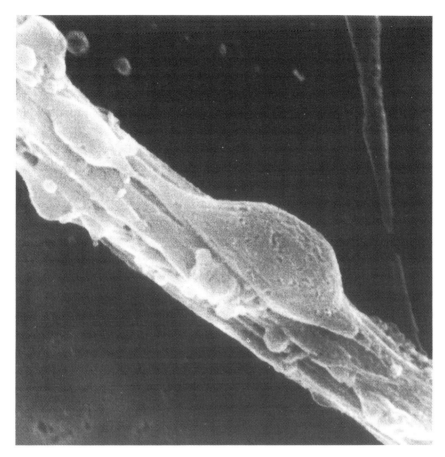

Fig. 32. A photomicrograph of a neuron migrating along a bundle of fibrils in culture. (Courtesy of Dr. Shinji Fushiki)

number might be quite small, and that the modulation and interaction of these few molecules provides the specificity needed to guide the cells, and subsequently their processes, to their destinations; others believe the number might be quite large, virtually one molecule for each step in this complex process.[40] Whatever the number, differential chemical affinities contribute importantly not only to the migratory movement of immature neuronal cells but to their aggregation and establishment of connections as well.

The importance of this process of migration cannot be overstated. In human beings and other primates, no neurons are actually generated in the cortex itself; all originate in other places, and therefore, their migration and siting in the cortex are essential to the establishment of a properly functioning brain. However, some cerebral neurons, presently thought to be relatively few in number, do not have their origin in the proliferative zones we have described, but arise elsewhere and migrate into the cerebrum. For example, the neurons that secrete the luteinizing hormone releasing hormone (LHRH)—which stimulates the production of the sex hormones testosterone and estrogen—originate in an oval area of thickened cells, the olfactory placode, lying outside the developing central hemispheres.[41]

Some of the complexity of brain development can be seen in the way in which the cortical architecture of the cerebrum develops. It evolves from the inside out. The most superficial neurons in the adult brain are the last to leave the proliferative zones and reach their destination by literally passing over the bodies of their predecessors. Immature neurons that have arrived via a particular glial process remain arranged in vertical columns; the late-arriving neurons slide past the already present ones and situate themselves closer to the outer surface of the brain than their predecessors. As a result, those cells that arrive first are displaced to deeper levels by the later arrivals. It has been argued, but not established firmly, that these columnar units, each derived from a different proliferative area, are the basic processing units of the brain,[42] they are our intellectual "chips," to draw an analogy with contemporary computers. While this migration proceeds, the cortical plate thickens. After the 18th week the ventricular zone has been reduced to a thin lining of the ventricular cavity itself. The cortical layer is now much thicker and stratified or layered.

Numerous factors can impinge upon the achievement of proper brain architecture and upon the occurrence of neurological abnormality. Several features about these untoward events seem especially pertinent here.

First, examples are known of the large-scale elimination of early formed neural pathways without the death of the parent cells through the selective loss of the longer stimulus-conducting extensions of the neuron. If ionizing radiation enhanced or impaired this loss, it could lead to the removal or reorganization of some, possibly many, neural pathways without cell death.

Second, with respect to the normally occurring reduction in neuronal cell number, it is possible that this diminution adjusts each neuronal population to the size or functional needs of its field of activity, or eliminates neurons whose extensions have grown to the wrong cell or cells. Selective neuronal death appears to be an important mechanism for the elimination of developmental errors.

Third, although most early cell deaths in the embryonic and fetal nervous system appear fortuitous, some evidence suggests that certain ones are not. Early death of these cells could be their inherited fate.

Their function might be to control the size of specific neuronal lineages, and once this is achieved they are removed.

Finally, as our knowledge of the embryology of the brain increases and recognition grows of the functional diversity that exists among neurons, it becomes more plausible to assume that the loss of small numbers of neuronal cells could have an effect out of proportion to the actual number of cells that are killed or have their differentiation impaired. For example, the establishment of certain axonal pathways is initiated by specialized cells, known as pioneer neurons, which provide the guidance matrix through which other cells establish their connections.[43] While the death or failure of a pioneer cell to differentiate might not always preclude establishment of a particular pathway—since the function of the pioneer could be taken on by another cell—in certain instances, at least, the pathway will not develop if the pioneer cell is prevented from differentiating. Moreover, in some animal species, where the geometry of a neuronal pathway might be quite complex (involving numerous sharp angles), still other specialized cells contribute to the delineation of the pathway. These cells have been called "landmark" cells, and their presence and proximity apparently signals the pioneers to change course.

Anatomically, the left and right sides of the brain are not identical. This asymmetry develops before birth. It has been demonstrated, for instance, that asymmetry of the temporal speech region arises in utero and, no less significantly, that the smaller homologous area on the right side of the brain develops earlier than the larger left-side region that will, later in life, process speech signals.[44] Thus the separation of cerebral function begins early and is associated with microscopically (and occasionally gross) anatomic differences between regions of the brain. Different areas of the cortex differ substantially in their layering or lamination, in the size and densities of the cells they contain, and in the distribution of cell types; moreover, there is much evidence that structurally distinct areas have different connectional patterns and functions.

Summary of the Developmental Features of the Brain

The various events just described are summarized in terms of when they occur in gestation in Table 3. However, one can itemize more simply the cardinal features of the development of the human brain and its associated structures, such as the eye and ear, and how this development differs from that of most other organs or organ systems. Briefly:

1. The brain is the most complex organ of the body, one with a very involved architecture in which different functions are localized in different structures.

Table 3. Summary of the week in gestation when selected, important developmental events involving the human central nervous system occur (modified from Williams, 1989).

Gestational age in weeks	Important developmental events
3rd	Neural folds close to form neural tube Cranial and cervical flexures appear
4th	Paired optic vesicles evert Closure of neural tube
5th	Choroid plexus develops and cerebrospinal fluid fills neural tube Beginnings of the cerebellum and cerebellar nuclei appear Thinning of the roof of the 4th ventricle allows cerebrospinal fluid to flow out
6th	Neural retina develops Olfactory nerves grow to base of brain Hippocampus and olfactory apparatus appear
7th	Neocortical primordia appear Olfactory bulb everts Formation of pigmented retinal epithelium and ciliary body
8th–11th	Cortical plate appears in neocortex First synapses in the molecular and subplate regions of the neocortex Neurons migrate from the proliferative zones Optic nerve pathways form Proliferative zone of third ventricle is exhausted Cortical plate of cerebellum appears Cortical plate of the hippocampus appears Sylvian and hippocampal fissures form
12th–15th	Corpus callosum forms Migration of neurons to neocortex in full swing Purkinje cell migration complete, inward migration of external granule cells begins
16th–20th	Germinal zones of lateral ventricles are depleted Prominence of subventricular germinal zone and first wave of glial migration Active phase of natural nerve cell death
20th–24th	Neuronal migration to neocortex complete Radial glial cells migrate into cortex as astrocytes Primary gyri and sulci form Myelination begins
25th–term	Granule cell migration continues Glial proliferation continues Robust growth of dendrites and axons, and synaptogenesis

2. These functions depend on the disposition and interconnection of specialized structures and cells. Developmentally, normal structure and function hinge on an orderly sequence of events, each of which must occur correctly in time and space.
3. The neurons of the central nervous system are not self-renewing. Their potential to divide is lost during the populating of the various layers of the brain and culminates in cells that no longer have the capacity to reproduce. Neuronal loss is irreparable, since repair through further cell division cannot occur.
4. The human brain is an exceptionally plastic organ, capable of remodeling itself to adjust to new stimuli, but it is also clear that some cortical functions are rigidly encoded, or alternatively stated, the development of cortical function is a complex process some portion of which is conservative and not readily amenable to change.

Developmental Abnormalities of the Brain

Malformations of the human brain and its supporting structures can occur either in the establishment of its major functional subdivisions or during the differentiation and growth of the cerebral cortex or brain mantle, as it is sometimes called. Among the former are such malformations as anencephaly, a fatal defect in which the cerebrum and cerebellum fail to develop properly and the bony covering of the brain does not form, or encephalomeningocele, a protrusion of portions of the brain through a defect in the skull. Both represent errors in the formation and elevation of the neural folds, and a subsequent failure of the closure of the neural tube. Defects can also arise from the inability of the primitive tube to produce two cerebral vesicles. The phenomena concerned—closure of the neural tube and its division into two vesicles—take place rapidly, probably within a few days in humans. These events occur early in development, sometime in the 4th to 6th week after fertilization,[45] and cannot be produced by exposure to ionizing radiation later in pregnancy. It is probable too that many of the defects that affect those portions of the brain other than the cerebrum and cerebellum, such as the brain stem, involve vital centers where loss or damage is incompatible with life.

Disturbances in the production of neurons and their migration from the proliferative zones to the cerebral cortical area give rise to the second class of malformations. Many of these stem from failures in the normal interactions of cells, neural and nonneural, that occur in the development of the human brain. Normal interaction hinges upon appropriate positioning of neurons, establishment of cell shapes, and formation of synaptic connections. Since damage varies as a function of when the insult occurs, defects that arise at this time can differ substantially in severity. Moreover, the sensitive period is not only later but undoubtedly much longer than that for most other malformations, certainly weeks instead of days. The longer sensitive period and the limited capacity for repair must be important reasons why these malformations are generally more common than those that arise during organ formation.

Because the cells of the different structures of the brain are produced at different times, if radiation-related damage were to occur when a certain cell type is being produced, even a brief insult could lead to damage to a particular region of the brain and bring about a permanent functional or behavioral abnormality. Several generations of neurons originate in the same restricted location in the proliferative zone, migrate along the same glial pathways, and accumulate in the same cortical region. If the guideline, the glial fiber, is destroyed or the surface properties of the cells (glial, neuronal, or both) are changed, neurons can aggregate at intermediate spots along their migratory route rather than at their normal sites. This gives rise to misplaced or ectopic gray matter.

The critical period for abnormalities of the cerebral cortex occurs when the primitive brain cells begin to differentiate into neurons and the cortical plate forms. Loss of these cortical cells can lead to convolutional abnormalities that could contribute to the origin of functional and behavioral abnormalities. Disorders such as dyslexia, the severe uncorrectable inability to recognize certain combinations of letters or words encountered in otherwise intelligent individuals, appear to be due to aberrations in specific cortical areas.[44]

Abnormalities that involve hearing, seeing, or smelling could arise from either maldevelopment of the eyes or ears, for example, or in the processing of the stimuli, the signals, transmitted from these organs to the brain, or both. Failures in the appropriate processing of these signals could be the result of a defect in the nerves from the eye or ear, or in the various cortical areas involved in auditory, olfactory, and visual function. Defects in the optic tract, for instance, could be manifested as aberrations in the field of vision, with their nature and extent dependent upon the location and severity of the damage. Olfactory impairment could produce a loss of smell, or an inability to perceive specific classes of odors.

Too little is known about the cellular and molecular events involved in the development of the cortex, as yet, to do more than speculate on the origin of the effects that have been observed among the prenatally exposed atomic bomb survivors. Thus far the most interesting insights have come either from autopsy examinations or from the use of magnetic resonance imaging, a recently introduced means of visualizing the living brain. These sources of information have revealed the following.

Among four deceased survivors, who were prenatally exposed and have been autopsied, two were mentally retarded and two were not. All four of their mothers were present in one or the other of these cities at the time of the bombing but only one received a dose in excess of 0.01 Gy. In the two cases with normal intelligence, the brains were of normal weight and the architecture appeared normal on visual and microscopic inspection. Both of the mentally retarded individuals, however, had brain weights substantially below normal. One had a brain weighing 840 g and the other 1,000 g, whereas the normal weight is about 1,450 g. Multiple slices through the larger brain, that of a female exposed in the 31st postovulatory week, revealed the usual pattern of gray and white matter and no evidence of swelling from the accumulation of fluid in the spaces between the brain cells, which could have increased brain weight. She died at age 20 of heart failure.

The other mentally retarded individual, a male with the smaller brain, died at age 16 of acute meningitis of probable viral origin. If he had been carried to the normal termination of a pregnancy, he would have been exposed in the 12th postovulatory week, but given his birth weight (1,950 g), he was undoubtedly premature. His weight suggests that he was actually exposed at about the 8th or 9th week following ovulation, since a full-term infant would weigh about 3,200 g. The estimated dose to his mother's uterus was approximately 1.2 Gy. Both of his eyes were abnormally small, and within each the retina was conspicuously underdeveloped. Posterior subcapsular opacities of the kind previously described were present in both eyes. He was severely myopic, and exhibited the purposeless constant, lateral scanning motion of the eyes known as nystagmus. Sections across the cerebrum revealed massive amounts of gray matter around the lateral ventricles, especially in the vicinity of the caudate nucleus.[46] Microscopic examination of these misplaced gray areas revealed an abortive laminar arrangement of nerve cells, imitating the usual arrangement of the cortical neurons. The cerebellum and the curved ridges on the floor of the inferior horn of the lateral ventricle, known as the hippocampi, were normal visually and on microscopic study. However, the two protuberances on the under surface of the brain, the mamillary bodies, lying beneath the corpus callosum (the network of nerves that provides communication between the two halves of the brain), were absent.[47] These structures are presumed to play a role in memory, since mamillary lesions are frequently associated with amnesia.

Misplaced gray matter was not observed in any of the other three autopsied cases, including the second mentally retarded individual.

Although the number of individuals that have been studied using magnetic resonance imaging is also small, several different anomalies of development have been seen.[48] These correlate well with the embryological events transpiring at the time of their exposure. Among the two survivors exposed in the 8th or 9th postovulatory week, the findings on neuroimaging are similar to those seen at autopsy in the case just described. There has been a failure of a significant number of neurons to migrate from the proliferative zone to their proper functional sites, and one of these individuals exhibits an underdeveloped area in the left temporal region. In both instances, the ventricles are somewhat enlarged. However, in these two cases, unlike the boy who was autopsied, the mamillary bodies are present and appear to be of normal size.

Ectopic gray matter has been seen in other instances of mental retardation not related to exposure to ionizing radiation but its prevalence among mentally retarded individuals is not reliably known. The limited data that are available suggest that the nature of the migratory error may be different in the two instances. In the cases described above, the failure occurs on both sides of the developing brain; whereas in nonradiation-related mental retardation it commonly involves only one side, although bilateral cases are known, and the ectopic area is often just beneath the cortex rather than around the ventricles. Ectopic gray matter is not invariably associated with mental retardation. Neuroimaging of individuals with the inherited Fragile X syndrome, where varying degrees of mental retardation commonly occur, has not revealed this defect. Among some 27 individuals who have been studied, just eight were found to be abnormal. Seven of these individuals exhibited only a mild enlargement of the ventricles, but in one case a moderate, generalized dilation was seen. Autopsy studies have, however, disclosed abnormalities in dendritic spine morphology—very thin, long tortuous spines with prominent heads and irregular dilatations were noted.[49] This suggests a developmental error occurring after migration was completed.

Individuals exposed in the 12th to 13th week, after completion of the initial wave of neuronal migration and late in the second, exhibit no conspicuous areas of misplaced neurons, but do show a faulty brain architecture. The prominently rounded elevations of the brain, the gyri, are enlarged. These elevations are separated by furrows or trenches, called sulci, and the latter furrows are shallower than normal. One of the cases studied at this time exhibited a corpus callosum that was markedly smaller than normal and a poorly developed cingulate gyrus (the prominence lying immediately above the corpus callosum), suggesting an aberration in the development of the band of association fibers that passes over the corpus callosum. Interestingly, experimental studies have shown this band of association fibers to be especially sensitive to ionizing radiation.[50] In both of these instances, the cistern involved in the recirculation of cerebrospinal fluid lying immediately behind and between the two lobes of the cerebellum, known as the cisterna magna, was markedly enlarged. While the size of the cisterna magna varies widely in the normal population, mega cisterna magna is thought to have an incidence of about 0.4% based on computed tomography, and thus the occurrence of two such similar anomalies is unusual.

Still later in development, at the 15th week, neither migrational errors nor conspicuous changes in brain architecture were seen. Presumably the functional impairment that exists must be related to the connections that occur between neurons, possibly similar to the abnormality in the Fragile X syndrome described above. Experimental evidence shows that exposure at this time in the development of the brain in other primates leads to a diminished number of connections between neuronal cells.[51] If all of these connections have functional significance, then the diminution must compromise performance in some manner.

The medical literature contains descriptions of other embryos or fetuses that have been exposed to ionizing radiation that supplement the findings just described. Shirley Driscoll and her colleagues, for example, have studied two fetuses, one a male exposed in the 16th or 17th week of pregnancy and the other a female exposed at 22 weeks to radium therapy in the course of treatment of maternal squamous cell carcinoma of the cervix of the uterus.[52] Both were alive at the time of the surgical removal of the diseased uterus and its adnexa, a day following the cessation of treatment in the first instance and six days later in the second. The doses were much larger than those of the prenatally exposed survivors. They were estimated to be about 4.3 Gy at the center of the fetal head and 7.7 Gy at the nearest point inside the cranium in the 16–17 week fetus, and about 16 Gy in the second fetus, and it is highly improbable that either fetus would have survived. In both cases, the brain incurred the greatest damage, but then it was also closest to the source of ionizing radiation. Neuronal cell loss was found to be selective. The primitive postmitotic migratory cells were promptly killed by the radiation. Damage to the cerebellum was less extensive but still noticeable, particularly in the older fetus. Extensive changes were seen in other organs, notably the bone marrow and lymph nodes.

Although the observations just described are more informative than the simple determination of the frequency of mental retardation as a function of dose, they do not indicate what cellular or molecular events are impaired. However, when coupled with recent experimental findings, they are provocative. It has been argued that each cortical neuron has not only a designated date of birth, but a definite functional address. Since neuronal cells arise largely in specific proliferative, circumventricular zones, proper function implies migration. Although the latter process extends over weeks, individual cells move and reach their destination in a matter of days at most. As previously stated, the process by which immature neuronal cells move from the sites of birth to those of their normal function is an active, timed phenomenon dependent on an interaction between their shapes, surface membranes, and guidance cells. Any damage to these membranes, however transitory, could impair migration. While there is as yet little direct evidence of the effects of low doses of irradiation on the membranal properties of either neurons or the radial glial cells that guide them,[53] very low doses of irradiation, 0.01 Gy or so, produce transitory changes in the cellular membrane of hematopoietic stem cells, which are also migratory.[54] In a process as exquisitely timed as neuronal migration, a similar delay in movement could culminate in dysfunctional cells through the failure to achieve their normal functional sites.

Pasko Rakic has argued that the cortex is a collection of developmental columns each arising from a specific proliferative unit, and quite possibly from a single cell, although this is not certain.[42] Substantial data can be mustered to support his contention. For example, 30 years ago, Vernon Mountcastle showed that the neurons within a single column in that portion of the cortex involving the processing of sensory perceptions arising elsewhere (the somatosensory cortex) are responsive to a specific type and field of stimulation.[55] Other sensory and association areas in the cortex are known to behave similarly. It is thought that those columns, stimulated by a single cluster of cells in the thalamic nucleus, serve as basic processing modules. If this perception is correct, the loss of a few cells, conceivably even a single cell, could result in the loss or compromise of specific somatosensory or association abilities if that loss occurs in the formative periods for these processing modules.

Some insight into the nature of the developmental effects to be anticipated from exposure to ionizing radiation might come from the effects on the developing brain seen in embryos and fetuses exposed to toxic chemicals. In Minamata, Japan, where the bay and its marine life were contaminated by methyl-mercury, 23 of 359 children born between 1955 and 1959 showed symptoms of cerebral palsy, a number 10 to 60 times higher than expected. Fetal exposure reduced brain weight in severely poisoned children to one-half or less of normal. Extensive areas of brain atrophy have been seen at autopsy and through the use of computed tomography, and abnormal cells have been seen distributed throughout the brain. Recent studies focusing on "normal" individuals suggest subclinical levels of impairment of brain function similar to those that have been observed in Hiroshima and Nagasaki. However, there are differences between the two exposures in terms of their effects on the central nervous system. Methyl mercury's effect on the developing cerebellum appears greater than has been seen among the prenatally exposed survivors in Hiroshima and Nagasaki, and the clearest cases of mental retardation occur at the lower rather than the higher exposures to methyl mercury. But this may be misleading. At the higher exposures the effects on motor development are so profound that it is impossible to assess cortical function adequately. Generally, these individuals cannot speak, and the loss of volitional motor control is so marked they are virtually uneducable.

Severe, permanent central nervous system damage leading to behavioral and other neurological disorders was also seen in Iraq, where methyl mercury contaminated seed grain was used as food. These incapacitating consequences were often observed in children of mothers whose most common symptom of methyl mercury poisoning during pregnancy was a mild, transient series of abnormal skin sensations such as prickling or burning. This suggests that the embryo and fetus are more sensitive to methyl mercury than the mother, but methyl mercury accumulates to higher concentrations in the blood and tissues of the embryo and fetus than in those of the mother, and the greater apparent sensitivity may merely reflect greater exposure. It is difficult to distinguish between these alternatives, since the initial exposure is rarely accurately known, and later, better measurements are not easily extrapolated to earlier exposure since methyl mercury is slowly cleared from the body and clearance rates may vary from one individual to another.

The fetal alcohol syndrome offers another possible paradigm. Abnormalities of the central nervous system, particularly mental retardation and small head size, are the most pronounced effects of heavy intrauterine exposure to alcohol. The average IQ of individuals with fetal alcohol syndrome is about 65, although scores may vary from 16 to 105; also, the severity of the mental retardation correlates with the severity of the abnormality in the features of the individual.[56] Investigators have noted that areas of ectopic gray matter in the frontal and temporal white regions of the cerebral hemispheres and heterotopia involving the covering of the brain, both evidences of abnormal cell migration, are common among infants with fetal alcohol syndrome.[57] However, the peculiar meningeal heterotopias sometimes observed in fetal alcohol syndrome have not been reported in either humans or other primates exposed to ionizing radiation.

It is thought that the teratogenic effect of alcohol, insofar as abnormalities of the central nervous system are concerned, is initiated during the first trimester, but this has not been well established. Given the commonly chronic nature of the exposure, it is not surprising that the sensitive period is poorly known. However, the evidence that is available suggests that the vulnerable periods for alcohol use and exposure to ionizing radiation may be similar. This belief rests in part on the knowledge that alcohol diminishes the incorporation of sialic acid, which is essentially a carbohydrate, into membrane-bound compounds, such as N-CAM, and inhibits the gap junctional intercellular communication referred to earlier. As has been stated, N-CAM plays an important role in neuronal cell migration recognition and adhesion, and it is believed that a reduction in the manufacture of N-CAM inhibits migration, leading to a premature aggregation of the neuronal cells, and thus to ectopic gray matter.

Ectopic gray matter is commonly seen in rodents following prenatal exposure to X-irradiation, and has also been described in beagles exposed to gamma irradiation from a cobalt-60 source at a stage in brain development roughly equivalent to that at which ectopias are seen in the prenatally exposed in Hiroshima and Nagasaki. Sam Hicks and his colleagues reported such occurrences in the rat more than three decades ago, and argued that they arose from surviving cells that retained the capacity to divide but did so in an abnormal environment.[58] More recently, Alejandro Donoso and Stata Norton found that rats exposed to 1.25 Gy of X-irradiation on gestational day 15 developed ectopic areas lying beneath the corpus callosum and adjacent to the caudate nucleus, and thus similarly situated as the ectopias previously described among the prenatally exposed in Hiroshima.[59] Presumably these isolated islands of neuronal cells arise through faulty repair of radiation-related damage to the wall of the lateral ventricles, leading to an encirclement of cells still capable of dividing. Subsequent divisions of these cells result in immature neurons migrating in all directions.

Physical Growth and Development

From the outset of the studies in Hiroshima and Nagasaki, concern has existed over the possible effect of prenatal exposure on subsequent physical growth and development. To determine whether this has or has not occurred, a variety of anthropometric, endocrinologic, and roentgenographic studies have been undertaken. Almost uniformly, these studies have shown a retardation of physical growth with increasing dose, particularly at the higher doses. This retardation is more than a delay; it is permanent. This fact has been recently reconfirmed using repeated measurements of stature from ages 10 to 18 and the DS86 dosimetry.[60] The retardation is most pronounced when exposure occurred in the first trimester of gestation.

One of the more informative studies involved roentgenographic examination of the age at closure of some 28 epiphyseal centers of bone formation in the fingers and wrist.[61] Five hundred and fifty-six boys and girls from the two cities, selected primarily on the basis of the number of hand and wrist roentgenograms available, were studied. Normally, closure of these centers occurs later in males than in females (about one and a half to two years in Japanese children, depending upon the center). However, in this study it was found that as dose (T65) increased, mean age at closure was delayed in females, but accelerated in males.

Radiation, therefore, tended to diminish the normal difference between the sexes, suggesting a "neutering" effect. Since hormonal changes in adolescence play an important role in the closure of the centers, this seems to imply an endocrinological disturbance. Although these authors did not conclude that radiation was responsible for the changes described, citing other possible factors such as nutrition, the data warrant reanalysis since their study used not only the older dosimetry but the doses attributed to the individuals were the shielded kerma of the mothers (the DS86 system suggests these doses would have been too high by about a factor of 2).

Ocular Damage

As noted earlier, two different radiation-related ocular effects have been reported among the survivors of the bombing of Hiroshima and Nagasaki and could occur among the prenatally exposed—the occurrence of radiation opacities and polychromatic granules. One ophthalmic survey of the atomic bomb survivors describes the examinations of 309 children of some 464 individuals exposed in utero who were invited to participate in the study.[62] Only one individual, a male, was observed to have any degree of opacification. He was exposed in the 6th postovulatory week, and although not judged to be mentally retarded clinically, he did poorly in school and had an IQ (Koga) of 65. Secondary lens fibers are present at this time, and presumably the fate of damaged germinative epithelial cells would be similar to that thought to occur among the postnatally exposed survivors. At least one other prenatally exposed individual, now deceased, the autopsied boy described earlier, is known to have had radiation opacities bilaterally. As to polychromatic plaques, of 47 individuals exposed prior to the 8th postovulatory week, 3 (6.2%) had polychromatic changes in one or both eyes, 14 (18.2%) of 77 exposed in the 8th through 15th week were affected, and 40 (21.6%) of 185 exposed after the 15th week exhibited posterior subcapsular changes. These results are consonant with the embryological development of the eye, and suggest no special period of vulnerability once organogenesis is completed.

Chromosomal Abnormalities

Studies of chromosomal abnormalities in the white cells of the peripheral blood of the survivors exposed in utero began in the early 1960s. These early studies focused on whether or not such abnormalities were present, and not upon the dose–response relationship, since individual estimates of dose were not available. Commonly, they involved the prenatally exposed children of mothers who had presumably received substantial doses, and a suitable comparison group. For example, one of these studies involved 38 children whose mothers were thought to have received more than 1 Gy (T65 dose), and 48 controls.[63] Among the former, the frequency of cells with complex chromosomal rearrangements was 0.52%; whereas among the latter, the frequency was only 0.04%. Expressed in terms of the frequency of children with chromosomal abnormalities rather than the frequency of abnormal cells, the percentages were 39 and 4, respectively. These findings led the investigators to conclude that both lymphocyte precursors and mature, immunologically competent lymphocytes had been affected by ionizing radiation.

Subsequent investigations have centered on those prenatally exposed survivors who are participants in the Adult Health Study and have employed more recent technical developments that permit a better characterization of chromosomal abnormalities than was possible at the time of the study cited above. The purpose was to define the dose–response relationship and the frequency of the various abnormalities that occur. These newer studies have, for example, shown that the loss or gain of an autosomal chromosome is, with singular exception, related to the size of the chromosome—the smaller ones are lost or gained more frequently than their larger counterparts. This is true for both males and females. The important exceptions involve chromosomes 13 and 21 and the sex chromosomes, where the loss or gain occurs far more frequently than size alone would predict. Why this should be is not clear.

Somatic Mutations

As yet there have been no systematic studies of the occurrence of somatic mutations among the prenatally exposed similar to those that have been described previously for the postnatally exposed survivors. However, even before the advent of these newer techniques, studies had been initiated that sought to find evidence of radiation-related somatic mutation. Possibly the earliest involved the prenatally exposed, and more specifically pigmentary changes in the iris of the eye. Normally the latter structure, which gives color to the eyes, has two layers—an inner or posterior layer and an outer or anterior one, which gradually atrophies or wastes away as the eye develops. These are separated by another layer containing numerous small blood vessels and the muscle that controls the size of the pupil.[64] These are suspended in a supportive structure of connective tissue or stroma. The inner layer is always heavily pigmented (save in albinos), but the vessel or stromal layer is not necessarily so. In blue-eyed individuals, for instance, little or no pigment forms in this region, and the eyes appear blue (or gray if the stromal layer is especially thick) because of the light reflected from the heavily pigmented inner layer. But if pigment does occur in the anterior cells that comprise the vessel layer, the eyes will have a color ranging from hazel to dark brown depending upon the extensiveness of the pigmentation. Embryologically, these pigmented regions develop in mid-gestation, that is, sometime between the fourth and the seventh month after fertilization, and if there were a radiation-vulnerable period this would be it.

Limited studies elsewhere had suggested that the frequency of an anomaly of the iris known as segmental heterochromia was increased following prenatal exposure.[65] Segmental iris heterochromia describes a developmental condition in which a sharply defined, wedge-shaped segment of the iris, usually of only one eye, exists that differs from the remainder in color. Commonly it is blue, for example, where the bulk of the iris is brown. Although the etiology of this defect was and still is unclear, and several different possibilities could be envisaged, some investigators held that it arose as a result of a somatic mutation in one of the genes in the cell at the point of the wedge (adjacent to the pupil). Subsequent divisions of this cell and its descendants were thought to produce the differently colored segment of the iris. To the extent that this argument is true, the mutation, possibly a deletion, presumably arose in a cell of the anterior layer, exposing the pigment in the deeper one and thus giving the segment a blue appearance. However, studies of the prevalence of segmental iris heterochromia among the prenatally exposed in Nagasaki failed to confirm this supposition. There was no demonstrable increase in prevalence with increasing dose; indeed, no cases of this anomaly were seen, although other irregularly pigmented areas were.[66]

Cancer among the Prenatally Exposed

Given that cancer is the predominant risk confronted by survivors exposed postnatally to atomic radiation, and that among individuals so exposed this risk is related to the age of the survivor at the time of exposure, it is of interest to know what the risk of malignancy is among the prenatally exposed. The first study to expose this was published in 1970; it was based on just those prenatally exposed survivors who could be identified through birth records in the two cities, some 1,250 individuals, and focused on childhood cancers occurring in the first 10 years following birth. No increase in mortality from these malignancies was observed.[67] Subsequently, in 1976, the period of observation was extended to cover all malignancies manifesting themselves within the entire preadult period. Again, no increase was observed, but the sample was small, and death certificates were not available on many of the children who died in the first year following the bombings because of the confused state of the nation's vital records in the immediate postwar period, particularly in 1945–1946. Accordingly, efforts were made to expand the study group through the roster of survivors maintained by the Commission and through the national censuses of 1950 and 1960. Even this expanded sample, however, failed to show an increase in childhood malignancies—only two cases were identified, one of liver cancer and the other a malignant embryonic tumor of the kidney, known as Wilms' tumor. Importantly, but unexpectedly, no cases of leukemia were observed.

These findings were at variance with the risk of childhood malignancy based on other, retrospective studies of the effects of exposure to low doses of medical irradiation. While the disparity between these studies has lessened with the recent downward revision of the doses of the survivors and the upward revision of the doses received in medical irradiation, the differences that do exist have sparked a lively debate. Proponents of a greater relative risk among individuals prenatally exposed to medical irradiation than has been seen in the Japanese survivors frequently contend that the ascertainment of cases of leukemia was incomplete in the early years of the studies in Hiroshima and Nagasaki and that this accounts for the disparity between the effect of medical irradiation and exposure to the bombings. While this might have been true, given the state of record keeping at that time, there is no evidence to support this conjecture. Moreover, a readily recognized increase in leukemia did not begin to appear among other survivors until two to three years after exposure, and by this time the Japanese system of vital records had been largely rebuilt. It has also been suggested that immune competence among the prenatally exposed was depressed or lost as a result of exposure, and that immunologically compromised survivors might have died more frequently of infectious diseases and thus failed to survive to develop leukemia. This conjecture too cannot be supported since no increase in such deaths was observed among the prenatally exposed.

Other investigators argue, however, that considerable caution must be exercised in interpreting the medical irradiation studies, especially in light of the many possible confounding or biasing factors. In a similar way, the fact that early postnatal irradiation in Japan showed carcinogenic effects only many years later (agreeing with other postnatal radiation effects), whereas the medical series showed their effects soon after birth, raises important radiobiological problems if both findings are the result of irradiation. The doubts are increased by the absence of serious effects in animal experimental data, and the constancy of relative risk values for many cancer sites in the medical series is at variance with other data, which suggest site-specific effects. Finally, differentiation of the blood-forming stem cells, where leukemia presumably arises, does not occur during the first trimester of fetal development, so that the excess reported in the medical series following first trimester irradiation is difficult to understand.

While the previous remarks concern childhood cancers, the evidence is mounting that the cancers of later years, the so-called adult ones, are increased in frequency among the prenatally exposed survivors in Japan.[68] It appears that not only do these latter cancers occur at earlier ages among individuals exposed to 0.30 Gy or more, but the incidence continues to increase, and the crude cumulative incidence rate, 40 years after the atomic bombings, is 3.5-fold greater in the |0.30 Gy group (see Figure 33). In the years from 1950–1984, based on the absorbed dose to the mother's uterus, and the augmented sample, the relative risk of cancer at 1 Gy is 3.77 with a 95% confidence interval, a measure of the uncertainty in this estimate, of 1.14–13.48. For the dose group receiving more than 0.01 Gy the average excess risk per 10^4 PYGy (person-year-gray) is 6.57 (0.47–14.49) and the estimated attributable risk is 40.9% (2.9–90.2%). Risks derived from only those cases identified through birth records are somewhat lower, but not significantly. The relative risk at 1 Gy in this group is 3.19 rather than 3.77, for example. It warrants noting that these risks include those individuals who are

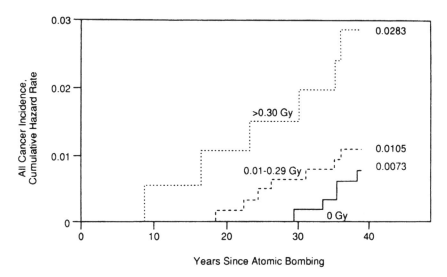

Fig. 33. The frequency of cancer among the in utero exposed by dose (Gy). (Adapted from Yoshimoto, Kato, and Schull, 1988)

known to have cancer but have not as yet died from their malignancy or some other cause.

When these findings are compared with those based on survivors who were 0 to 9 years old at the time of the bombings, who have higher relative risks at a given time since exposure than adults, the risks among the prenatally exposed are slightly higher, except for leukemia. The difference with regard to the latter malignancy remains enigmatic. Nonetheless, the results with respect to adult cancers, when viewed in the perspective of fetal doses, suggest that susceptibility to radiation-induced cancer may be slightly higher in pre- than in postnatally exposed survivors (at least those exposed as adults). This conclusion must be guardedly drawn, however, since only now are the prenatally exposed survivors reaching those ages in life when the natural rate of deaths ascribable to cancer increases dramatically. It will be a number of years before the full impact of exposure can be assessed with the accuracy and reliability this issue warrants, since the sample is small and in the short-term, chance variation in the occurrence of cancer cases or deaths could obscure the dose–response relationship.

Fertility of the Prenatally Exposed

Once the prenatally exposed survivors reached those ages at which marriage and child-bearing occur, it was natural that they should be concerned about the risks they confronted. There was no precedent for anticipating what these might be, although it seemed unlikely they would be less than those of the survivors exposed postnatally, and significant long-term effects of radiation on the rate of marriage or reproductivity have not been seen in the latter group. It could be argued, nonetheless, that exposure during the formation of the ovaries and testes might have more dire consequences than exposure at later ages, or that damage to endocrine organs, such as the pituitary, whose hormones are important in sexual maturation, might have similar such effects. Among some primates, the squirrel monkey for instance, it is known that prenatal radiation exposure of the female germ cells is much more damaging than postnatal exposure. However, studies of human fetal ovaries grown in culture have not revealed them to be notably radiosensitive.

Whether it is possible to unequivocally establish a radiation-related biological effect on human fertility is moot, for the separation of biological from other factors determining marriageability and child-bearing, particularly in contracepting societies, is fraught with many pitfalls. Nevertheless, to provide some insight into the possible effects of exposure, whether biologically or socially mediated, William Blot and co-workers have studied the frequency of marriage and births among members of the In Utero Mortality Sample, some 2,756 persons.[69] They found the marriage rate to be lower among persons exposed prenatally to 1 Gy or more of radiation; however, this difference is partly ascribable to the lesser marriageability of persons with mental retardation who are common in this particular exposure group. No consistent relationship was observed between dose and three reproductive indices—the number of childless marriages, the number of births, or the interval of time between marriage and first birth. The latter measure is often used by reproductive biologists, since it is thought to be sensitive to the loss of a conception early in pregnancy. Still another survey failed to find evidence that exposure in utero to higher doses of atomic radiation induced complete sterility.[70]

The Future

Future studies of the prenatally exposed will undoubtedly focus on the relationship of exposure to ionizing radiation and aging, including that of the central nervous system, radiation carcinogenesis, and reproductivity. Each of these areas of endeavor offers unique opportunities.

Aging

The prenatally exposed survivors are unusual in many respects, not the least of which is the fact that they are the only group of survivors whose life experience subsequent to exposure can be followed from birth to death and, as such, can provide unique insights into the effect of exposure on aging. Magnetic resonance imaging studies could be especially informative, providing quantitative measurements on brain atrophy during aging, and they could also be used to detect the occurrence of many diseases associated with advancing age. Obviously, such studies must have direction and recent experimental investigations could provide this. For example, whole-body gamma radiation of beagles exposed at 55 days following conception (corresponding to about the 33rd or 34th week in the human case) and at two days after birth has revealed effects on the developing kidney at doses as low as 0.16 Gy.[71] These effects involve a significant increase in the total number of immature, abnormally formed glomeruli, the tufts of capillaries at the end of the tubules within the kidney, and a hardening of the connective tissue extending into the glomerulus similar to that seen in progressive kidney failure. While it is difficult to say what one should expect to see clinically in the human, partial renal loss could result in hypertension or possibly a less concentrated urine than normal due to excessive glomerular filtration. It would be worthwhile also to look for radiation-related damage to the kidney during its early formation, given the apparent sensitivity of this organ to irradiation, the earlier reports of urinary incontinence among the prenatally exposed, and the finding of a dose-related increase in the frequency of proteinuria, which reflects a dysfunctional kidney among these individuals.[72]

Impairment of Brain Development

Numerous events are involved in the process that brings forth a functional brain, any one of which is potentially susceptible to radiation damage and could lead to a different result. Patently, there is a need to confirm and extend the findings on cerebral cortical impairment following prenatal exposure to ionizing radiation described in the earlier paragraphs. To do so, however, will entail not only more neurologically focused clinical examinations, including the various techniques now available to image the brain of the still living prenatally exposed atomic bomb survivors, but a concerted effort, national and international, to identify groups of individuals prenatally exposed to ionizing radiation of possible importance, such as the prenatally exposed in the Chernobyl accident. These studies could have value well beyond the immediate assessment of the risk of prenatal exposure to irradiation, and could contribute to a deeper understanding of human embryonic and fetal development, to a clearer appreciation of the diversity among individuals in the age at achievement of specific embryonic or fetal landmarks, and to a sharper definition of the developmental ages most vulnerable to exposure to chemical or physical teratogens.

As yet, among the prenatally exposed survivors there have been no studies directed toward the effect of irradiation on specific, cortical functions. Nevertheless, many of these functions can be investigated with a surprising degree of precision, and the time at which cortical neurogenesis is initiated in these areas, and its duration, is often reasonably well known. Particularly appealing are the various aspects of visual function. Some 30% or so of the cortex appears to be involved in the processing of visual stimuli, and the mechanisms through which this processing occurs are better understood than for any other cortical area. Similarly, it has long been known that the brain emits weak electrical activity, which can be measured. Its recording is not painful or invasive, and other studies have revealed measurable radiation-related alterations in the normal record of these electrical potentials, known as the electroencephalogram.[73] This electrical activity can be mapped, and these maps might reveal areas of impaired function that would not be revealed by clinical neurological examinations.

Future investigations of possible central nervous system impairment should examine not only damage to the cerebrum but to the cerebellum and brain stem, to the extent that effects on the latter can be dissociated from the former. Here, such methods of visualizing the living brain as magnetic resonance imaging or positron emission tomography (PET) might reveal subclinical evidence of radiation-related central nervous system damage. If there are metabolic differences between neuronal and nonneuronal cells, PET scans, since they can measure some metabolic functions, could give evidence of impaired migration or of seemingly histologically normal but nonfunctional sites in the brain. Individuals with intractable seizures, for example, often have regions of the brain that show no functional activity yet appear normal grossly. Since the risks, if any, associated with PET scans are still unclear, their use as a screening procedure would be unwarranted. However, recent advances in magnetic resonance imaging, which does not involve exposure to ionizing radiation, have made possible some functional measurements of brain activity, and further developments can be anticipated.

Members of the prenatally exposed clinical sample are still examined biennially at the Radiation Effects Research Foundation. This examination emphasizes general health but a search should also be made for evidence of central nervous system damage. The neurological examination itself does not now but should include tests of motor control and development,[74] and the electroencephalographic measurements previously mentioned. Some cognitive tests, such as word association, learning ability, and memory and intelligence, should also be included. But there is the opportunity to do more. In the light of experimental findings on other primates, careful studies of auditory and visual acuity, olfaction, and taste should be contemplated. These might include audiometric assessment of the left and right ears, and conventional appraisal of visual acuity, including detailed examination of color vision. In addition, smell and taste could be evaluated through exposure to a battery of tastes or aromas at different concentrations.[75] Special neurons in the nose are capable of detecting a myriad of odors and communicating to the brain what we smell. These neurons, unlike most others, when they age and die are replaced with new ones derived from a population of progenitor cells in the nasal epithelium. These new neurons will acquire chemosensitivity and form synaptic connections in the olfactory bulb of the brain. This process occurs continuously, but despite this constant change the sense of smell is normally quite stable. But would this stability also be true in the survivors? Evidence of an earlier loss in hearing or in vision that normally accompanies aging should also be sought, since a lesser initial number of neuronal cells could lead to earlier manifestation of an aging central nervous system.

Still other neurophysiological effects, with different critical periods, can be envisaged. Recently, for example, it has been shown that mice irradiated prenatally on day 18 (corresponding to about the 33rd week of gestation in the human) suffer a significant loss in spatial memory.[76] The integrity of spatial memory appears to depend upon the proper development of the hippocampus—an anatomically distinct part of the cortex lying beneath the cerebral hemispheres. This structure arises relatively late in the development of the human brain but damage to it results in a recognized cognitive defect characterized by severe amnesia and includes deficits in learning mazes. Moreover, there is evidence that associates a reduction in the pyramidal cells within the hippocampus with memory impairment. Here simple pencil and paper maze tests might be informative.

Radiation carcinogenesis

As the years unfold and these survivors move deeper into the time in life when cancers commonly occur, undoubtedly more will develop a malignancy. This event, unfortunate though it be, should make the risk estimates more secure and the time course of manifestation of radiation-related cancers more obvious. At present, the number of cancer cases that have occurred is too small for the risk and its change with time to be well estimated. There is also the possibility that this group of survivors might contribute disproportionately to our understanding of the molecular and cellular events involved in carcinogenesis. Their tumors will be occurring when more tools are available to describe and understand these events than was true of those survivors who have already succumbed. Whether this does or does not come to pass, the survivors should profit from new and better means to alleviate or intervene in the process of carcinogenesis.

Reproductivity

The studies of reproductivity reviewed earlier warrant repeating for at least two reasons. First, the follow-up period is now longer, bordering on the complete reproductive lifetime of the individuals concerned, and second, the earlier investigations were based on the old dosimetry, using as the estimate of the relevant dose the mother's shielded kerma. Although it is unlikely that the findings will change qualitatively, they might change quantitatively, and better estimates of risk would surely follow from the use of the individual-specific doses absorbed by the gonads that are now possible.

8

The Survivors' Children

At the outset of the Commission's activities, public and scientific concern over the possible genetic effects of exposure to atomic radiation was at least as great as that over cancer, and possibly greater. To most prospective parents the thought of producing a seriously malformed infant is deeply disturbing, but it is even more distressing when coupled with the belief that the abnormality might have arisen through an avoidable exposure to ionizing radiation. As a result, no other human population has been scrutinized for evidence of transmissible mutations more closely, continuously, or thoroughly than the children of the survivors of the bombings of these two cities. A major portion of the resources of the Commission and its successor, the Foundation has been devoted to studies of the genetic effects since their inception in Hiroshima and Nagasaki in 1947.

A number of surveillance strategies have been employed to look for newly arisen mutations; these include a search for alterations in the frequency of certain population characteristics, such as the occurrence of life-threatening or socially handicapping congenital defects and premature death, chromosomal changes, or changes in the biochemical structure or activity of a variety of enzymes and proteins normally present in the blood. Diverse as these alternatives are, their aims are the same—to estimate the probability of mutation following exposure to ionizing radiation and to determine the public health implications of an increase in the number of mutations measured. These various studies, however painstaking and thorough, serve these ends unequally well since some—such as the protein and enzyme studies—measure the direct product of genes, and others—such as the study of socially handicapping defects—albeit of great public health importance, examine characteristics considerably removed from the level at which genes act.

These facts notwithstanding, the data accumulated provide the clearest picture available on transmitted genetic damage following the exposure of human beings to ionizing radiation.

Pregnancy Outcome

As previously stated (see Chapter 3), the first steps toward a continuous surveillance of the children conceived and born in Hiroshima and Nagasaki after the bombings were taken in 1946 and a full-scale program was initiated in the late spring of 1948.[1] To reiterate briefly, at that time certain dietary staples were rationed in Japan, and had been throughout much of the war, but in the interests of health the ration regulations made special provision for pregnant women. Several lines of evidence revealed that most pregnant women did request these special rations. The initial system of surveillance was tied, therefore, to the process wherein pregnant women registered for the supplementary rations. Through this means, it was possible to identify more than 90% of the pregnancies in these cities that persisted for at least 20 weeks of gestation (the earliest stage in pregnancy at which the additional rations were made available), and to examine the outcome at birth and thereafter. [1]

The indicators of possible genetic effects that could be drawn from the physical examination of newborns were gender, birth weight and prematurity, presence of compromising or life-threatening malformations, occurrence of death during the neonatal period (usually defined as the first 30 days following birth), and growth and development at age 8–10 months. All of these events are influenced to a greater or lesser extent by circumstances other than exposure to irradiation. Collection of observations on these indicators and the factors that could alter their occurrence was tedious, time-consuming and costly. It was mandatory, therefore, that the observations be as complete as circumstances permitted. To this end, the data were continuously monitored for accuracy. Ambiguous or uncertain observations were verified and an annual report of progress was submitted to the Academy's supervisory Committee on Atomic Casualties.

A comprehensive analysis of the data accumulated through 1953 suggested that this extensive clinical program had reached its logical conclusion, since few further births could be expected to those survivors then in their reproductive years, and it seemed unlikely at that time that the studies would continue long enough for many of the younger survivors to reach the age of reproduction.[2] Borderline findings on the effect of radiation on the gender of an infant and his or her survival, however, prompted a continuation of the collection of data on gender and mortality. The study of mortality that followed, known as the F_1 Mortality

Study, the surveillance of deaths and their causes among the children of the survivors, continues.[3] It will be recalled that this study focuses on three age- and gender-matched groups in each of the two cities, namely:

1. all infants liveborn between May 1946 and December 1958, one or both of whose parents were within 2,000 m of the hypocenter at the time of the bombings (the "proximally exposed"),
2. an age- and gender-matched group randomly drawn from the remaining births in the two cities during this same period, where one parent was exposed at 2,500 m or beyond (the "distally exposed") and the other was either similarly exposed or not exposed at all, and
3. an age- and gender-matched group randomly drawn from the births in these cities during these years, where neither parent was exposed.

Recently, as has been stated, these groups have been enlarged, using the family registers, to include births to parents in the proximal and distal groups in the period from January 1959 through December 1983. Presently, the study embraces approximately 75,000 individuals, and includes virtually all of the children born to proximally exposed parents, irrespective of their age at exposure.

As a part of the initial clinical program, 76,617 pregnancies were studied,[1,4] but not all of these can be analyzed (see Table 4). Some 3,264, for example, were reported to the Commission by the attending midwife

Table 4. The two major genetic samples and the bases for the exclusion of some of the cases initially ascertained.

Item	Clinical sample[a]	Mortality sample[b]	
		Original	Extension
Total cases initially	76,617	54,243	23,298
Exclusions	6,911	3,714	1,599
Unregistered births	3,264	—	—
Stillbirths	—	694	—
No household census	—	799	—
Unknown exposure status	1,595	2,191	1,599
No T65DR dose	1,228		
No DS86 dose	367		
Incomplete observations	2,052	30	
Analyzable cases	69,706	50,529	21,699
		(72,228)	

[a]Sources: Schull et al., 1981 and Otake et al., 1989.
[b]Sources: Kato et al., 1965 and Yoshimoto et al., 1991.

but the pregnancy had not been registered with the municipal authorities. Although the midwives were encouraged to report all pregnancy terminations without regard to whether they were or were not registered, as noted before (see Chapter 3), it is unclear how complete the reporting of unregistered births was. But since the unregistered births that were reported appear to be a selected set, they have never been included in any analysis of the clinical data. In some other instances the information available on the mother, the father, or both was inadequate to determine exposure status or the dose both parents received. Finally, despite repeated checking, occasionally the data on a termination were incomplete in other ways, such as in recording the age of the father or the birthweight of the infant.

The findings on the 70,073 terminations that can be analyzed can be conveniently summarized under the broad rubric "untoward pregnancy outcome." The latter includes pregnancies that ended in a child with a major congenital defect, that was stillborn, that died during the first week of life, or some combination of these events. Among the 70,073 pregnancies, there were 3,332 untoward pregnancy outcomes. To the extent that these various outcomes are hereditary (genetic) in origin, they should increase in proportion to the radiation dose of the parents because of the induction of transmissible mutations with harmful effects. When the untoward outcomes are distributed by parental exposure, no statistically significant association emerges, although at face value there is a small increase as the combined parental dose grows higher.[5] This also holds true if malformations, stillbirths, or neonatal deaths are analyzed separately. Earlier analyses did not show an increase in the occurrence of any specific congenital malformation, including Down syndrome (trisomy of chromosome 21). The most common of these specific abnormalities, exclusive of congenital heart disease, were anencephaly, cleft palate, club foot, cleft lip with or without cleft palate, polydactyly (an additional finger), and syndactyly (the fusion of two fingers). These seven abnormalities, common in all human populations, accounted for 445 of the 594 malformed infants born to parents who were not related to each other. It warrants noting, parenthetically, that except for anencephaly these defects are repairable surgically.

The absence of a statistically significant effect of ionizing radiation on the frequency of untoward pregnancy outcome should not be construed as evidence that mutations were not induced by parental exposure to atomic radiation. At least two reasons argue otherwise. First, mutations have been seen in every animal and plant species studied under suitable experimental conditions, and it would be contrary to all biological evidence to presume that human genes are not mutable when exposed to ionizing radiation. Second, the magnitude of a difference between two or more groups that can be detected statistically depends upon the number of observations made and on the "natural" frequency of the event under scrutiny, as well as the difference between the groups resulting from exposure. One can ask, therefore, how adequate this study has been, or to pose the question differently, how large a difference would have had to exist to be demonstrable with a study of 70,000 infants, only half of whom had one or more exposed parents? Suffice it to say that a clinical study of the kind described would be able to detect a doubling of the rate of major congenital malformations if such occurred, and a 1.8-fold increase in the stillbirth or neonatal death rate. [5]

Since major congenital defects of the kind to which we refer, that is, those recognizable at or shortly after birth, normally occur in about one out of every 100 pregnancies that persist for at least seven months of gestation, this says that if the risk had been changed to two in 100 as a result of parental exposure, that fact would have been recognized. However, the frequency of detected congenital defects depends upon the clinical tools available and the length of time the children are studied. Examinations conducted within days after the birth of an infant would not detect most cases of mental and motor retardation nor would they be likely to identify those congenital defects of the heart that do not involve cyanosis (a bluing of the child) and are commonly not detected until the infant becomes older. Nevertheless, there is no reason to believe that the malformations that can be readily recognized in the newly born infant are more or less heritable in origin than those that would be seen a few years later in life, and hence would respond to ionizing radiation differently from those defects that cannot be diagnosed until the individuals are older.

The distribution of "untoward pregnancy outcomes" by parental exposure when the cities and genders are combined, and the results of an analysis that examines the frequency of these events as parental exposure and a variety of other variables known to influence their occurrence change, have been described in detail elsewhere.[5] Briefly, as stated earlier, 70,073 pregnancy terminations were available for analysis, but of these, parental gonadal doses could not be computed using the DS86 dosimetry on 14,770. To avoid the loss of information inherent in the pregnancy terminations where DS86 doses were lacking on the parents, two analyses were undertaken—one based on the 55,303 pregnancies in which the DS86 dose was known, and another based on 69,706 pregnancies. The latter group included those instances in which the parental DS86 doses were known and 14,403 pregnancies where DS86 doses were not available but a tentative dose could be assigned pending completion of efforts to provide DS86 doses (no dose could be assigned, even tentatively, to 367 of the 70,073 cases). The increase in frequency of untoward pregnancy outcome in the larger sample (as measured by the slope of the dose–response relationship after adjustment for extraneous sources of variation) was 26 cases per 1,000,000 pregnancies to parents who had received 0.01 Gy or more of gonadal exposure. The public health burden these additional cases would impose must be seen in the context of the number of untoward pregnancy outcomes that would be expected had the parents not been exposed to atomic radiation. This number is about 38,500 per 1,000,000, bringing the total in a population exposed to 10 mGy to about 38,526. The smaller sample, where the doses were more reliable, suggested a slightly greater increase—42 cases per 1,000,000. There is a large statistical error associated with both of these estimates, however, and the actual number of excess cases could be as small as none or possibly as large as 110 per 1,000,000, or a total in an exposed population of at most 38,610 cases. These numbers suggest that if a woman conceived after exposure to 10 mGy and if that pregnancy terminated in an untoward outcome, the likelihood that this event was due to irradiation would be 110/38,610, or about three chances in 1,000. Thus, the odds are overwhelming—997 chances in 1,000—that it was due to some cause other than exposure to ionizing radiation.

Although these estimates are the best that can be deduced from the data, there remains considerable statistical uncertainty. A better approach than accepting these values literally is to ask what do these data suggest is the lower limit of the dose that would double the risk. To obtain this range one can reason as follows.

The excess relative risk is merely the ratio of the increase in untoward pregnancy outcomes per unit dose equivalent (here a sievert) to the naturally occurring frequency of such an outcome without exposure to atomic-bomb radiation. Since the increase expressed per Sv is 0.00264 and the background rate is 0.03856, this ratio is 0.0685. Now we ask, what is the sampling error of this ratio? This proves to be 0.0726, and hence the plausible upper limit of the ratio (commonly calculated as the value which would be exceeded only 1 time in 20 by chance alone) is about 0.19 (taking the ratio plus 1.64 times its sampling error). Since the lowest plausible dose that would double the risk of "untoward pregnancy outcome" is associated with the highest plausible excess risk suggested by the data and since the doubling of the risk is by definition that dose at which the radiation-related contribution is precisely equal to the background, the lower limit to the dose, D, that would double the risk suggested by these data, is about 0.2 Sv (or 0.03856 = 0.19D). Or to return to the original aim—to determine what range of doses would be unlikely to double the risk—we can exclude doses below 0.2 Sv.[6]

Mortality after Birth

When originally constructed, the F_1 mortality cohort consisted of 52,621 live, single births for which the exposure of both parents was known. To this group has subsequently been added 22,984 births in the years from January 1959 through December 1983, of which 11,196 were to parents of known exposure status. Most of these latter parents were too young at the time of their exposure to have had children in the years from 1946 through 1958 when the original three groups were selected. Deaths within the study population are identified through examination of the obligatory family register previously described. Altogether the original cohorts and the later extensions consist of 76,817 livebirths, but of these 3,589 have had to be excluded either because they did not have Japanese citizenship (and hence could not be followed through the *koseki* system) or because the information available on parental exposure was inadequate to compute a dose. Among the 72,228 remaining individuals, the exposures of both parents can be estimated directly in 67,586 cases, and indirectly in 4,642 instances.

The results of two analyses of these data will now be described: one that focuses on the occurrence of death attributed to causes other than cancer, and the other on cancer.[7,8] Attention centers on those 67,586 children whose parents have directly estimated DS86 doses. Among these children, there were 3,852 deaths. Of these, 2,766 were attributed to diseases other than cancer, 115 to cancer, 28 to benign or unspecified tumors, 584 to accidents or suicide, and 359 to unknown causes. Of those deaths without known cause, 322 occurred in the first year of life, and could have been due to infantile diarrhea, respiratory diseases, or other infections, but this is not certain.

When the frequency of noncancer deaths, exclusive of those ascribed to accident, suicide, or unknown cause, is related to parental exposure considered jointly, the deaths increase slightly, but not statistically significantly so, as exposure increases. This is true whether or not one tries to adjust the data to take into account nonradiation-related sources of variation that contribute to early death. Although these latter sources of variability affect the estimate of the dose–response relationship, their effects are small when the errors inherent in the estimates of this relationship are considered. The excess relative risk of death at all ages due to diseases other than cancer at 1 Sv is 0.030 (standard error: ±0.046). If this risk is computed on the basis of only those deaths occurring before age 20 (some 73% of the sample had attained this age in 1985), the excess relative risk is somewhat higher (0.038) but not dramatically so.

Given the relationship of mutations (somatic as well as germinal) to cancer, it is interesting to note that of the 2,881 deaths attributed to disease (that is excluding deaths from unknown causes, accidents or suicide, or unknown tumors) occurring in the F_1 study group between 1946 and 1985, 115 were ascribed to cancer.[8] The most common of these cancers was leukemia; indeed, 44 of the 115 cancer deaths were due to this malignancy and 30 of these 44 resulted in death before the age of 20. It is not surprising that this cancer should be the predominant one since it is the most frequent fatal cancer found in children, worldwide. *However, no clear trend in the occurrence of either leukemia or other cancers before the age of 20, or after, with increasing parental dose exists as yet.* The excess relative risk at 1 Sv, when age at death is ignored, is -0.220 (standard error: ± 0.340) for all cancers except leukemia and -0.117 (standard error: ± 0.934) for leukemia. And for cancer deaths prior to age 20, the excess relative risks are -0.224 (standard error: ± 0.310) and 0.021 (standard error: ± 1.516), respectively. Although these risks are not statistically significant, as has been stated, if the estimated values are taken literally, the risk of cancer appears to decrease with increasing parental exposure, which seems improbable and suggests the negative values are due to chance variation.

Again, as in the instance of untoward pregnancy outcomes, doses can be estimated at which the added risk of cancer or of noncancer deaths would be doubled. However, to provide an estimate of the doubling dose for noncancer deaths that is independent of the one for untoward pregnancy outcome, those deaths occurring in the first 14 days of life in the years 1948 through 1953 must be excluded, since these deaths are already included in the estimate for untoward pregnancy outcomes. When this is done, and account is taken of the fraction of such deaths ascribable to mutation in the previous generation, the lower bounds (90%) for the doses that would double the rate of occurrence of cancer and noncancer deaths are 0.07–0.15 and 0.81–1.32, respectively.

It is important to reiterate that although others have alleged an association of leukemia with preconception exposure, most recently in the context of the studies of leukemia in the vicinity of the Sellafield nuclear processing facility in England,[9] a significant increase in leukemia with parental exposure has not been seen in Hiroshima and Nagasaki. Some Sellafield investigators have focused on the risk associated with exposure within six months before conception and suggest that the apparent increase in leukemia cases they observe is due to transmitted preconception-induced germinal mutation in exposed fathers. The data described thus far involve conceptions at any time following exposure. However, when the Japanese data are restricted to those study members conceived in the same postexposure period as that studied at Sellafield, no cases of leukemia have been seen among the exposed, and only one case among some 1,100 children born to parents who received a total dose of less than 0.01 Gy. Limited though these numbers are, when they are analyzed in the context of the cases to be expected assuming the Sellafield findings to be attributable to paternal exposure, the two sets of observations are clearly incompatible.[10] This incompatibility obtains whether one considers the total dose accumulated before conception or merely that in the six months preceding conception.

Could this disparity between the findings at Sellafield and in Hiroshima and Nagasaki have arisen because cases of leukemia might have been missed in the latter two cities? Obviously, one cannot assert categorically that no cases went undetected, but this is unlikely for two reasons. First, if deaths ascribed to blood disorders other than leukemia (where one might expect to find misdiagnosed cases) are distributed by parental dose, no increase in the frequency of such deaths with increasing dose is seen. Second, it could be argued that deaths attributable to leukemia in early life could have been erroneously ascribed to "unknown causes." Three hundred and fifty-nine children in the sample of 67,586 who died in the first year of life had their deaths attributed to unknown causes. When these cases are examined in the context of parental dose, the frequency of such deaths does not increase with dose, and is, in fact, lower at all doses than the frequency seen among the children of parents receiving less than 0.01 Gy. Moreover, only three of these 359 children were conceived within six months after the bombing and in each instance the father's estimated dose was less than 0.005 Gy. Given these findings, the explanation for the apparent increase in leukemia in the vicinity of the Sellafield facility must be sought in factors other than irradiation. If this contention is correct, what then could be the explanation?

A variety of reasons have been advanced for the differences that exist between the Sellafield study and that of the children of the atomic-bomb survivors. Among these are the failure in the Sellafield study to consider maternal exposure, differences in the methods of measurement of doses used in the two studies, possible differences in the genetic susceptibility between the Japanese and United Kingdom populations, the exposure to other chemical mutagens in an occupational environment, and the possibility, favored by the Sellafield investigators, that the disparity reflects the long-term, chronic occupational dose as contrasted with the instantaneous one in the case of the atomic-bomb survivors.[11] While the contribution that these differences could make singly or in aggregate is debatable, at present the weight of the evidence suggests that some other factor might account for the Sellafield findings. Four particularly telling observations leading to this conclusion exist. First, there is the finding in the United Kingdom that the frequency of childhood leukemia is also increased in the vicinity of what have been called "phantom" sites, that is, locations where nuclear facilities were to be built but were never constructed, or if constructed, the rise in leukemia occurred before construction took place. Second, efforts to repeat the Sellafield study in other areas in the United Kingdom (Dounreay), Canada (Ontario), or the United States have failed to yield similar results.[12]

Indeed, these studies are statistically incompatible with the one at Sellafield. Third, a study of the fathers of 1,024 cases of leukemia and 237 cases of non-Hodgkin's lymphoma diagnosed before the age of 25 among individuals born in Scotland since nuclear operations began in 1958 and the fathers of 3,783 randomly chosen controls failed to find a significant excess of leukemia and non-Hodgkin's lymphoma at any radiation level in any preconception period. Finally, if the risk of mutation is as great as the Sellafield data suggest when taken at face value, then one would expect to find evidence of other mutations such as those culminating in an untoward pregnancy outcome. This has not been observed, but the data are limited.[13]

Sentinel Phenotypes

One potentially useful mutational surveillance strategy involves the search for changes in the rate of occurrence of isolated cases (within the family) of certain phenotypes, so-called sentinel ones, which have a high probability of being due to dominant mutation, that is, a mutation that will express itself if present in only a single copy. Among such phenotypes are aniridia (the absence of the iris of the eye), chondrodystrophy (a form of disproportionate dwarfism), epiloia (an epileptic disorder associated with mental retardation and characteristic changes in the skin in the area around the nose and mouth), neurofibromatosis (the "elephant man" syndrome, a condition in which many fibrous tumors develop in association with nerve fibers), retinoblastoma (a cancer of the retina), and possibly neuroblastoma (an embryonic tumor of nerve cells). Some of these are, or have been until recently, invariably fatal tumors, and most are readily diagnosable. Here, we also include among surveys of sentinel phenotypes those clinical surveys that have sought to establish the existence of changes in the frequency of Down, Klinefelter, and Turner syndromes, all of which are ascribable to specific chromosomal abnormalities.

Insofar as the phenotypes enumerated earlier are concerned, the only ones that have been under constant scrutiny in Hiroshima and Nagasaki have been those associated with childhood malignancies—leukemia, neuroblastoma, retinoblastoma, Wilms' tumor (an embryonic tumor of the kidney), and the like. The tumor registries to which reference was made earlier make possible the identification of these occurrences as well as malignancies among cancer-prone diseases such as those associated with immunodeficiency or neurofibromatosis. During the years these registries have existed only one case of retinoblastoma has been reported (neither parent exposed) and two cases of Wilms' tumor (neither parent exposed in one instance, both parents exposed in the other). Patently, there is no persuasive evidence of a relationship between their occurrence and parental exposure to radiation, but the numbers upon which this assumption is based are abysmally small.

Chromosomal Abnormalities

As described earlier (Chapter 3), experimental evidence suggests that exposure to ionizing radiation results in an increased frequency of the failure of chromosomes to separate. This event is termed nondisjunction and results in one of the two daughter cells arising from cell division lacking a chromosome, whereas the other daughter cell has one more chromosome than normal. If, for example, chromosome 21 is involved in nondisjunction, then the fertilization of an ovum with two chromosome 21s by a sperm with one 21 will give rise to an individual with three rather than the normal two chromosome 21s. Individuals with one, two, or a few chromosomes more or less than the normal number are said to be aneuploids. And in the case of chromosome 21, the individual is generally moderately to severely mentally retarded and characteristically has a flat-bridged nose and a fold to the eyelids that gives the individual a somewhat mongoloid appearance. This disorder was first described in the 19th century by an English physician, Langdon Down, and now bears his name. Still other forms of aneuploidy arise. Commonly, these involve the X or the Y chromosomes, which are instrumental in the determination of an individual's gender. In general, abnormalities of the X or Y chromosome are less life-threatening than abnormalities of the other chromosomes, 1 through 22, known as autosomes. In addition to aneuploidy, chromosomal rearrangements, where material from one chromosome is transferred to another, are possible. Rearrangements are either balanced, if the total chromosomal material is unchanged or unbalanced, if chromosomal material is lost or gained.

Two separate studies of Down syndrome and one of the syndromes associated with sex chromosomal aneuploidy have been made. Neither of the studies of Down syndrome found evidence of a relationship between maternal irradiation and this disorder.[14,15] The study of sex chromosomal aneuploidy conducted in Hiroshima was based on a search for sex chromatin abnormalities in cells taken from the lining of the mouth. This is a simple screening technique based on the knowledge that in a resting cell (one not undergoing division) physiologically inactive X chromosomes will appear as characteristically dark-staining bodies near the nuclear membrane. Thus, in the cells lining the mouth of a normal male, who has only one X chromosome, which is active, these bodies are not seen; whereas in similar cells of a normal female, with two X chromosomes, only one of which is active, one body is seen. However, if extra

X chromosomes are present for whatever reason, more than one of these Barr bodies, as they are termed, will be apparent.

This survey revealed no cases of Turner's syndrome—individuals who appear to be females on the basis of their external genitalia but have only one X chromosome—among some 2,660 females examined, but three cases of Klinefelter's syndrome—individuals who appear to be males but actually have more than one X-chromosome—were seen among the 4,481 males examined.[16] None of the three, however, were conceived by parents exposed to the bombing of Hiroshima or Nagasaki. It must be emphasized that the sample studied, junior and senior high school students, is not necessarily a representative one, and since individual estimates of dose were not available, these data have limited value in the estimation of radiation-induced rates of chromosomal abnormality. However, they do provide indirect evidence of the difficulties inherent in the sentinel phenotype approach. No less than two of these three males would not have been detected clinically because of their normal mental status and the virtual absence of the usual physical abnormalities, such as reduction in axillary hair or enlargement of the breasts, associated with this syndrome.

To offset the limitations of the studies just described, a more systematic cytogenetic investigation of the children of exposed parents was begun in 1967, the subjects being drawn from the cohorts established for the F_1 mortality study.[17] Since children younger than 12 are not enrolled in the cytogenetic study (venipuncture is viewed as too psychologically traumatic to younger children), the survey will not yield adequate data on the frequency of chromosomal abnormalities, such as unbalanced autosomal rearrangements and autosomal trisomies, where death commonly occurs early in life, that is before the age of 12. The data on sex chromosomal abnormalities and balanced autosomal rearrangements should, however, be relatively unbiased, since these abnormalities are not associated with early mortality as stated previously. Awa has recently reported the frequency of sex chromosome abnormalities and autosomal structural rearrangements among 8,322 children of exposed parents and 7,976 children of "controls" (parents exposed beyond 2,499 m at the time of the bombings).[17] He notes that "among the children born to exposed parents, 19 individuals (0.23%) exhibited sex chromosome abnormalities and 23 (0.28%, exhibited autosomal structural rearrangements; whereas among the children of unexposed parents, 24 (0.30%) and 27 (0.34%), respectively, were observed to exhibit these abnormalities."

At face value, these findings would seem to suggest that sex chromosome abnormalities actually decrease with exposure, but this simple comparison of the exposed with the unexposed is misleading. As Table 5 indicates, when the frequency of sex chromosome abnormalities is examined in terms of the combined gonadal dose received by the par-

Table 5. The frequency of sex-chromosome aneuploids among the children of the survivors of the atomic bombing of Hiroshima and Nagasaki as a function of the combined parental doses, based on the DS86. (Adapted from Neel et al., 1990, Table 3)

Combined parental dose (Sv)	Number of children	Mean dose (Sv)	Number of aneuploids	Percent
.0	8,225	0	24	0.29
.001–.050	1,346	.024	0	—
.051–.100	951	.073	2	0.21
.101–.500	2,693	.263	9	0.33
.501–1.00	1,531	.719	3	0.20
1.01–1.50	686	1.227	2	0.29
1.50–2.00	295	1.716	1	0.34
2.00–2.50	157	2.228	0	—
2.50+	331	3.674	2	0.60
Unknown	83		0	
Total	16,298		43	0.26

ents, the frequency increases slightly, albeit not statistically significantly so, with combined parental dose. The rate of increase in sex chromosome abnormalities expressed as the change in percent per sievert is 0.044, with a standard error of 0.069, and the estimated background rate is 0.252 (±0.044). Stated somewhat differently, these numbers imply that among 10,000 children whose parents were not exposed one would expect to encounter 25 with a sex chromosomal abnormality, whereas if their parents had been exposed to 1 Sv that number would rise to about 29. No similar analysis is possible for the autosomal structural rearrangements, since so few of these appear to be mutations—most were also present in one or the other of the parents of the child with the chromosomal abnormality.

The most commonly encountered sex chromosomal anomalies were males with Klinefelter's syndrome (XXY) and females with an additional X chromosome (XXX). Both of these disorders are associated with infertility and a variety of clinical signs that can differ in their severity, ranging from normality or near normality to an obvious physical defect.

Among the balanced rearrangements involving different autosomal chromosomes, the majority were either reciprocal translocations, Robertsonian translocations, or pericentric inversions—inversions involving the region of the chromosome astride the centromere. Most of these rearrangements (90%) could not be ascribed to newly arising chromosomal mutations, as stated above, since the same defect was often found in the lymphocytes of one or the other of the two parents. While family studies were not possible on all of the children with balanced structural changes, because one of the parents had died, lived outside the contact area, or refused to participate, they did occur in 27 out of 43 cases, and among these 27 only two appeared to be new mutations, one in the proximal group and one in the distal.

Biochemical Studies

The advent of electrophoretic techniques for the identification of abnormal protein molecules created a new approach to an assessment of the genetic effects of atomic-bomb exposure. It rests on the recognition that proteins carry an electrical charge—whether this charge is positive, negative, or is absent depends upon the specific amino acids that make up the protein. Arne Tiselius, a Swedish physical chemist, showed that this fact could be used to separate mixtures of differently charged proteins. He observed that if a mixture of proteins is placed in an electrical field, where temperature and other important physical variables are held constant, the constituent proteins, if differently charged, will move toward either the positively or negatively charged end of the field at different speeds and thus can be separated. However, the technique developed by Tiselius—known as free, moving-boundary electrophoresis—required a costly and elaborate machine, and allowed one to study only one specimen at a time. It was impractical, therefore, as a survey tool where thousands of specimens need to be studied.

However, in 1955, Oliver Smithies pioneered an inexpensive alternative, one using a gel made from potato starch, that permitted the investigator to study several specimens simultaneously.[18] Over time, his procedure has undergone further refinements, and today, hundreds of proteins can be scrutinized on the basis of their electrical charge. These different proteins, when appropriately stained, will appear as distinct bands on the gel. Unfortunately, similarly charged proteins will not separate, since they will move at the same speed (they are said to have the same electrophoretic mobility). But since a starch gel is like a mesh whose holes are all of the same size (or nearly so), a molecule too big to pass through these holes will not migrate whatever its electrical charge; it is blocked by its own size. Thus, unlike a fluid medium, a starch gel separates proteins not only on the basis of their charge but also according to their molecular size. Despite the sensitivity of electrophoresis, a substantial fraction of proteins that are known to be different in their amino acid composition cannot be separated either through free or starch gel electrophoresis.

Genes determine the structure of proteins, and they do so through the determination of the particular amino acids, life's building blocks, that make up the protein. The genes, composed of DNA (deoxyribonucleic acid), contain the coded information for each protein. Many proteins are enzymes, though proteins serve many other roles as well. The location wherein a gene resides on a chromosome is known as a "locus" ("loci," in plural). Each amino acid within a protein is represented in the gene by specific combinations of three fundamental substances called nucleotides. This set of three is termed a codon, and all genes that specify protein structure are made up of some number of codons. If a gene is changed as a result of mutation, that is, if one or more of the codons are altered, the protein it specifies may also be changed. If this results in an alteration in the charge of the protein, then electrophoresis might show that the protein has been altered structurally. Such alterations are called electrophoretic variants or structural mutations. If there is no alteration in charge, and this would be expected on theoretical grounds almost two times out of three, or if the alteration does not produce a band that can be visualized electrophoretically, electrophoresis will be uninformative. Similarly, if the mutation results in a failure to specify a protein at all, electrophoresis will be uninformative, since there will be no protein to migrate. The latter mutations are described as deficiency variants or "null" mutations.

Detection of these deficiency variants or null mutants has been difficult biochemically. They are recognized through a diminution in enzyme activity when the latter is measured under standardized conditions. Assessing enzymatic activity manually is tedious and prone to error. It was not, therefore, a promising avenue of investigation until another technological innovation occurred. At the Oak Ridge National Laboratory a machine was developed—the Centrifugal Fast Analyzer—that could automatically load, mix, and analyze 30–40 specimens at a time. The strategy employed is ingenious but simple in principle. Specimens are placed in wells in a clear circular disk, next to these specimens in a communicating well is added a predetermined amount of any substance on which the enzyme acts, and then the disk is spun. The centrifugal force generated in the spinning shoves the specimen and reactive material together into still another well, a reaction chamber, mounted above a light so the speed of the reaction can be measured photometrically. All of this is controlled by a small computer. This technique has made it possible to analyze not only a large number of specimens, but numerous enzymes as well. However, it has some limitations. First, it can be applied only to those enzymes whose activity is measurable spectrophotometrically, that is, by a change in color or density. If no such test exists then the activity of the enzyme cannot be assayed by this method. Second, if the variability that normally occurs between specimens from different individuals is too large, the diminution in activity associated with a new null mutant may not be separable from the background variability. This further restricts the applicability of the automated method.

This biochemical approach, like the cytogenetic one, is free of many of the ambiguities inherent in the study of population characteristics or sentinel phenotypes, and has, as a result, been vigorously pursued. After a pilot study extending from 1972 through 1975, a full-scale investigation employing electrophoretic techniques was undertaken in Hiroshima and Nagasaki in 1976. The subjects were drawn from the groups of children born to the proximally and distally exposed parents identified for the mortality study previously described. The same blood sample served the needs of this biochemical program and the cytogenetic one. Most of the proteins easily studied in blood specimens are enzymes, found either in the red blood cells or in white cells, but some nonenzymic proteins in the serum can also be studied electrophoretically. Accordingly, each child was examined for rare electrophoretic variants of 28 proteins of the blood plasma and red cells, and since 1979, a subset of the children has been examined for deficiency variants of 10 of the red-cell enzymes.

A rare electrophoretic variant is defined in this context as one with a frequency of less than 2% in the population and an "enzyme deficiency" or "low activity" variant as one resulting in an enzyme activity level three standard deviations below the mean (or less than 66% of normal enzymic activity). When either variant is encountered, its occurrence is first verified—the possibility of a technical error is excluded—and then blood samples from both parents are examined for the presence of a similar variant. If the variant is not found in one or the other parent, and if an error in assigning parentage is improbable, it presumably represents a new mutation. To establish parentage, since a priori the probability that the putative parents might not be the real parents is very much larger, several orders of magnitude, than the probability of a new mutation, some 11 different red-cell antigenic systems and the major histocompatibility phenotypes, the HLA system, were used to search for evidence that the putative parents were not the actual parents of the child. While such testing does not prove parentage, it can only exclude falsely identified parents, the battery used was sufficiently large that the a priori probability of failing to detect a falsely identified parent was approximately the same as the a priori probability of a new mutation.

James V. Neel and his colleagues have estimated that they have information on the equivalent of 667,404 locus tests on 13,052 children born to parents whose average combined gonadal dose is about 0.47 Sv.[19] Three probable mutations were seen—one, a slowly migrating variant of the enzyme glutamate pyruvate transaminase, another a slowly migrating variant of phosphoglucomutase-2, and the final one a variant of nucleoside phosphorylase. Three mutants have also been seen in the equivalent of 466,881 locus tests on 10,609 children whose parents, one or both, were exposed beyond 2,499 m and who received less than 10 mSv. These mutants involved the proteins known as haptoglobin, 6-phosphogluconate dehydrogenase, and adenosine deaminase. The mutation rates in the two groups of children are almost identical; the values are 0.60×10^{-5} mutations per generation in those who are the offspring of parents receiving more than 0.01 Sv of gonadal exposure, and 0.64×10^{-5} in those whose parents received less than 0.01 Sv. The confidence intervals for these two estimates, that is, the probable range in which the "true" value lies, are 0.2–1.5 and $0.1–1.9 \times 10^{-5}$, respectively.

Chiyoko Satoh and her coworkers have reported the results of the 122,270 determinations, distributed over nine enzymes, to assess the impact of parental exposure on the frequency of "deficiency variants."[20] One probable mutant in 60,529 locus tests on children whose parents, one or both, received more than 0.01 Sv of radiation has been seen, but none among the 61,741 tests on the children of distally exposed parents. The one apparent mutant involves the enzyme triosephosphate isomerase. Thus, after more than 1,256,000 biochemical tests, when the results of the studies of structural and activity variants are combined, four mutants have been seen among the children of parents receiving more than 0.01 Sv, and three among those whose parents received less than 0.01 Sv.

It was recognized at the outset that this approach, although it would entail a massive amount of work, would have little discriminatory power if the information at our disposal was correct. This can be illustrated by the following simple example: if one assumes that (a) the induced mutation rate per codon per 0.01 Sv is about 1 in 100,000,000, (b) 200 codons are needed to specify the structure of the average enzyme, and (c) 28 enzymes can be tested on (d) 20,000 individuals whose parents received an average exposure of 0.20 Sv, then less than 11 new mutations would be expected.[21] Of these only a third or so would be detected electrophoretically, for reasons given previously. These three or four potentially recognizable variants would be hidden, as it were, in 1,120,000 tests. While the numbers cited are merely approximations to the actual observations, they cannot be wrong by much and if, perchance, the "true" mutation rate is lower than suggested, the expected number of mutants is smaller still.

The Growth and Development of These Children

It will be recalled that at the time of the initial, clinical studies of these children, measurements were obtained to characterize their growth and development in the first nine months or so of life. Four measurements were made—weight, height, and head and chest circumference. Analysis of these in 1956 did not reveal any systematic impairment of growth and development with increasing parental exposure.[1] Although doses could not then be estimated for individual parents, it was possible, through knowledge of their distance from the hypocenter and the occurrence of symptoms of acute radiation sickness, to order their exposure in a manner reflecting the presumed doses.

Later, when these children were 6 to 17 years old, Toshiyuki Furusho and Masanori Otake reassessed their growth and development using the measurements of stature, weight, sitting height, and chest circumference routinely obtained each year in the public schools in Hiroshima.[22] At this time, T65 doses were available for most of their parents. Again, no evidence emerged of a radiation-related retardation in their development.

Neither of these sets of data has been reanalyzed with the new dosimetry, and it is debatable whether such an analysis would be worthwhile. The absence of any trend of impairment with dose in the first year of life or later makes it highly improbable that one would be found now. Moreover, since so little is known about the role of genetic factors in childhood growth and development it is unlikely that these observations could be couched in meaningful genetic terms. It should be reassuring, nonetheless, that these empirical measurements of development fail to suggest damage that could be compromising to the children of the survivors.

The Sex Ratio

When the genetic studies began, it was believed that a person's gender was simply determined. Individuals inheriting an X chromosome from their father and one from their mother were destined to be females; whereas those individuals who inherited a Y chromosome from their father and an X from their mother would be males. Thus, females would have two X chromosomes and males only one. These notions suggested, in turn, that when mutations induced in the X chromosome by ionizing radiation are incompatible with survival (are lethal), their expression would be manifested differently in the two genders and would depend, partly, upon whether the X chromosome was inherited from the mother or the father. More specifically, since a father transmits his X chromosome to his daughters exclusively, if a lethal mutation were present on the X chromosome in the father's sperm, it would find expression only in his daughters. Whereas, since mothers transmit their X chromosomes equally to their sons and daughters, a lethal mutation might find expression in either sex. If the mutation were dominant, that is, expressed itself if only one copy was present, the two sexes would be affected equally often; however, if the mutation was recessive (normally requiring two copies for expression), since the male has only one X chromosome, it would invariably manifest itself in males, but in females manifestation of the new mutant would occur only if the second X chromosome fortuitously carried a functionally similar gene. It follows from these thoughts that since the likelihood of a mutation would increase as dose increased, if the father were exposed, more female embryos would be lost, and at birth the relative proportion of males would be greater than would be true if the father were not exposed (see Figure 34). If, on the other hand, the mother were exposed, more male embryos would be lost, and at birth the relative proportion of males would decrease. If both parents were exposed, the resulting sex ratio or proportion of male births, would be related to the individual parental doses and the frequency of dominant versus recessive lethal mutations. As can be seen, this theory of sex determination made fairly specific predictions that could be compared with the actual observations that were accumulating.

When the data from the initial study were examined, it appeared that the proportion of male births was, in fact, declining with dose when the mother was exposed, and increasing, albeit modestly, with increasing paternal dose. The rate of change with dose was not, however, statis-

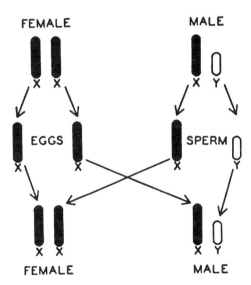

Fig. 34. A schematic representation of the distribution of sex-linked lethal genes in man.

tically significant, although in the direction predicted by theory. It was for this reason that when the clinical phase of the studies ended, data on the sex ratio continued to be collected on the supposition that the rate of change might become statistically significant with further information. To this end, observations on the frequency of male births were continued through 1966. However, when these additional observations were analyzed, the results did not support the earlier findings; indeed, the modest changes seen were opposite to those predicted by theory.

Today, the earlier arguments regarding sex determination are known to have been overly simplistic. First, they did not take into account the occurrence of X chromosome aneuploids, such as the Klinefelter and Turner syndromes, which might confuse the determination of the gender of an individual as revealed by a clinical examination shortly after birth. The first of these aneuploids was discovered in 1959 (in unrelated studies) and soon many others were identified. As a result of these discoveries, it is known that it is possible for some females to have only one X chromosome (or even as many as five), and for some males to have two or more. Moreover, these individuals with abnormal numbers of X chromosomes are more frequent in most populations than one would expect new sex-linked lethal mutations to be, following exposure to ionizing radiation at least at the dose of the average survivor. Second, it is now known that in females only one of the two X chromosomes within a cell is functionally active. This inactivation of one of the X chromosomes, known as Lyonization (after its discoverer, the English geneticist Mary Lyons), makes the prediction of the behavior of a potentially lethal gene on the X chromosome more difficult, and particularly if the inactivation is not random (and it does not appear to be). Given these developments, most human geneticists no longer accept the simple, early arguments, and contend that prediction of the effects of lethal mutations on the proportion of male births is not possible. Thus, no effort will be made to estimate the possible doubling dose for X-linked lethal mutations in the paragraphs that follow.

An Overall Estimate of Genetic Risk

As has been seen, despite the monumental investment in time and labor that has been made, no unequivocal evidence of radiation-related genetic damage emerges. One could argue, therefore, that there is nothing further to do, and that it would be presumptuous, if not misleading, to attempt to generate individual estimates of the dose at which a doubling of the genetic risk occurs. This view, however, seems an abandonment of scientific responsibility. The enormous effort invested in these studies demands that every iota of information that can be extracted from these data be extracted, and this includes the best possible estimate (and lower bound) of the doubling dose, accepting the nonsignificant differences that are seen as real, not merely random fluctuations.

Conventionally, two parameters have been used to characterize the risk of radiation-related mutation. These are the probability of a mutation at a specific genetic locus per unit of dose to the gonads, and the "doubling dose." The latter, as previously noted, is that dose at which there would be induced as many radiation-related mutations as would occur spontaneously in a generation, so that the overall rate of mutation is actually doubled. Each of these measures has its uses and its advocates. Experimentalists favor the first, since it reflects the nature of the data they collect and the problems they face. Population biologists, however, favor the doubling dose. This is especially true of human population geneticists, since few estimates exist of locus-specific rates of mutation, either induced or spontaneously occurring. Accordingly, we turn to the estimation of the doubling dose suggested by these various sets of data and the uncertainties inherent in this estimate.

Theoretically, at least, an estimate of the doubling dose could be obtained for each of the different measures of mutational damage, i.e., untoward pregnancy outcome, cancer and noncancer mortality, sex chromosome aneuploids, biochemical studies, and so on. These would undoubtedly differ for a variety of reasons including chance, the different samples employed, the sensitivity of the specific measurement, the validity of the assumptions inherent in passing from the observations to the estimate, and the fact that the doubling dose may actually differ for these outcomes. The preferred estimate, therefore, would be some combination of the individual estimates that takes into account their differences. But how are they best combined, since they could be combined in a variety of different ways—on the basis of their presumed genetic specificity, the sizes of the samples involved, or in some other manner?

The process of estimating the doubling dose for each of the measures of genetic damage previously described can be illustrated using the data on untoward pregnancy outcomes. Recall that the increase in such pregnancy outcomes is 0.00264 per sievert, the background rate is 0.03856, and the excess relative risk at 1 Sv is 0.0685.[5] We have seen that the lowest plausible dose that would double the risk of an untoward pregnancy outcome derived from the excess relative risk is about 0.20 Sv. But this is not actually the genetic doubling dose, since it is known that not all untoward pregnancy outcomes are genetic in origin; some could arise, for example, from exposure during pregnancy to a variety of possible drugs or environmental agents that impair the normal growth and development of the embryo or fetus. Such exposures could lead to early death or the occurrence of a "specific" congenital malformation, as in the much publicized case of thalidomide three decades ago.

To derive a genetic doubling dose, it is necessary to postulate what fraction of untoward pregnancy outcomes is attributable to spontaneous mutation in the preceding generation. This value is not known with certainty; however, from a variety of lines of evidence it can be estimated to lie between 0.33 and 0.53%.[6] These values suggest a genetic doubling dose in the range of 0.63 to 1.01 Sv. Note that had the findings with the smaller sample been the basis of this calculation—that is, the sample excluding those cases where an ad hoc dose was used—the doubling dose would lie between 0.50 Sv and 0.80 Sv. The difference between these two ranges is small when the uncertainties in the estimates are taken into account.

Elsewhere, Neel and his associates have argued that the simplest overall estimate is the one that adds together the individual estimates.[6] They justify this cumulation "on the grounds that each regression involves the relationship to radiation exposure, in the same cohort, of specific, independently determined events." The individual regressions they derived from the various sets of data are given in Table 6.

The lower estimate of the doubling dose associated with exposure to acute doses of low-LET radiation derived from these values is 1.69 Sv (.00632/.00375), and the upper estimate is 2.23 Sv (.00835/.00375). It should be noted, however, that these limits reflect only the biological uncertainties. This range, 1.69 to 2.23 Sv, does not consider the additional errors inherent in the estimation procedure itself, and if this could be adequately done, the range of uncertainty would be even larger. Be this as it may, based on the experiences in Hiroshima and Nagasaki, the doubling dose for acute gonadal radiation appears to be about 2.0 Sv. And for chronic exposure, assuming a reduction in effectiveness with a low dose rate of two (which is consonant with experimental studies of mutagenesis), the doubling dose is about 4.0 Sv.

Table 6. Summary of the regression of various indicators of genetic damage on parental radiation exposure and of the impact of spontaneous mutation on each indicator.

Trait	Regression/combined parental Sv	Contribution of spontaneous mutation
Untoward pregnancy outcome	+.00264	.0033–.0053
F_1 noncancer mortality	+.00076	.0033–.0053
Protein mutations	−.00001	
Sex-chromosome aneuploids	+.00044	.0030
F_1 cancer mortality	−.00008	.00002–.00005
Total	.00375	.00632–.00835

What is to be made of these numbers? First, it warrants noting that these estimates are higher than the 1 Gy that has been used to guide regulatory action. However, the estimate of 1 Gy was based largely on experimental studies of mutations arising in mice following exposure to ionizing radiation, commonly X-rays or gamma rays derived from the radioactive isotope cobalt-60. Little cognizance was taken in this estimate of the human data, limited though they were. Moreover, it rested almost exclusively on one large series of studies, those of William Russell and his colleagues at the Oak Ridge National Laboratory, using a particular strain of mice and a technique known as the specific-locus test, involving seven recessive genes. The doubling dose derived from these data was not only lower than that from *Drosophila* on which an extensive body of data existed, but inconsistent with other, but smaller, rodent studies. This led some investigators, including Russell himself, to suggest that the specific loci used in the Oak Ridge studies might be unusually sensitive to irradiation and therefore not a suitable basis for estimating the rate of mutation at all loci. Nonetheless, the experimental and human data appeared to conflict. Recently, James Neel and Susan Lewis have reexamined the earlier experimental data and those that have accumulated since Russell's studies, focusing on those observations they contend are most appropriate to the human situation.[23] When this is done, a doubling dose is obtained for chronic exposure of the mouse of about 4.0 Gy. They conclude, therefore, that the data lead to a "somewhat higher estimate of the doubling dose of chronic radiation for the mouse than the Committee's (Committee on the Biological Effects of Ionizing Radiation), and lessens the apparent conflict between the two sets of data that the Committee identifies in its report."

Second, it must be noted that the human data are based upon instantaneous exposure, and experimental evidence shows that doses accumulated slowly over time, such as would occur in most occupational settings, produce fewer mutations for the same total dose than acute exposure. How much smaller the number may be is debatable. We have assumed chronic exposure to low-LET radiation to be only half as effective as acute exposure—a value consistent with present genetic knowledge—but if in fact chronic exposures are even less effective, the doubling dose would be higher than estimated here.

Finally, if these newer estimates are correct, then in the past there has been an overestimation of the risk, but given the many uncertainties to which we have alluded, this error was prudent.

As has been stated before, at the outset of the Commission's activities, public concern over the possible genetic effects of exposure to atomic radiation was at least as great as that over cancer, and possibly greater. Over time, however, this emphasis has slowly shifted to more interest in cancer. This undoubtedly reflects the failure to find unequivocal evidence of genetic damage on the one hand, and the more dramatic findings on cancer, on the other. Understandably, but unfortunately, this has led to a failure to recognize that the absence of demonstrable radiation-related findings is no less noteworthy. But such negative findings must be more guardedly interpreted, and have a tentativeness some find disturbing. They cannot be interpreted as implying that *absolutely no effect* has occurred. The latter conclusion could be legitimately drawn only if it were possible to examine, in the necessary detail, every single pertinent event within the population of interest. Epidemiological studies deal, however, with samples, not populations, since statistically the population under investigation is viewed as indefinitely or infinitely large. The sample available to an investigator might or might not represent the larger population from which it is drawn, or irrespective of its size, it could be too small to reveal those changes that occur very infrequently. And in this sense, the genetic findings can be seen as "inconclusive," since although an effect was not found, it cannot be concluded that none exists. However, as has been indicated previously, the extent to which they are "inconclusive" can be estimated—one can compute the size of an effect that could have existed but gone unrecognized. But ultimately, of course, the acceptance of these findings should rest on the comprehensiveness of the study design and the care and thoroughness with which that design was implemented.

The Commission's findings can be seen as reassuring in at least two different ways. First, where genetic theory suggested no effect would be demonstrable, "no effect" was found. The study does not then contradict current theory, and as a result makes more plausible extrapolations based upon previously accumulated information and understanding. This would not necessarily be true of a single study that did not find an effect, particularly if it were based upon a small sample. However, in the present instance, there is not merely one but many mutually reinforcing "negative" findings, and all are based upon samples of considerable size. Second, the findings should be seen as reassuring to the public, since they argue forcefully against the fears of a devastating genetic effect. More immediately, they do not support the notion of the birth of an epidemic of malformed infants, genetic or otherwise in origin.[24]

Evolutionists have long recognized that mutations provide the means through which new genetic combinations come into existence in a species. If a new mutation, either alone or in combination with pre-existing genes, enhances the survival, reproductivity, or both, of the individual who carries it, the mutant gene will be favored, and in time might replace its predecessor. This process of replacing one gene or one genetic combination by another has been called "natural selection." However, most mutations, if not neutral, are deleterious. Thus, it is also recognized that there must be a dynamic balance between the rate at which genes mutate and the capacity to select advantageous combinations. If the mutation rate is too high, no genetic combination however good will be secure for long, and an orderly improvement in the species would be impossible. As human beings, we have done much to blunt the effects of natural selection through the manipulation of the environments in which we live, but we too are bound by these inexorable rules. Precisely how this balance between new mutations, on the one hand, and the opportunity to protect the genetic advances of the past, on the other, is achieved or maintained is not fully understood. However, human beings have been subject to exposure to ionizing radiation since our species arose, as have all organisms, and it is reasonable to suppose that the same evolutionary processes described have shaped the human capacity to respond to radiation damage.

One out of every two or three human conceptuses, for example, is aborted spontaneously.[25] If abortion is a random event, one would expect that the number of surviving detectable mutations is only 50% or so of those that are present at conception. However, if mutations are generally harmful, as the evidence supports, abortions are likely to be nonrandom, and the number of mutations surviving long enough to be detected must be even smaller. A variety of observations indicate that ionizing radiation is a poor point mutagen—that is, a mutagen that alters the composition of a gene; more commonly, it leads to the loss of genetic material, and this loss in turn often leads to premature death. As a result, human populations may purge some portion of newly arisen mutations through early embryonic death and thus lessen their public health impact. Whether such a purging occurred in Hiroshima and Nagasaki is debatable, and can no longer be tested rigorously. Even under optimal conditions, it would be difficult to show an increase of a few percent, or even as much as 10%, in the frequency of occurrence of spontaneous abortions. Yet an increase of this amount could readily account for the higher "doubling dose" that has been estimated. As a public health issue, only those mutational events that are potentially transmissible are important and those that culminate in an abortion, often before the woman is even aware of her pregnancy, would impose little burden, seemingly, even of a psychological nature, on the members of the society involved.

The Future

Virtually everyone is aware, at least vaguely, of the existence and importance of deoxyribonucleic acid or DNA. Newspapers, magazines, and news broadcasts are constantly informing us of this or that "breakthrough" in the understanding of DNA's role in development, the extent of its variability between individuals or groups of individuals, and even its utility in identifying the culprit in a sexual assault. But this has not always been so. When the genetic studies in Hiroshima and Nagasaki began, little was known biochemically about how genetic information was transmitted from parent to offspring and translated within a cell. It was, however, clear that the vehicles of this information were the chromosomes, and it was suspected that one of the constituent components of the chromosome, deoxyribonucleic acid (DNA), was a central player. It was also known that DNA itself consisted of four nucleotides—adenine, cytosine, guanine, and thymidine—and several elements, most notably phosphorus. Enough data had accumulated by 1953 to permit James Watson and Francis Crick to propose a structural model of DNA, the now famous double helix, one that implicitly defined a series of crucial experiments. These experiments soon confirmed the validity of the structure they had postulated, and focused attention on an effort to determine precisely how the sequence of nucleotides in DNA specified which amino acids were to be made within a cell.

Subsequently, Marshall Nirenberg was able to demonstrate that three successive nucleotides are enough to define a particular amino acid; these nucleotide triplets are the codons described earlier. Since there were four nucleotides involved and three positions, this meant that 4^3 or 64 different combinations could occur. This was more than the twenty-odd amino acids known, and implied that some amino acids must be represented (or coded) in the DNA in more than one manner. Quickly, each of the different ways in which the various amino acids could be coded was deciphered. This left unsolved, however, the mechanism that communicated this information to the organelles (ribosomes) within a cell, where the enzymes and other proteins are actually made. It was soon discovered that DNA served as a template (or model) on which a closely related molecule, ribonucleic acid (RNA), was constructed. The precise composition of the RNA that was formed was determined by the structure of the DNA that served as the model. This process is called transcription and the pieces of RNA that are formed are termed messengers (generally designated as mRNA). These messengers "tell" the ribosomes, located in the cytoplasm of the cell, what protein to make, and what its structural composition is to be. They are said to be "translated" within the ribosome.

The process of transcripting and translating the genetic information is actually more complicated than this simple summary would suggest. First, within the segment of DNA that corresponds to a gene, not all of the nucleotides are translated, only those that constitute what are called exons. Other portions of the gene, the introns, do not code information for protein synthesis but have some other, as yet unknown function. Second, large regions of DNA are not translated at all; they appear to be redundant or "junk." Some of these segments have structures very similar to those of active genes, enough so that they are termed pseudogenes. These functionally inactive copies of genes are thought to be relics of evolution.

Since it is known that DNA is the substance that carries the code that determines which proteins our bodies make, when they should be made, and at what rate, if we wish to have a measure of mutation uncomplicated by other processes or events, DNA should be the object of study. Yet until quite recently, the means to do this were not available. They are now, but the procedures are costly, time-consuming, and not presently applicable to the screening of a large number of individuals. This will surely change, and the consequences can be foreseen to some extent. With singular exceptions, such as the red blood cell, which has no nucleus, DNA is present in all of our cells. It is amenable to study either from small snippets of skin or, more conveniently still, from the white blood cells present in a sample of blood. In either event, these cells can be grown in culture until sufficient quantities are available to provide the amount of DNA present techniques require. This amount continually diminishes as molecular biology evolves. Newly developed techniques, such as the one known as the polymerase chain reaction (PCR), which makes possible the amplification of very small amounts of DNA, can be applied to single cells.

At least two different biochemical approaches to future studies can be foreseen. First, and most exciting, is DNA sequencing. This procedure entails the determination of the biochemical structure of all of the DNA within a given cell, or at least selected portions. Presumably, through this technique every change in the molecule could be recognized. This is indeed the aim of the oft-publicized Human Genome Project. Second, the information could be obtained, less efficiently but more simply, through the study of pieces of DNA, so-called restriction fragments. There is a class of enzymes, the endonucleases, that cut DNA molecules. A particular endonuclease will always cut the molecule at precisely the same place. The fragments that result can be examined structurally using electrophoresis and radioactive labeling techniques. A hundred or so such endonucleases are already known, with no end in sight. Thus, in theory at least, the DNA molecule could be cut into a very large number of fragments, each of which would represent many genes. These pieces could then be studied to determine whether they are structurally normal or abnormal. Again, however, this is a costly and labor-intensive technique at present, and until some degree of automation or a simpler alternative is developed, it would be difficult to use on the scale needed to assess the mutational damage that has resulted from the atomic bombings.

On the supposition that sooner or later these techniques will become practical on the scale required, the Foundation has launched a program to collect and establish permanent cell lines of peripheral B-lymphocytes, specific white cells, of a group consisting of 1,000 children of survivors and their parents (500 trios—the child and its two parents—selected because the parents had relatively high radiation exposures, and 500 controls).[26] Since the genetic effects, if any, associated with the manipulation of the cells to establish permanent lines are unknown, intact white cells from these survivors and their children are also being preserved as a basis of reference to the original DNA or RNA if this should be needed. These cells and cell lines can be stored indefinitely in liquid nitrogen, and will therefore be available for study as new developments occur. Even now, the utility of a new biochemical technique, known as denaturing gradient gel electrophoresis is being examined. This method is suitable for large-scale studies, and has other advantages as well: notably, it is inexpensive, since it does not require the use of costly radioactive isotopes to reveal genetic changes. Moreover, it appears capable of detecting not only nucleotide substitutions but those mutations that lead to the loss of genetic information, currently presumed to be the most common genetic consequence of exposure to ionizing radiation.

There is still another study opportunity, one unique to Japan. It has been the tradition among Japanese parents to keep indefinitely a piece, an inch or so long, of the umbilical cord from the birth of their children. DNA could be recovered successfully from these snippets, if samples can be obtained, and where warranted, interesting portions of the DNA amplified by the polymerase chain reaction. This could be a promising avenue for the study of those children who were born alive, but died before the age at which they would have been eligible to enroll in the biochemical studies we have cited. These children would have been more likely to have inherited mutations that were incompatible with survival, and that manifested themselves early in life, than the children seen during the biochemical studies. Similarly, it should be possible using these techniques to search for transmitted genetic damage in the preserved samples from the 717 infants autopsied in Hiroshima during the period 1948–1953.

Although the techniques and opportunities just described are exceptionally promising, they will not address directly the question of the public health impact that ensues from radiation-related mutational damage. For example, no less than 14 copies of the gene responsible for the enzyme argininosuccinate synthetase appear to exist. These pseudogenes are located on different chromosomes and most are not active. What then would happen if a change occurred in one of the nonactive copies? It is hard to believe that such a change would have important health implications if the affected copy continued to be functionally inactive. The situation with regard to argininosuccinate synthetase is not unusual; many similar instances of pseudogenes are known. Obviously, still other studies will be needed to assess this aspect of the total issue.

Measures of mortality or the frequency of congenital malformations have immediate meaning to everyone; the occurrence of a new mutant of glutamate pyruvate transaminase, for example, does not. To the geneticist, however, events of the latter kind have a precision in the measurement of the frequency of mutation not to be found in changes in the frequencies of variables where the role of genetic factors is still uncertain. Mortality surveillances, such as the one in Hiroshima and Nagasaki are inexpensive, but biochemical studies are not. How, then, are these seemingly conflicting cost and pertinacity differences to be reconciled? Undoubtedly, with time, technological developments will narrow the cost difference. Even now, the feasibility of such techniques as two-dimensional gel electrophoresis, DNA sequencing, and various uses of restriction fragments or denaturing gradient gel electrophoresis to assess the frequency of newly arisen mutations is being explored. These alternatives, more elegant than those used in the past to assess mutagenesis in this population, must still confront the matter of their relevance to the concerns of the public, and these center on the health and well-being of our children. It would seem important in the public health perspective, therefore, that future studies focus primarily on genes with known functions and not upon the massive amount of DNA that is apparently nonfunctional or involved in regulatory functions that cannot now be discerned. This is not to imply there are not interesting biological questions associated with the apparently nonfunctional portions of the genome, such as whether mutations arise in these areas at different rates than elsewhere, but merely that it will be difficult to extrapolate this information to matters of public health. Ultimately, it is information to guide personal conduct that the public seeks and the provision of this information provides the greatest justification for the enormous investment in time and money that has been made in the genetic studies of the survivors and their children.

9

Summary

To summarize simply and briefly, the search for the late effects of exposure to ionizing radiation on the health of the survivors and their children initiated in 1948 has revealed or suggested the following:

Significant Radiation-Related Increase

Malignant tumors: leukemia, cancers of the breast (female), colon, liver, lung, ovary, skin (nonmelanoma), stomach, and thyroid
Lenticular opacities
Small head size, mental retardation, diminished IQ and school performance, increased frequency of seizures (prenatally exposed)
Retarded growth and development (among survivors exposed at young age or prenatally)
Chromosome abnormalities in lymphocytes
Somatic mutation in erythrocytes and lymphocytes

Suggestive Radiation-Related Increase

Malignant tumors: cancers of the esophagus, and urinary bladder, malignant lymphoma, salivary gland tumors, and, possibly, multiple myeloma
Adult-type malignancies among the prenatally exposed
Impairment of neuromuscular development among the survivors exposed in utero
Parathyroid disease
Mortality from diseases other than malignant tumors, specifically cardiovascular disease and liver cirrhosis, at higher doses
Specific (humoral or cell-mediated) changes in immunologic competence

No Radiation-Related Increase Seen to Date

Malignant tumors: chronic lymphocytic leukemia, osteosarcoma
Acceleration of aging
Sterility or infertility among the prenatally or postnatally exposed
F_1: congenital abnormalities, mortality, including childhood cancer, chromosome aberrations, and in biochemically identifiable genes

Undoubtedly the list of effects will grow as the surveillance of the survivors continues, and some of those effects that are now only "suggestive" will become more certain, or at least less ambiguous.

Cancer has involved, and will continue to involve, more survivors than any other health-related late effect of exposure to atomic radiation. With the exception of leukemia, this increased risk does not generally appear until the survivors reach those ages in life at which the natural occurrence of cancer begins to rise sharply. And for the nonleukemic malignancies, the increase in mortality or incident cases is linear with dose. Although this increased risk is evident at all ages at the time of the bombings, it has been most pronounced among those survivors exposed in the first two decades of life. However, their risk now appears to be declining, and significantly, among those survivors exposed prior to the age of 10. Further study should permit a better delineation of this apparently changing risk than has been possible thus far.

The attributable risk of a site-specific cancer among the survivors exposed to 0.01 Gy or more ranges from a few percent to as high as 59% in the case of leukemia at 1 Gy.[1] Many of the solid tumors have attributable risks of 10 to 25%, as judged by the recent data on cancer incidence. But, it should be noted that most of these estimates of risk for specific sites have relatively large statistical errors associated with them, and it may be that the variability seen is ascribable, in part or wholly, to random variation, and not due to inherent differences in the sensitivity of different tissue sites.

Although the number of survivors involved is small, the most poignant of the effects seen is severe mental retardation and diminished cortical function among the prenatally exposed. Their handicap, which can manifest itself in a variety of ways, persists from the moment of birth throughout a lifetime.

Finally, no epidemic of congenital abnormalities occurred following the bombing nor is there evidence that the health or development of the children of the survivors has been measurably impaired. Estimates of the genetic "doubling dose" suggest that earlier appraisals might have been overly pessimistic. Currently, it appears that the "doubling dose" following acute exposure, such as occurred in Hiroshima and Nagasaki, is in the neighborhood of 2 Gy, and for chronic exposure approximately 4 Gy.

10

Epilogue

Most of us will not receive doses of ionizing radiation in our lifetime commensurate with those received by many of the atomic bomb survivors and will not face their carcinogenic, mutagenic, and teratogenic risks. However, numerous individuals have been exposed under a wide enough variety of circumstances to make members of the public concerned and apprehensive. Hundreds of thousands have been exposed in our nation's nuclear weapons complex, in the atmospheric testing of nuclear weapons, in the construction or refitting of nuclear powered submarines, in the generation of nuclear power, or because they live near one of the weapons complex installations, such as that at Fernald, Ohio or Savannah River, Georgia. While it is generally believed that the dose these individuals received is small, this is not invariably true. In some instances, such as the release of radioactive iodine at the Hanford Facility in Washington in the early years of its operation, the absorbed dose to the thyroid of infants and children was appreciable and greater than that to adults living in the same areas because of the size and activity of their thyroid gland, where iodine is normally sequestered. Manufacturing safeguards used in the past are worrisome too, since these have often led to the contamination of surface and ground water around some of the facilities involved in the making of nuclear weapons. The residents of areas surrounding these facilities are understandably fearful that this contamination might affect their health.

Isolated cases of high exposure due to accidents of various sorts have also occurred, and will undoubtedly occur in the future. Some of these accidents have involved nuclear power reactors, as at Chernobyl in the Ukraine,[1] or an explosion at a nuclear processing facility, as in the accident involving the faulty cooling of an underground tank for radioactive waste storage that occurred at a nuclear weapons production complex, known as Kyshtym, in the Soviet Union on September 29, 1957.[2] Overheating of the tank caused a chemical explosion with the release into the atmosphere of a large amount of radioactive material and its dispersal over a wide area. Still other accidents have happened, often unpredictably but involving unsuspecting, innocent people. The situation a few years ago in Ciudad Juarez in Mexico is typical.[3] A cobalt-60 source, used in hospitals for radiation therapy, was improperly disposed, and fell into the hands of a junk dealer. Unaware of the danger, the driver of the truck transporting the source to a steel manufacturing facility parked it on an open street in a crowded neighborhood. Some 300–500 individuals, passersby and children playing on or near the truck, were exposed, some to doses of known biologic consequence. Another recent accident in Goiânia in Brazil was even more disastrous.[4] A cesium-137 source was dismantled in a residential area. Out of ignorance of the hazard, some 240 individuals were exposed, and 54 had to be hospitalized. Four later died. An accident in El Salvador in February 1989 led to three poorly trained men being seriously injured, one fatally, as a result of faulty and slipshod maintenance of a cobalt-60 irradiator used in the sterilization of prepackaged medical products.[5] A protective metal shroud designed to prevent the product boxes from interfering with the movement of the radioactive source was never installed. The boxes jammed the source in the open position, and one of the operators, despite instruction to the contrary, entered the facility to release them. When he could not do this alone, he called the other two men to help. Within hours all three were acutely ill. World-wide, over 160 similar irradiators are in use, many in countries with poor or virtually nonexistent standards of radiation protection. Finally, it has recently been discovered that in Taiwan, cobalt-60 contaminated reinforcing rods were used in the construction of a number of large buildings, and that several thousand individuals have been needlessly exposed, some to doses as high as 1.5 Gy.[6] It is not yet clear how the contamination occurred, but it is tempting to speculate that the situation may have been analogous to that in Ciudad Juarez.

It is reasonable, given the diversity of the situations under which these exposures occurred, to wonder what pertinence the experiences of the atomic bomb survivors have to the assessment of the biological effects of exposure to the much lower doses to which the average person may be subject and the regulatory limits designed to protect the public. As is frequently true when the data are limited, there is no consensus about the probable size of the effects of exposure to low doses of ionizing radiation and different schools of scientific thought exist. One extreme argues that the effects at low doses might be relatively greater than those at high. If this is true, the experiences of the survivors are a poor guide, since the risks will be underestimated. Still another extreme claims that low doses may, in fact, be beneficial through the stimulation of the body's response mechanisms and that to extrapolate the findings in Hiroshima and Nagasaki will overestimate the risk. Neither of these points of view is widely accepted nor is either rigorously defensible since, as already noted, the average survivor received a dose that was intermediate, that is, in the range 0.20 to 2 Gy, and many received doses no higher than those others might experience under more widely prevailing situations. But were the circumstances attending the exposure of the survivors so unique as to limit the lessons to be learned, as some have argued?

In August 1945, Hiroshima and Nagasaki were not typical of other Japanese cities either before or after the war in a number of respects. First, both communities, but particularly Hiroshima, were significantly involved in the nation's war effort. Though a less secretive area than its neighboring city, Kure to the southeast, Hiroshima was not and had not been an open area for a decade or more before the bombing. Second, as would be anticipated in a nation at war, there were fewer young men living in the cities than normally expected; those of military age were in Japan's armed forces. Third, the systematic bombing of the country that began after the fall of the Marianas had led to the evacuation of many preschool- and school-aged children and pregnant women to the countryside. Fourth, the progressively worsening economic conditions and the demands of the military reduced the food available to the general population, and compromised nutritionally related aspects of health to some degree. Finally, Japan's isolation from the remainder of the world for a half decade or more had limited access to such therapeutic agents as the sulfonamides and antibiotics, which if available could have diminished mortality.

Doubtless, there were other differences between these cities and peacetime ones, and these differences have been cited to suggest that the findings on the survivors have little relevance to the nuclear problems that the world now confronts.[7] But this is an extreme position not shared by most radiation biologists. Nonetheless, as has been repeatedly implied, there is a legitimate basis for concern about the importance of the experiences of the survivors of the bombing of Hiroshima and Nagasaki as a basis for extrapolating the risk likely to occur in other populations under different exposure circumstances. Clearly, such extrapolations cannot be made without error, but this does not mean that the experiences of the survivors are misleading. It only means that the conclusions to be drawn must be guarded scientifically, and stated in language that makes clear the strengths and the limitations of the information at our disposal. However, such extrapolations do assume that a causal relationship between exposure to ionizing radiation and subsequent ill-health extends throughout the range of possible exposures, or alternatively stated, that any dose of ionizing radiation, however small, is ultimately prejudicial to health. Is this assumption a legitimate one? Put somewhat differently, how does one establish in an epidemiological study that an association of one event, exposure to ionizing radiation, with another, the occurrence of cancer, is evidence of causality?

Some years ago, in an effort to make inferences regarding causality more objective, and less sensitive to individual interpretation, Sir A. Bradford Hill, an eminent English statistician and epidemiologist, proposed a series of criteria that should be met before causality is inferred. Specifically, he identified seven bases for judgment. These are the *strength of the association* (the size of the relative risk), *consistency* (the findings can be or have been replicated by other investigators), *dose response* (the relative effect increases with increasing exposure), *temporality* (the cause must precede the effect), *biological plausibility* (the findings should be consistent with experimental or other research), *coherence* (the observed association should be consonant with temporal patterns of disease), and finally, *specificity* (the association should be related to one disease or biologic process, if that process can culminate in more than one disease). We may ask, then, does the often-seen association of an increase in cancer following exposure to ionizing radiation meet these criteria of causality? At moderate to high doses of ionizing radiation these conditions are met, certainly with respect to the occurrence of a variety of cancers. Doses of 0.20 Gy or more demonstrably increase the risk of cancer, as has been previously seen. But at lower doses the evidence is much less clear-cut, and this makes the assumption of a causal relationship at these dose levels more tenuous. Yet, ultimately, the public's concern over the hazards of exposure to ionizing radiation involves those exposures where current evidence is least compelling. To obtain some insight into the possible magnitude of the risk and to provide guidance to regulatory agencies charged with minimizing that risk it is necessary to project the risk from the data where the evidence of causality is most persuasive.

It might seem logical and more desirable to try to measure directly the cancers that occur at low doses or low dose rates rather than to project the expected number from experiences at intermediate to high doses. This has been tried repeatedly. However, these investigations have many limitations—the sample sizes must be astronomically large, the follow-up period long, it is difficult to obtain person-specific estimates of dose that are trustworthy, and so forth. Moreover, as the size of the study sample increases, the likelihood that an association between radiation and its presumed effects is solely a function of dose decreases, since exposure to other potential toxicants increases. Furthermore, since the effects expected are small, they could easily be "lost" in the much higher naturally occurring rates of cancer or obscured by variation among individuals and studies in smoking habits, differences in nutrition, and other life-style factors associated with cancer. It is not surprising, therefore, that the results of these studies have been inconsistent and that most, but not all, have not disclosed an increase in cancer.

A recent study of employees of the Oak Ridge National Laboratory, for example, suggests an increase in cancer mortality, and that ascribed to the occurrence of leukemia in particular.[8] The latter increase is not dose-related, however, and is based on expected mortality using national statistics, which might inappropriately represent the expectations among a largely east Tennessee population. Two larger studies of nuclear workers, one embracing 35,933 Americans,[9] and the second 95,217 persons included in the United Kingdom's National Registry for Radiation Workers,[10] have reported findings consistent with the studies in Japan and lifetime risk estimates that accord reasonably well with those recommended by the International Commission on Radiological Protection. The British study, for example, finds external radiation and mortality from all cancers to be weakly but positively correlated; whereas the correlation was stronger for multiple myeloma and was statistically significant for leukemia. Most of the increase in the leukemias was accounted for by the chronic myelogenous types,[10] and this too is consistent with the Japanese experience, given the ages of the labor force involved. Possibly, future analyses of these two groups or some of the other presently contemplated industry-wide studies of nuclear power workers or defense reactor employees will prove more informative, since they will embrace hundreds of thousands of workers. But one should not be overly optimistic since, as previously stated, in such very large samples the likelihood that the comparison groups will differ in ways other than exposure increases dramatically. Moreover, exposure to other known carcinogenic agents in the workplace is difficult to evaluate, and so too are individual medical exposures to ionizing radiation that could equal or exceed the doses occupationally incurred by the same individuals. Nonetheless, irrespective of the outcome, these are important studies, since they provide a means to evaluate the reliability of current projections and to establish upper bounds of risk that can be contrasted with existing occupational and public safety standards to determine their adequacy.

Projections of Risk at Low Doses and Low Dose Rates

All epidemiological studies are unique in one way or another; they are generally based on situations that might or might not be representative over a longer time span or applicable to other circumstances. It is not known with certainty how reliable an individual study may be as a basis for predicting future events. This uncertainty can be diminished or resolved through the repetition of the study elsewhere to see if the same results are obtained. In the present instance—the experiences of the atomic bomb survivors—this cannot occur. Other studies of exposed individuals do not fully meet this need for repetition, since often the exposure happened because of an illness—whose effects could be interwoven with those of radiation—or the age and gender composition of the study group is not representative of the general population, or the period of surveillance has not been long enough to reveal all of the effects that might occur. This is not to imply that these studies are valueless, but merely that their findings have to be carefully interpreted with due cognizance of the nature of the populations involved.

The Projection of Risk of Cancer to a Lifetime

No one of the major groups, upon which our understanding of radiation-related health effects rests, has been followed for the lifetime of all of the members, and thus if lifetime risks of malignancy, say, are to be determined, they must be projected from the data that are available. These projections depend on a variety of assumptions, some of which are of uncertain validity. For example, it is known that over the period of the studies in Hiroshima and Nagasaki the risk of a malignancy is higher among individuals exposed in the first decade of life than at later ages, but it is not known whether this elevated risk will persist throughout their lives. There is some evidence that it might not. But to project risk to a population that includes young children, some assumption must be made about the future pattern of the risk at these younger ages, and this assumption might or might not be correct. For this and other reasons, risk projection is a tenuous enterprise, an uncertain art, and is likely to remain so until the biological processes that transform a normal cell into a malignant one, and its subsequent growth into a tumor, are better understood. However, to be realistic, this information might not be forthcoming shortly, and when it is available, it might not provide the insight needed to model cancer case occurrence more reliably.

As just stated, risk projection hinges on a variety of assumptions, not the least of which are the presumed dose–response relationship, the effects of a low dose rate or fractionated exposure, the choice of the projection model, the validity of using risks based upon one population (the Japanese atomic bomb survivors for instance) to describe those in another population or group, the changing nature of the exposure to other carcinogens in most populations, and the possible, indeed probable, existence of individual differences in response to ionizing radiation. A critical consideration of each of these areas of uncertainty lies beyond our intent; however, it is impossible to form a judgment about the acceptability of the risks that have been estimated without some discussion of how the estimation proceeds. The following will be limited to only three aspects of the projection process: (1) the choice of the projection model, (2) the use of radiation-related estimates of risk based on the Japanese to project the lifetime experience in other populations, and (3) the issue of low doses and low dose rates, since most exposed populations of regulatory concern will receive their doses in repeated small exposures rather than in one single event. These are not the only concerns associated with risk projection, but they are central to the process and can serve to illustrate the complexity of the issues.

Choice of the Projection Model

Conventionally, two models have been used to project lifetime risks, although other models may, in some instances, be preferable. The common ones are known as the additive (sometimes termed the absolute) and the multiplicative (or relative) models. One of the past justifications for their use has been the presumption that these two, when considered jointly, establish the upper and lower limits for the lifetime risk. Recent studies suggest that this presumption might not be correct, and that intermediate models, combining features of these two common ones, could be better.[11]

Conceptually, the two models are simple. The additive model says that at each age in life the number of cancer deaths (or cases) attributable to a specific exposure is the same, or put somewhat differently, that *the number of excess cases is independent of the naturally occurring rate of cancer deaths (or cases)*. By "naturally occurring" is meant the rate of cancer given no exposure to ionizing radiation above the "normal" levels that individuals receive through cosmic radiation, the occurrence of radioactive elements in the soil, and the like. If, for example, at age 5 among a specified number of individuals all exposed to the same dose, 10 additional cancer deaths occur because of exposure, then among the same number of individuals at age 70 and the same dose, this model assumes that 10 additional cancer deaths will also occur. Note that this implies that both the relative and the attributable risk—the contribution of radiation-related cancer deaths to all cancer deaths—will decline with time, since cancer deaths will be more common at age 70 than at age 5.

The multiplicative model assumes that the *excess relative risk*, the contribution of radiation-related cancer deaths to all cancer deaths in any population of individuals at the same dose, is *independent of the naturally occurring or baseline rate*. This implies that the number of radiation-related cancer deaths will increase with age, but the attributable risk will remain the same, since the proportion of deaths due to radiation exposure is a constant multiple of the background rate. The difference between these two models is shown graphically in Figure 35, where we assume, for simplicity, a population exposed shortly after birth and distribute the same number of ensuing cancer deaths by age at occurrence, assuming a latency of 10 years. Note particularly that the excess relative risk (the ratio of the additional cases of cancer to the baseline) declines with advancing age under the additive model but does not do so under the multiplicative one.

Since the two models make different projections with respect to time, as this figure illustrates, their applicability is open to scrutiny through the examination of the cancer experience revealed thus far in Hiroshima and Nagasaki. If among the survivors the relative risk and the excess deaths per 10^4 PYGy of cancers other than leukemia are computed as functions of age at exposure and age at time of death, the relative risk does not increase significantly for ages at time of bombing (ATB) of 20 or greater, and certainly 30 or more (see Table 7a), but the number of excess deaths continues to rise (see Table 7b). These observations suggest that for exposures occurring after the age of 20 a multiplicative risk projection is a better descriptor of future deaths from non-leukemic cancer than an additive one. This is not true for exposure in the

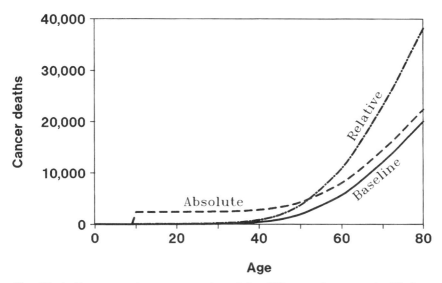

Fig. 35. A diagrammatic representation of the difference between the lifetime risk of death from cancer at various ages under the relative and absolute projection models.

Table 7a. Relative risk at 1 gray by age at time of bombing (ATB) and age at death for leukemia and all cancers except leukemia. The latent period for leukemia is assumed to be 2 years and that for all other cancers to be 10. (Adapted from Shimizu, Kato, and Schull, 1988, Table 6)

| Age ATB | Total | Age at time of death |||||||
		<20	20–29	30–39	40–49	50–59	60–69	70+
Leukemia									
<10	17.05	44.16	3.41	8.64	0.95			
10–19	4.76	54.74		2.45	1.02	0.82		
20–29	5.06		5.33	3.54	43.09	1.02	0.82	
30–39	3.99				24.05	10.58	1.47	3.89
40–49	2.55				0.83	3.82	0.82	3.10
50+	6.50					15.63	5.18	6.90
Total	4.92	46.47	9.81	4.75	5.68	3.98	1.70	4.40
All cancers except leukemia									
<10	2.32		5.89	1.96	1.86			
10–19	1.65			1.66	1.59	1.68		
20–29	1.65				2.09	1.74	1.37	
30–39	1.26					1.11	1.23	1.48
40–49	1.24						1.13	1.33
50+	1.11							1.15
Total	1.29		2.22	1.60	1.58	1.39	1.13	1.29

first two decades of life, where the relative risk has been falling, significantly among those survivors exposed before the age of 10, but the number of excess deaths continues to rise. These trends do not accord with either of the models. However, when one examines the data for leukemia, the excess mortality has declined with time, but remains significantly elevated in 1981–1985 in Hiroshima. This implies that for leukemia the constant additive risk projection model is a better descriptor of the data than the constant relative risk model. But as previously stated and as Tables 7a and 7b show, neither model describes the events satisfactorily among those survivors exposed before the age of 20. This is a matter of concern, since these age groups will usually make up 25–30% of the general population and the estimation of their risk has a profound effect upon lifetime projections for a general population.

More realistic projections might be based on models that allow the excess relative or absolute risk to vary with time rather than be constant. Indeed, this notion prompted the National Academy of Science's Committee on the Biological Effects of Ionizing Radiation (BEIR), in its latest

Table 7b. Excess deaths (per 10^4 PYGy) by age at time of bombing (ATB) and age at death for leukemia and all cancers except leukemia. The latent period for leukemia is assumed to be 2 years and that for all other cancers to be 10. (Adapted from Shimizu, Kato, and Schull, 1988, Table 7)

Age ATB	Age at time of death							
	Total	<20	20–29	30–39	40–49	50–59	60–69	70+
Leukemia								
<10	2.93	6.71	0.93	1.27	−0.01			
10–19	1.19	3.95		0.56	0.02	−0.06		
20–29	2.13		3.93	1.52	4.84	0.01	−0.28	
30–39	2.54				3.18	2.26	1.09	3.89
40–49	2.11				−0.35	3.07	−0.24	3.50
50+	4.56					4.31	3.84	5.12
Total	2.29	6.48	2.17	1.16	1.88	1.54	1.09	4.24
All cancers except leukemia								
<10	2.29		1.32	2.85	5.16			
10–19	4.66			2.00	5.84	13.91		
20–29	9.38				9.40	15.71	14.33	
30–39	9.31					3.16	11.00	41.01
40–49	14.52						7.31	37.30
50+							7.89	17.21
Total	7.41		0.54	1.98	5.35	9.62	6.85	30.53

effort to assess the lifetime risk of cancer following exposure, to advocate the use of a modified multiplicative model rather than either of the two simple models.[12] This modification incorporates time since exposure as a variable in the projections, and assumes that the expression of cancer may not only vary between different age groups at the time of exposure but within a group as time elapses following exposure. Although, in principle, this accords with the experience of the Japanese survivors and would seem the proper pathway in projection, it is not without its own difficulties in application. For example, if the number of years of life lived following the diagnosis of a cancer that will prove fatal to the individual concerned continues to increase, as it has, estimates of risk of death based on time since exposure and the limited observations presently available could suggest a diminished risk with time where a diminution does not exist, or a distribution of time to expression following exposure that is flawed. Few of the Japanese exposed at the age of 50 or more are still alive, and most of these individuals who died of cancer died when the means for medical intervention were limited.

Therefore, the distribution of their deaths as a function of time since exposure could be different from that of individuals exposed at age 20 because of changing medical practices rather than an inherent difference in their response to radiation. That is, individuals who were young at the time of their exposure, and who would normally have more years of life to expect, stand to profit more from recent or future advances in cancer therapy than those individuals who were older in the summer of 1945 and had fewer years of life remaining. Similarly, the modeling of time since exposure is sensitive to late reporting of death, and if the latter is age dependent, the distribution of time since exposure could be flawed more seriously for some age ATB groups than others. Some late reporting is inevitable due to the way that data on death accumulates in the Life Span Study sample, and this will undoubtedly continue to be so as long as the number of survivors remains large. These concerns also apply to the use of constant risk models, but the effects of some can be mitigated by a suitable change in the method of computing risks. For example, late reporting could be dealt with by incorporating the date of the last known vital status in the computation of the person-years at risk. It also warrants noting that risks estimated from registry-based cancer incidence data will be freer of this concern, and this fact argues for greater use of incidence data.

Some of these ambiguities could be erased through suitably designed experiments. Most, if not all mammalian species develop cancer, and the response to ionizing radiation of many of these species, in particular dogs (beagles specifically), mice, and rats, has been studied. Heretofore, these investigations have focused more on delineating the steps in the carcinogenic process than on the projection of risk. Indeed, the experimental strains studied were often selected because the occurrence of a particular cancer was naturally high, and such animals might not, therefore, represent the human species, where cancer is less common. But recently, John Storer, Toby Mitchell, and Michael Fry, in a review of the experimental data on radiation carcinogenesis, have argued that the constant relative risk model describes these data well for many, albeit not all, cancer sites.[13] More of this could be done if the basic similarities and dissimilarities among different species in the occurrence of cancer were more clearly established and understood. The National Radiobiology Archives may contribute to this end. A broad synthesis of the experimental data, if coupled with such understanding, might provide a better means to utilize the considerable body of experimental information that exists in addressing everyday problems arising in radiation protection. But, undoubtedly, some problems in the projection of risk would remain. For example, if the constant relative risk model is indeed correct, what risk would one choose for projection when the relative risk varies, as it appears to do?

The Use of Risk Estimates Derived from One Population to Project the Lifetime Experience of Another Population

Radiation-related risk coefficients are inevitably estimated in the context of a particular population, and the latter will have its own background mortality rates from all causes of death, and its own age- and gender-specific cancer rates. These can differ substantially. To illustrate these differences, consider Table 8, where age-adjusted cancer death rates in 1988 are given for eight of the most common malignancies in Japan, the United Kingdom, and the United States. It will be noted that cancers of the liver and stomach are more frequent in Japan, at all ages, than in the United Kingdom or the United States, and cancers of the breast, colon, and lung are more common in the latter two countries than in Japan. Similarly, some forms of leukemia, such as chronic lymphocytic leukemia, are relatively rare in Japan but are common in the United States and Western Europe. In the United States, for example, chronic lymphocytic leukemia annually accounts for more than one-third of all leukemias occurring among adults. Still other studies, in particular those of thyroid cancer among children exposed to radiation

Table 8. Age-adjusted death rates from cancer at eight specific sites per 100,000 population per year in Japan, the United Kingdom, and the United States in 1988. (Abstracted from the World Health Statistics Annual 1989 and 1990)

Site	Sex	Japan	United Kingdom	United States
Bladder	M	3.1	13.2	5.5
	F	1.4	5.8	2.6
Breast*	M	—	—	—
	F	9.0	52.3	33.5
Colon	M	11.3	21.1	19.6
	F	11.1	24.3	19.6
Leukemia	M	5.4	7.8	8.1
	F	3.8	6.5	6.2
Liver	M	20.0	1.9	2.0
	F	6.3	0.8	0.8
Lung	M	40.6	100.6	73.5
	F	14.6	41.9	35.9
Rectum	M	8.51	2.8	3.4
	F	5.8	10.0	2.9
Stomach	M	50.0	22.7	6.8
	F	29.0	14.7	4.4

*Females only.

in the treatment of ringworm of the scalp, suggest that blacks in the United States may be less sensitive to ionizing radiation than whites, and that among whites, ethnic groups may differ in sensitivity. Given these observations, what use can be made of risk coefficients obtained from one population, the Japanese, for predicting lifetime risks in any other population—Americans, for example? And, more specifically, which risk estimate, the relative or absolute one, should be used?

UNSCEAR, in its 1988 assessment of radiation-related malignancy, focused its projections on Japan, arguing that the best projection should be obtained by applying the risk estimates to the closest possible expected rates of death (from cancer and from all other causes), namely, those from the same country as the exposed.[14] Nevertheless, baseline risks in the Japanese exposed to the atomic bombs were not the same as those of current Japanese, nor even those of individuals living in Hiroshima and Nagasaki today. Baseline mortality rates have been varying with time in Japan—the frequency of stomach cancer has been declining and that of breast cancer rising, for instance, and there are regional differences in these rates, as in every other country.

It is also true that exposure to carcinogens other than ionizing radiation is changing not only in Japan but in most populations, and so it is difficult to know how accurate lifetime risk projections might be. One possible way to assess the size of this potential source of error is to compare the lifetime risk estimates based on the same risk coefficients and different baseline mortality patterns. This has been done by UNSCEAR, where the excess deaths and excess relative risk coefficients derived from the experiences of the atomic bomb survivors have been applied to (a) the Japanese population using the Japanese 1980 national mortality patterns, the nearest available representative rates for those coefficients; (b) the United Kingdom, a nation of longer term industrialization; and (c) Puerto Rico, a population with high infant and infectious disease mortality and low cancer rates, and in these respects similar to many preindustrialized or newly industrializing nations.

Across these three populations there is virtually no difference in risk projected by the additive model. Even for the multiplicative model, the maximum difference, using Japan as the basis of comparison, is only 22%.[15] This suggests that the lifetime risk projections are not particularly sensitive to differences in overall or cancer-specific mortality within the range of contemporary large national populations. Thus, risk projections would seem to have rather broad generality and applicability. Much greater differences may apply, though, to specific cancers with large variation in risk from country to country, such as cancer of the female breast, esophagus, large bowel, lung, and stomach. Figure 36 shows the baseline and radiation-induced lifetime mortality risks of cancer for various sites among 100,000 males and females theoretically exposed to

1 Sv at age 20. As one notices, the six-specific baseline lifetime risks are readily apparent. The differences between the relative and absolute excess risks are explained in part by the ratios of the U.S. age-adjusted rates to Japanese age-adjusted rates (Table 8). While the ratios of age-adjusted rates for the colon, lung, and bladder in the U.S. to those for Japan are greater than unity, the same ratio for stomach cancer is less than unity. Thus the absolute risk projection for the stomach site dominates the overall excess cancer deaths.

This conclusion about the broad applicability of the risks seen among the Japanese survivors applies to only one of the uncertainties in the extrapolation of risk projections to other populations, specifically the importance of differences in background rates. It is not known how much the risk coefficients themselves might vary between different ethnic groups or populations with differing exposures to other carcinogens. The latter exposures could act synergistically, that is, in a manner greater than the sum of their separate effects. The range of today's knowledge is limited essentially to data from Japan and from a variety of industrialized populations. Within this context, and given the statistical problems in estimation, the range of risk coefficients seen among the

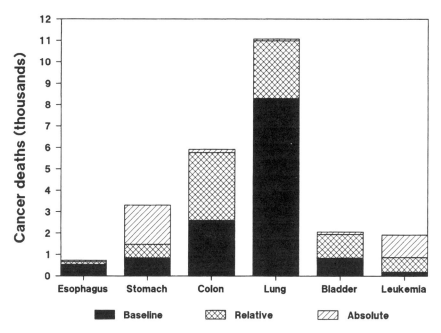

Fig. 36. The number of radiation-induced and spontaneously occurring (baseline) cancer deaths at 1 Sv among 100,000 deaths in males and females at six specific sites projected with the absolute and relative risk models.

Japanese survivors and among English patients with ankylosing spondylitis, for example, is small, well below a factor (multiple) of 10.

Low Doses and Dose Rates

As stated at the outset of this chapter, few individuals will ever be exposed to doses similar to those received by many of the survivors of the atomic bombings, but many may be exposed to doses above natural background. When this does occur, the dose will generally be small, under 0.10 Gy, and accumulated over some substantial time, possibly years. Since the exposures in Hiroshima and Nagasaki were instantaneous, how is allowance to be made in projections using the experiences of the survivors for differences not only in the total dose but in the rate at which the dose is accumulated? This issue is the most vexing of those in the projection of risk, and the one where the greatest uncertainties occur. Experimental studies, either with cells in culture or laboratory animals, indicate that a dose administered at a low dose rate is less injurious than the same dose at a high rate. It is presumed that this difference in outcome reflects the capacity of cells to repair some, if not all, radiation injury, provided the repair process is not overwhelmed by the amount of cellular damage. Whether these findings would obtain in the human case is uncertain, although it is known that human cells also have this repair capability. The epidemiological information that is available is limited, but it also suggests that at low dose rates the risk might be reduced.[16]

As pointed out before, the lowest specific absorbed dose at which unequivocal effects can be demonstrated among the atomic-bomb survivors is 0.20–0.49 Gy, a dose substantially above that of most regulatory interest. At lower doses, if risks are to be estimated, assumptions must of necessity be made about the shape of the dose–response curve and its attenuation or alteration at low dose rates. The latter, the degree of attenuation, is commonly termed the *dose-rate-reduction effectiveness factor* or the DRREF. Unfortunately, the proper factor is not known, but as a first approximation it is often taken to be the ratio of the estimate of the rate of increase in cancer with dose obtained by fitting a linear dose–response curve to the linear term when a linear–quadratic dose–response is used. When this is done with the cancer mortality experiences of the survivors of the bombings of Hiroshima and Nagasaki, the ratio that is obtained lies between 1 and 2, and for the solid tumor incidence data the ratio is very close to 1. This suggests that the lessening of cancer risk at low dose rates is not particularly great.

National and international committees, such as the National Academy of Science's Committee on the Biological Effects of Ionizing Radiation and UNSCEAR, charged with the prediction of risk, have merely noted the problem, provided a plausible range of dose-rate-reduction effectiveness factors, but have made no explicit numeric recommendations.[12,14] Although this is a defensible strategy, given the limited epidemiological and experimental evidence at hand, it is an unsatisfactory resolution of a difficult issue, and places the burden of judgment on the regulator or the exposed individual. Possibly with time, as information accumulates on those atomic bomb survivors exposed to doses of less than 0.20 Gy, of whom there is a very large number, a more satisfactory evaluation will be possible. At the moment, the prudent course, insofar as an exposed individual is concerned, is to assume there is no attenuation of risk at low dose rates. However, this assumption might create more apprehension than is warranted, and can also pose difficulties for regulatory agencies involved in the protection of the occupationally exposed as well as the general population.

Current Estimates of Lifetime Risk

To provide guidance to regulatory agencies, a number of national and international bodies, while cognizant of the uncertainties we have described, have attempted to project the lifetime risk of a malignancy to individuals exposed to 1 Gy. The data these agencies have used are essentially the same—the experiences of the atomic bomb survivors augmented to greater or lesser degree by other studies—but the projection methods have differed. Some have projected the risks for males and females separately, some have used the simple average of the sex-specific risks, and still others have used the geometric mean of these risks. Despite these methodological differences, the estimated lifetime risk for all fatal cancers is surprisingly similar, although the site-specific risks vary more (see Table 9). Depending upon how the estimates from the Japanese atomic-bomb survivors are projected to the end of life, and how they are modified to fit an American population with its different underlying cancer rates, the estimated lifetime risk, in percent, of cancer mortality associated with an acute radiation exposure to 1 Gy, averaged over a lifetable population of both sexes and all ages, varies between 7% and 10% (that is, between 712 and 1000 per 10,000 persons). This is in addition to the normal lifetime risk of death from cancer, which the National Cancer Institute estimates to be 22% among Americans. Assuming a dose and dose rate reduction factor of 2, the lifetime risk from 1 Gy delivered continuously or in very small increments over a long period of time is estimated to be half that for acute exposure, that is, between 3.5% and 5%, and that from a relatively low dose of 10 mGy is estimated to be between 0.035% and 0.05%, whether the exposure was

Table 9. A variety of estimates of the lifetime risk of a fatal radiogenic cancer (10^{-4} person-Gy). To avoid the impression of a greater precision than actually exists, all estimates have been rounded to the nearest integer.

Site	Acute exposure[1]			Chronic exposure[2]		
	UNSCEAR[3]	NCRP	U.S. EPA	UNSCEAR	NCRP[4]	U.S. EPA[5]
Bladder	39	62	50	20	30	25
Bone	—	—	2	—	5	1
Breast	60	55	46	30	20	46
Colon	79	254	196	40	85	98
Esophagus	34	20	18	17	3	09
Kidney	—	—	11	—	—	6
Leukemia	97	94	99	49	50	50
Liver	—	—	30	—	15	15
Lung	151	163	143	76	85	72
Ovary	31	27	33	16	10	17
Skin	—	—	2	—	2	1
Stomach	126	175	89	63	110	44
Thyroid	—	—	6	—	8	3
Remainder[6]	118	150	246	68	50	123
Total[7]	712	1000	972	356	500	509

[1]Note that the NCRP report does not provide explicit estimates for acute exposure but does provide relative probabilities of fatal cancers at the sites indicated and an estimate of the total probability of a fatal cancer based on the averaging of the risk of males and females and the values from two transfer models (Table 13.6, Report 115); the values given here are derived by multiplying these relative probabilites and the total probability.

[2]Note that UNSCEAR did not make separate estimates for chronic exposure; to provide a basis for comparison the values given here assume a DDREF of two as has been assumed by NCRP and the U.S. EPA.

[3]The UNSCEAR values are taken from Table 69, UNSCEAR 1988. These estimates are based on the 1982 Japanese population.

[4]The NCRP values are taken from Table 13.8, Report 115, NCRP 1993. These estimates are identical to those given in ICRP, 1991, and are based on the current United States population.

[5]The U.S. EPA values are taken from Tables 4 (acute exposure) and 6 (chronic exposure), U.S. EPA, 1994. These estimates are based on the assumption of a stationary population with the vital statistics of the United States in 1980.

[6]The remainder represents the risk of all cancers not specifically estimated.

[7]The total is not always the simple sum of the site-specific risks; in some instances, such as the UNSCEAR value, it was separately estimated.

acute or chronic. For a population of working age, the corresponding estimates would be somewhat lower because sensitivity to radiation carcinogenesis is thought to be greatest for exposure during childhood.

Since the uncertainties are many, it might seem that risk estimation and risk projection are futile exercises. This would be an unwarranted and unduly pessimistic position, since substantial progress has been made, and will undoubtedly continue to be made. The essential first steps have been taken. The hazard has been identified, the process of reconciling the uncertainties in our present knowledge has begun, and regulatory actions, albeit possibly imperfect ones, have been initiated. More data will accumulate and advances will occur in our understanding of the process of carcinogenesis, and with these will surely come the means to mitigate the effects that have been seen. Even now, it is clear that the risk of significant biological damage accruing from exposure to ionizing radiation of the magnitude of greatest regulatory concern is small, and from a public health perspective substantially less than many of the risks accepted daily. This is not to imply that one should not remain vigilant, nor that every practical effort should not be made to reduce exposures still further. Indeed, every effort should be made to keep exposures as low as reasonably achievable. The risk must be seen in a larger context, however, one that is reconcilable with the many useful applications of ionizing radiation in medicine and in the production of energy (see Table 10). There has never been a risk-free world, nor is one ever likely to exist, but it is possible to live in one in which the acceptability of each risk is judged after carefully weighing the benefits that accompany the risk. If these benefits are too small to justify the risk, prompt remedial action is the only sensible course.

Table 10. Exposures to man-made sources expressed as equivalent periods of exposure to natural sources. (Adapted from UNSCEAR, 1993, Table 5)

Source	Duration of exposure	Equivalent exposure to natural sources
Medical exposures	One year at the current rate	90 days
Nuclear weapons test	Completed practice	2.3 years
Nuclear power	Total practice to date	10 days
	One year at the current rate	1 day
Severe accidents	Events to date	20 days
Occupational exposures	One year at the current rate	8 hours

Perceptions of Risk

Until the accident at Three Mile Island in 1980, and the subsequent one at Chernobyl in 1986, most individuals either ignored the risk associated with nuclear energy or presumed it was small. These two events dramatically changed the public's perception of the safety of the nuclear power industry and the preparedness of governments, local and national, to cope with accidents when they occur.[17] What was accepted, almost blithely by most, is now seen as a hazard beyond entertaining or even discussing. This makes reasoned debate about the role of nuclear power in global energy production virtually impossible, particularly in the United States, where so much polarization of thought has occurred. Nevertheless, as concern mounts about global warming and the use of fossil fuels, the role, if any, nuclear power can play—reconciling national needs with the inherent risks—must be examined. But how is the necessary state of enlightenment to be achieved when the public perception of the risks of nuclear power is so ambivalent? Presumably some of this ambivalence stems from the notion that a nuclear reactor could explode like a nuclear bomb. It has been estimated that 40% of individuals believe this to be a possibility. But this cannot happen; the physics of the two situations are different. A nuclear bomb makes use of "fast" neutrons, since the explosive force depends upon achieving a series of fissions (each of which releases energy) so closely spaced together timewise that most, if not all, occur before the explosive material itself can blow apart. Reactors, however, depend upon controlled fissions, and utilize "slow," readily manageable neutrons. This control requires a moderator, a substance that will slow down or absorb the "fast," potentially uncontrollable neutrons. Uranium-238, the principal naturally occurring isotope of uranium, absorbs neutrons and is a moderator of the fast neutron flux if the fissionable material, uranium-235, is not too highly enriched. As a consequence, the uranium is rarely enriched to more than 3% (only about 0.7% of naturally occurring uranium is uranium-235).

Reactor accidents can and obviously have occurred. However, the consequence of these is very different from the detonation of a nuclear bomb. Heat is generated when atoms are split, and the danger in a power reactor arises primarily from the overheating of the nuclear fuel. To forestall this, a coolant, usually water, is continuously circulated through the reactor, and to guard against failures in the cooling system, layer after layer of containment walls and other devices are designed into the reactor. If all of these should fail, the most serious result would be a "meltdown." The overheated core could burn its way through the containment walls of the reactor and into the earth. If this occurred, some release of radioactive materials into the atmosphere is possible. This did happen in the Chernobyl accident. Safer reactors than most of those presently in use can be designed, largely through the minimization of the need for human intervention to prevent an accident. A variety of such reactors, widely described as "passively safe," have been proposed,[18] but none have actually been tested and it is moot whether they would be above any chance of a major accident.

But concerns of the public appear deeper than merely a confusion of a reactor with a nuclear bomb. As an illustration, some years ago members of the Oregon League of Women Voters were asked to rank their perceptions of the risk associated with 30 activities and technologies.[19] Nuclear power was perceived to carry a greater risk than that from the use of motor vehicles, yet more than 45,000 Americans die each year in traffic accidents whereas fatalities attributable to nuclear power are few.[20] Ironically, electric power was rated the most beneficial of these 30 activities and technologies, substantially higher than the availability of motor vehicles.

Similar surveys conducted in other countries have yielded similar results. For example, in France, respondents to a national survey in 1988 thought the danger associated with a nuclear power plant was greater than that in uranium mining, but mining, whether for uranium or other ores, causes more deaths each year than occur in all nuclear power facilities without regard to country.[21] Moreover, the risk associated with nuclear power plants is invariably seen as greater than that stemming from exposure to diagnostic or therapeutic X-rays, but again, in fact, the risk of a second malignancy following X-ray therapy for a primary cancer is substantially higher than the risk of radiation-related cancer among workers in nuclear facilities.

The elements that define one's perception of risk associated with a particular activity are still poorly known. But they must begin early in life. In the case of ionizing radiation, for instance, comic books routinely present radiation as the source of malevolent, often grossly disfigured individuals who seek world control for their own evil purposes. Titillating as these comics may be to the young, they present ionizing radiation in a way that seriously distorts the truth in minds unable to distinguish fact from fiction. Whether these initial impressions are lasting may be moot, but it is difficult to believe they are not without some effect. Nonetheless, individual notions do hinge on a variety of factors, including whether the risk is seen as voluntary or involuntary, chronic or catastrophic, controllable or not controllable, old or new, immediate or delayed, fatal or not fatal. It is clear too that this perception is colored by the depth of one's knowledge of the actual hazard. For example, surveys have shown that nuclear engineers—who surely must have a well-informed knowledge of the likelihood of an accident—are far less reluctant to live near a nuclear power facility than are environmentalists. Obviously in some manner the gap between the perception of risk and its actuality must be closed.

The Challenge

As energy consumption grows globally, and fears mount about the effects of continued use of fossil fuels on earth-warming and the occurrence of acid rain it is imprudent to temporize. A consensus must be reached on the role, if any, nuclear energy is to play in the future. To do so will require everyone to be more critical of what is read and heard, to examine the facts rather than to be passive vessels into which the thoughts and prejudices of others are poured. Neither the proclamations of nuclear power advocates, on the one hand, nor those of the environmentalists, on the other, should be seen as unprejudiced. These groups have their own agendas and can advocate these with equal intensity, trading on the concerns of an uninformed or poorly informed public. Constructive advocacy can be important, however, since it often draws the lines of debate more sharply and clearly. But ultimately, everyone must actively seek the information on which an informed judgment can be made. This is not an impossible task. Organizations such as the League of Women Voters have published highly readable accounts of how nuclear energy is produced, the functions of the various agencies that are involved in regulatory activities, the kind and nature of the accidents that can occur, and the like.[22] Similar primers exist to enlighten the public on the issues that attend the disposal of nuclear waste, another matter of common concern.[23] Citizen groups, such as the Hanford Health Information Network, can also be useful sources of information. An informed public could insist upon better standards of training of reactor operators, and of greater managerial and corporate responsibility. Moreover, since secrecy and continued governmental dissembling have contributed so much to the estrangement of the public, we must demand a greater openness in matters that concern the public weal, including the preparedness for accidents. But are we, as members of the public, prepared to invest the time and energy required to inform our discretion?

Other nations—Belgium, Britain, France, Germany, South Korea, and Taiwan among them—have reaffirmed their commitment to the development of nuclear energy. In Japan, where one might imagine the opposition to nuclear power would be greatest, the government has set for itself the goal of doubling its number of nuclear reactors from 40 to

80 to generate more than 43% of its energy needs through nuclear power by the year 2010.[24] Initial plans called for these reactors to be fast breeders, using recyclable plutonium, which is highly toxic and where a meltdown as a result of a reactor accident could conceivably occur. Delays in the development of fast breeder reactors have, however, thwarted these initial projections, and the plutonium is now to be employed to enrich conventional uranium fuel in a new advanced generation of boiling water reactors. The latter, ostensibly safer than their predecessors, are an outgrowth of collaborative research involving General Electric, Hitachi, and the Tokyo Electric Power Company (Toshiba). To make these developments palatable to a wary public, Japan's Science and Technology Agency proposes to launch a broad educational program and to support this with generous subsidies to those communities prepared to accept a reactor in their vicinity. Whether these steps will overcome public concern is arguable. However, siting a nuclear power reactor in Japan is a less democratic process than in the United States, and the opportunities for the public to protest bureaucratic decisions perceived as unwise are more limited. But leaving aside the wisdom of a specific decision, the nation does have a plan to diminish its dependence upon fossil fuels, whereas in the United States we have not begun a serious debate on the issue nor does one seem in the offing.

If our influence and prominence in world affairs is to continue, plausible alternatives to our energy needs must be developed or other nation's policies adopted. Conservation through the wider use of high-efficiency light bulbs, electronically controlled industrial motors, more heavily insulated refrigerators, and dispensing with needless gadgetry as well as the use of renewable sources of energy such as solar-powered air conditioners can obviously be helpful, but there is a limit to what these alone can achieve. Regardless of the choice or choices, years of commitment will be needed. It takes 6 to 12 years to build a nuclear power reactor and involves not only the labor of thousands of employees but several billion dollars as well. The research needed to make other alternatives economic could take even longer.

As a nation, do we have the capacity to plan carefully and to adhere single-mindedly to a course of action? Recent history suggests this is problematic. State and national legislative bodies have become cabals of self-interest and legislators casualties of their own hubris. The only avenue to a sound and widely acceptable nuclear policy is, therefore, through an informed and committed public and their representatives in government, willing to exercise its collective rights and responsibilities through the electoral process if need be. Thomas Jefferson said it best when almost two hundred years ago he wrote:

I know of no safe depository of the ultimate powers of society but the people themselves; and if we think them not enlightened enough to exercise their control with a wholesome discretion, the remedy is not to take it from them, but to inform their discretion.

A Chronology of Major Events*

16 July 1945   Trinity, the first detonation of an atomic weapon at a test site some 50 miles northwest of Alamogordo, New Mexico
6 August 1945   The atomic bombing of Hiroshima
9 August 1945   The atomic bombing of Nagasaki
14 August 1945   Japan announces its willingness to accept an unconditional surrender
15 August 1945   Cessation of hostilities
2 September 1945   The signing of the instruments of surrender aboard the U.S.S. Missouri
12 October 1945   The creation of the Joint Committee for the Investigation of the Effects of the Atomic Bomb in Japan under the direction of Colonel Ashley Oughterson. The recommendation for a long-term study of this Committee was transmitted to James Forrestal, Secretary of the Navy, in 1946,and subsequently to President Truman
26 November 1946   President Truman directs the National Academy of Sciences to initiate the studies of the long-term effects on the survivors of the atomic bombings
November 1946   Visit of the Academy's study feasibility committee comprised of Drs. Austin Brues, Paul Henshaw, James V. Neel, Melvin Bloch, and Frederick Ullrich
Early 1947   Creation of the Atomic Bomb Casualty Commission (ABCC); Dr. James V. Neel, a medical geneticist, appointed acting director
February 1948   **Initiation of the Genetics Studies in Hiroshima; to be followed in May 1948 with the beginning of the studies in Nagasaki**
March 1948   Dr. Carl Tessmer, a pathologist, succeeds Neel as the

*To distinguish administrative developments from scientific ones, the latter are in boldface.

director of the Commission, and Dr. Hiroshi Maki is appointed director of the Hiroshima Branch Laboratory of the Japanese National Institute of Health attached to ABCC

Summer 1949 **The first cases of lenticular opacities are seen by three Japanese ophthalmologists, Drs. Kinnosuke Hirose, Tadashi Fujino, and Hiroshi Ikui**

Fall 1949 **The beginning of the first ophthalmological survey that reveals the frequency of radiation-related lenticular opacities in Hiroshima and Nagasaki**

Fall 1949 **Dr. Takuso Yamawaki surmises that leukemia may be increased among the survivors**

October 1950   The occurrence of the national census of Japan on which most of the major samples used in the studies of ABCC and RERF are based

November 1950   Dr. H. Grant Taylor, a pediatrician, succeeds Tessmer as the director of the Commission

January 1951   The new clinical facilities in Hiroshima are occupied

Spring 1951 **The first survey of women who were pregnant at the time of their exposure to the atomic bombing of Nagasaki; this survey revealed not only an increase in the frequency of pregnancy loss but suggested mental retardation was common among those infants who were born alive**

October 1952   The exchange of Notes Verbale between the U.S. Embassy and the Ministry of Foreign Affairs establishes the Commission as an agency attached to the Embassy

December 1953   Dr. John Morton, a surgeon, succeeds Taylor as the director of the Commission

February 1954   The major clinically oriented Genetics Program is phased out

July 1954   Dr. Robert H. Holmes, a pathologist, appointed to succeed Morton as director

October 1955   The Francis Committee arrives in Japan to review the Commission's activities and to recommend the direction of future studies

March 1956   The Committee on Atomic Casualties of the National Academy of Sciences endorses the recommendations of the Francis Committee; the implementation of the Unified Study Program begins

April 1956   The beginning of the program, known as *Ichiban,* at the Oak Ridge National Laboratory that would culminate in the T65 dosimetry

Spring 1956 **Publication of the results of the clinical study of pregnancy terminations occurring in Hiroshima and Nagasaki**

Summer 1956   Dr. Masanori Nakaidzumi, formerly professor of radiology and dean of the Faculty of Medicine at Tokyo University, is appointed an Associate Director to interact with the Ministry of Health and Welfare

January 1957   Dr. R. Keith Cannan, head of the Academy's Division of Medical Sciences, assumes the role of interim director, pending the appointment of a permanent director

April 1957   The Japanese Diet passes the Atomic Bomb Survivors Medical Treatment Law; Dr. Isamu Nagai appointed Director of the Nagasaki Branch Laboratory of the Japanese National Institute of Health attached to ABCC

May 1957   Establishment of a Tumor Registry in Hiroshima within the City Medical Association, but with the technical support of the Commission

6 June 1957   Dr. George B. Darling, a human ecologist, appointed director, a position he held for fifteen years. Introduction of the T57 dosimetry, an interim system that was little used

May 1958   Establishment of the Nagasaki Tumor Registry

July 1958   The beginning of the implementation of the Francis Committee recommendations with the initiation of the first cycle of the Adult Health Study

August 1958   Initiation of the Life Span Study, and the beginning of the Child Health Study

1959   Tomin Harada and Morihiro Ishida publish the first data suggesting an increased frequency of nonleukemic malignant tumors, particularly among survivors exposed within 1 km of the hypocenter

January 1960   Initiation of the F_1 mortality surveillance, based on approximately 54,000 children born between 1 May 1946 and 31 December 1958

March 1961   Completion of the first Atomic-Bomb Hospital in Nagasaki with a grant-in-aid from the U.S. government; a similar facility was completed in Hiroshima in September 1961

February 1962   Installation of the first electronic computing system, an IBM 1401; this was replaced in June 1965 with an IBM 1440

1962   Dorothy Hollingsworth and her colleagues publish the first evidence that thyroid disease, and in particular thyroid carcinoma, was probably elevated among the more heavily exposed members of the Adult Health Study

1 January 1965   The Commission enters into a collaborative study of cardiovascular disease, known as the Ni-Hon-San Study, sponsored, in part, by the U.S. National Heart Institute; approval had occurred on 19 August 1964

1965   The introduction of the T65 dosimetry system; computation of the doses was not completed until the summer of 1967

1965 New cytogenetic techniques are introduced; these have been applied as a means to estimate the dose received by survivors as well as to estimate mutational damage

1967 Clifford Wanebo and his colleagues show that the incidence of breast cancer is increased among the more heavily exposed women in the Adult Health Study sample

20 May 1968   Passage of the Atom Bomb Survivor's Special Measures Law, which extended the benefits under the original Atomic Bomb Sufferer's Medical Treatment Law and broadened the definition of the qualifiers

January 1969   Initiation of the study of the body burdens of residents in the Nishiyama Reservoir area in Nagasaki, where the fallout had occurred

1972   The beginning of the pilot study on the feasibility of identifying electrophoretically recognizable mutations among the children of the survivors; a full-scale study was begun in 1975

1 January 1973   Dr. LeRoy R. Allen succeeds Darling as the Commission's director

1 April 1973   Establishment of the Hiroshima Tissue Registry with the assistance of funds from the U.S. National Cancer Institute; a similar registry was begun in Nagasaki in the following year

April 1973   Exchange of a second set of Notes Verbale between the U.S. Embassy and the Ministry of Foreign Affairs terminating the status of ABCC as an agency attached to the Embassy but recognizing the Commission as a U.S. governmental research facility in Japan. This was a necessary preliminary to the reorganization then under discussion

26 March 1975   A second major scientific review of the activities of the Commission under the chairmanship of James F. Crow

1 April 1975   Creation of the Radiation Effects Research Foundation and the dissolution of the Atomic Bomb Casualty Commission; the first Japanese chairman, Dr. Hisao Yamashita, a radiologist, is elected

1 April 1978   Dr. Masao Tamaki, also a radiologist, replaces Yamashita as the Foundation's chairman

1 April 1981   Dr. Itsuzo Shigematsu, an epidemiologist, succeeds Tamaki as the chairman

Early 1982   Creation of the joint U.S.–Japan research program for reassessment of the atomic bomb radiation dosimetry

September 1982   RERF occupies new facilities in Nagasaki

April 1983   First publication on solid tumor incidence as revealed by the Nagasaki tumor registry

March 1986   Approval of the use of the DS86 dosimetry

September 1987   Publication of the first of the periodic Life Span Study reports to use the DS86 dosimetry

March 1988   The first recognition that adult-type tumors are increased in frequency among the survivors prenatally exposed

June 1990   Publication of the reassessment of the genetics studies and the evidence supporting the notion of a lower genetic sensitivity than has been previously supposed

July 1990   Saeko Fujiwara and her colleagues establish that among the members of the Adult Health Study sample the frequency of hyperparathyroidism rises linearly with dose

1991   First comprehensive report on solid tumor incidence based upon the tumor and tissue registries in both Hiroshima and Nagasaki

Notes

Preface

1. See, e.g., Ibuse, 1953; Kumami, 1993; Lifton, 1967; Maruki and Maruki, 1985; Nagai, 1949; Omuta, 1966; Shirabe, 1982; Shohno, 1986; Soka Gakkai, 1978; and Hibakusha: Survivors of Hiroshima and Nagasaki.
2. Crease and Samios, 1991.
3. Although in the Japanese convention the family surname precedes the given name, to avoid ambiguity throughout the text both Japanese and non-Japanese names are indicated with the family name following the given one.

Chapter 1

1. The Joint Commission estimated 64,000 civilians in Hiroshima died within the first four months following the bombing and that another 72,000 were injured. Their figures for Nagasaki were 39,000 and 25,000, respectively (Oughterson and Warren, 1956, Table 4.2). The reliability of these numbers is uncertain. They do not include the dead among the military, conscripted Korean laborers, nor the transients found in every city. Much more recently, Japanese scientists attempting to take these other groups into account placed the number of dead in Hiroshima at 140,000 and in Nagasaki at 70,000 (Committee for the Compilation of Materials on Damage Caused by the Atomic Bombs in Hiroshima and Nagasaki, 1981, p. 113). These numbers too are subject to error and can be construed as a political statement, much as the numbers advanced by the Joint Commission can be. In the light of governmentally published—ostensibly national—figures appearing shortly after the cessation of hostilities, the numbers reported by the Committee for the Compilation of Materials on Damage Caused by the Atomic Bombs in Hiroshima and Nagasaki are much too high. The circumstances obtaining after the bombing precluded accurate counts, but the death toll was obviously awesome.

2. Trumbull, 1957, page 79; see also Shirabe, 1982. I first met Dr. Shirabe in 1950, when, at the urging of James Yamazaki, who was then the pediatrician-in-charge of the studies that were underway in Nagasaki, Dr. Shirabe arranged an informal meeting to discuss the personal experiences of those survivors of the bombing who were still members of the Medical School. Sixteen individuals were present. They had been faculty members, students, or nurses at the time of the bombing. With the aid of floor plans of the various university medical buildings and a drawing of the campus, each described what had occurred in the hours and days immediately following the disaster. Much of what is written here draws upon Dr. Shirabe's remarks at this time.
3. Arakatsu, 1953; Matsumae, 1982; Shimizu, 1982.
4. Penney, Samuels, and Scorgie, 1970. Later in 1945 a team of British observers from the Ministry of Home Security were sent to Hiroshima and Nagasaki to make an independent assessment of the blast damage. Their evaluation is commonly described as the "Report of the British Mission."

In the summer of 1985, to provide some evidence on the time variation in dynamic pressure, the Defense Nuclear Agency conducted a series of multiton explosions at the White Sands Missile Range in New Mexico. One of these, known as the Minor Scale event, involved a charge of 4,744 short tons of an ammonium nitrate–fuel oil mixture designed to produce a blast wave equivalent to that from an 8 kT nuclear explosion. Ten replica I-section poles were exposed at 250 ft intervals between 1250 ft and 3500 ft from ground zero. Using these new data led to the slightly higher yield estimate based upon blast of 13 ± 2 kT. As was noted at a meeting convened by the Defense Nuclear Agency in October 1986, this did not resolve the problem, since there remained a divergence in yield estimates—13 kT for blast, 15 kT for thermal neutrons, and 17 kT for other radiation effects. Accordingly, working figures of 15 ± 2 kT for Hiroshima and 21 kT for Nagasaki were recommended by the meeting participants. However, they also urged the completion of assembly/disassembly calculations at Los Alamos National Laboratory to give more evidence of the effect of the steel gun barrel on energy partition and neutron spectrum. This was done.
5. Oughterson was aided in his responsibilities by Drs. Stafford L. Warren, Shields Warren and Masao Tsuzuki representing the Manhattan District, the Naval Technical Mission to Japan, and the Japanese Imperial Government, respectively.
6. Oughterson et al., 1951; Oughterson and Warren, 1956; see also the report of the Committee for the Compilation of Materials on Damage Caused by the Atomic Bombs in Hiroshima and Nagasaki, 1981.
7. Langham, 1967.
8. It should be noted, however, that some evidence exists that for individuals receiving the same dose those who experienced epilation are apparently more prone to develop cancer later in life than those who did not experience this symptom of acute radiation sickness and they are also more prone to exhibit chromosomal abnormalities. These observations have been interpreted to mean that some individuals are more radiosensitive than others but the evidence is not compelling.
9. It is clear from the linkages that can be made that age was described in a

variety of ways, but often in the old Japanese fashion (*kazoedoshi*) wherein a person was one year old on the day born, and attained an additional year of age on the first of January of each succeeding year. Thus an individual born on December 31st would be two years old on the next day.

10. Numerous attempts have been made to estimate the distance or dose at which 50% of the exposed individuals died (see, e.g., Shirabe et al., 1953; Shirabe, 1982; Ishida and Matsubayashi, 1961; Masuyama, 1952; Hayakawa et al., 1986; Rotblat, 1986; Fujita, Kato and Schull, 1987). Each of these studies has its strengths and weaknesses; none is above fault finding. Often the data were collected under circumstances that precluded definition of a study group representative of the population immediately prior to the bombing, and, with rare exceptions, identification of individuals at risk of death had been obtained after the fact, through surviving members of a family or through friends. Furthermore, for those studies removed in time from the actual bombings, migration from these cities, if related to exposure, could obscure the true distribution of survivors, and thus the estimate of the $LD_{50/60}$.

Intuitively, this quantity would seem to be a simple one to estimate, but it is not. Formidable difficulties arise. Estimates of the $LD_{50/60}$ based on retrospective reconstructions of the populations of these cities at the time of the bombing are plagued by three virtually insurmountable problems. First, there are no wholly reliable estimates of the number of individuals (nor their age or gender) who were exposed at a given distance or dose and hence of the number at risk of death, save in selected circumstances. Even in those rare instances in which these parameters are known, the population at risk may not be representative of the general population in age, gender, or health status. Second, it is exceptionally difficult, if not impossible, to separate deaths due to exposure to ionizing radiation from those ascribable to other bomb-related causes, such as fire or the blast itself. Finally, other factors, for example, the devastating typhoon on September 17, 1945, and the lack of adequate medical care, undoubtedly contributed to the 60-day mortality rates. The contributions these made to mortality, and hence to the obfuscation of the estimation of the $LD_{50/60}$, can only be conjectured.

Although no consensus exists as yet, these studies suggest that this critical dose, when expressed in terms of the radiation absorbed by the bone marrow is around 2.5 to 3 Gy, and that a dose of about 5 Gy would have killed 95% of those exposed. It warrants noting that this estimate is contingent upon the circumstances obtaining in these cities at these specific times, and may poorly describe the prospects of survival in a nuclear accident such as Chernobyl, where heroic measures made possible by the relatively small number of individuals exposed to life-threatening doses undoubtedly kept alive individuals who would have died in Hiroshima or Nagasaki. This has been true of other nuclear accidents too.

Chapter 2

1. The National Academy of Sciences was established in 1863 during the presidency of Abraham Lincoln. It began as essentially an honorary society

of American scientists, and continues in this fashion, in part. The National Research Council was founded in 1916 as a result of an executive order by President Woodrow Wilson. It was organized under the aegis of the Academy. Its stated aim is to mobilize the nation's science and engineering talent to support pressing national needs, which it does through providing advice on matters involving governmental policy. It is, in effect, the working arm of the Academy.

2. Letter from James Forrestal, Secretary of the Navy, to President Harry S. Truman dated 18 November 1946.
3. Beebe, 1979. This is an especially worthwhile article for readers interested in the history of the studies in Japan, and no less importantly those interested in the lessons to be learned from these studies that are applicable to all large-scale epidemiological investigations. See also Putnam, 1994.
4. This imposing name, which might appear callous, was coined by Paul Henshaw allegedly on the spur of the moment to meet SCAP's administrative needs when the interim fact-finding committee arrived in Japan. For an entertaining account of the committee's activities and some of the attending events see Brues, 1968. With the creation of a permanent organization in 1947, this name was continued. It has been said that Henshaw chose this title because the acronym ABCC would be memorable, as indeed it has proven to be. Its Japanese name, however, was Genbaku Shôgai Chôsa Iinkai, which was less catchy. Administratively, ABCC was an agency of the NAS–NRC although technically a dependency of the Supreme Commander of the Allied Powers prior to the formal cessation of hostilities in the autumn of 1951. Its first director was James V. Neel who, as one of the few officers in the Army Medical Corps with training in genetics, had accompanied the civilian consultants, Austin Brues and Paul Henshaw, on their fact-finding mission.

Originally, responsibility for the U.S. involvement in the studies in Hiroshima and Nagasaki rested in the Division of Medical Sciences of the National Research Council. This Division came into existence in the earliest days of the NRC. However, in 1973, a general reorganization of the Council occurred and the Division of Medical Sciences along with the biology components in several other units within the Council were incorporated into a new Assembly of Life Sciences. In 1982, a further reorganization took place and the Assembly of Life Sciences, somewhat altered in its composition, was renamed the Commission on Life Sciences. Responsibility for the U.S. contribution to the Radiation Effects Research Foundation now rests in the National Research Council's Board on Radiation Effects Research (BRER). The latter was established in 1981 to coordinate the Council's interests in the biological effects of all forms of radiation. Initially, it was a part of the Assembly of Life Sciences, but with the reorganization of the Council in 1982 it became an administrative part of the Commission on Life Sciences.
5. For one person's reactions to these cultural disparities see Schull, 1990.
6. Once the Occupation of Japan ended, the exclusionary policies of the Australians no longer prevailed, but the housing situation remained labile if not uncertain. The Notes Verbale that established ABCC as an agency of the

U.S. Embassy ensured the Commission's personnel access to military housing in the area of Hiroshima. However, with the cessation of the Korean War and the reduction in U.S. military personnel in Japan, the need for substantial military housing waned and the military authorities began to release surplus housing to the Japanese. This meant that the accommodations at Nijimura, an Occupation-constructed housing project some 25 miles from Hiroshima, where the bulk of the Commission's foreign staff resided, would in time be released. This forced the Commission in 1958 and 1959 to seek housing in Hiroshima itself. This had several merits. First, it brought the staff into closer contact with the citizens of the city, and second, it meant an end to the long commute that had been necessary. However, suitable housing was still scarce, and this put the Commission's needs into competition with local ones. Nonetheless a series of houses were leased at terms that were manageable but not especially attractive. A more economically favorable solution would have been to purchase houses, which could have been done at fairly reasonable prices. However, under the Eisenhower administration a limit had been placed upon federal purchases of property outside of the United States. While leasing solved the immediate problem, in time, as land and house values rose, the owners saw the arrangements less equitable to them and sought either to not renew the leases or to substantially increase the rentals. Fortunately, when this occurred, more houses and apartments were available in the city, and the Foundation continues to identify suitable accommodations for its foreign employees.

7. See Eisenbud, 1990, page 112.
8. See Willard Machle's Report on Field Operations of the Atomic Bomb Casualty Commission addressed to Detlev Bronk, and dated 3 January 1951. Pamphlet Box M, Archives of the RERF Library. The tenor of this report is surprisingly pessimistic, even for the time. Machle introduced his suggestion that consideration be given to the discontinuance of the studies in Japan with the statement "With the likelihood that atomic bombing of the United States may occur in the near future, thought may be given to the abandonment of the present project and the preparation of plans for study of populations elsewhere." He seems to have been persuaded that the conflict in Korea was merely a prelude to World War III. Indeed, he began his remarks on future policy with the statement "Newer elements such as the national emergency and the beginning of world war III require new definition of affirmation of policy."
9. Taylor, 1991; for other personal accounts of the activities of the Joint Commission and the fact finding group from the National Research Council see Liebow, 1965, Brues, 1968, and Neel, 1994.
10. The fact that this decision was not a unilateral one is commonly ignored in Japan today, and little recognition is given to the fact that many of the American physicians were not in sympathy with it. For an exception to this statement, however, see the newspaper article entitled "ABCC's Conscience" that appeared in the Nagasaki issue of the Mainichi Shinbun on 9 August 1954.
11. See letter of 3 February 1956 to Dr. R. Keith Cannan from Robert H. Holmes, Director of ABCC, ABCC–RERF Archives.

12. The old shibboleth that the Commission saw the survivors as "guinea pigs," since they were not treated, was hurled about anew. Whether this was merely a convenient psittacism, or was the expression of deeply held beliefs is debatable. It had a simplistic emotional appeal that made the charge immune to rational arguments. Moreover it summarized, in a sense, Japanese cultural expectations of an apology and some form of remuneration (here treatment) when one perceives oneself to have been wronged.
13. See Atomic Bomb Casualty Commission Annual Report, 1 July 1974–31 March 1975 for further details. It should be noted that these numbers represent the operating costs in Japan alone, and do not include the expenses the National Academy of Sciences incurred in the administration of the contract, the travel of U.S. and other foreign personnel to and from Japan, or the salaries of individuals assigned to the Commission from the United States Public Health Service. Interestingly, the operating budget in Japan for the 9 months immediately preceding the formation of the Radiation Effects Research Foundation had risen to $8.68 million, largely as a consequence of an 18.6% increase in staff salaries, which then constituted some 89% of the annual operating expenses.
14. When established in 1975, the Act of Endowment provided for only four "permanent" directors, but this number was broadened to six in 1981.
15. See, e.g., Vogel, 1988.
16. Dr. Matsuzaka was a survivor of the atomic bombing of Hiroshima and Vice Chairman of the Hiroshima A-bomb Casualty Council. His "Notes," which are one-sided and erroneous in part, nonetheless provide an insight into the sentiments toward ABCC harbored by some Japanese survivors. His admission cited here is all the more unusual given his sharply critical attitude toward the Commission and the Foundation and their research activities.

Chapter 3

1. Snell et al., 1949.
2. Yamasawa, 1953.
3. At the time of registration, the mother was issued a maternal-and-child handbook. Holders of this handbook were not only entitled to a special ration of rice but, in addition, a special clothing ration ticket, a special allocation of sugar, fruit, soap, salt, bean paste, absorbent cotton, tissue paper, and diaper cover, and, with the birth of the child, a special baby clothing ration ticket. As Japan's economy recovered, rationing and other restrictions were gradually lifted, and by the time of the termination of the genetics study the benefits associated with the handbook were limited to additional rice.
4. Most midwives then in practice were licensed under a system that required a candidate for admission to a school of midwifery to have completed the equivalent of about eight years of education. Although not required, some candidates had also completed Japanese middle school, the equal of four more years of study. The training in midwifery itself covered two years. A candidate either spent both of these in a school of midwifery, or one year in

school and the other obtaining experience in the obstetrics-gynecology department of a hospital, or in association with a senior midwife. However, before a midwife could begin her practice, a prefectural examination had to be passed for licensure. Not uncommonly, a licensed midwife married soon after completing school, and often did not return to the practice of midwifery until some years later when her own children were grown or at least of an age at which they could be responsible for themselves.

5. In the beginning, each midwife was given 15 yen for reporting a normal birth and 50 for an abnormal one; this soon rose to 20 and 100 yen, respectively, and by 1954 they were receiving 50 yen (then about 14 cents) for each normal termination and up to 200 (55 cents) for a stillbirth or a neonatal death reported within 12 hours of its occurrence. In addition, the teaching fund of the midwives' association was receiving 50 yen for each stillborn child, or one dying in the neonatal period whose body was obtained for autopsy. Eventually, this supplementary contribution was made directly to the midwife through whose intercession the body was obtained. When the clinical phase of the genetic studies ended in early 1954, each midwife received a small termination payment, the amount determined by the number of births she reported, as a token of the Commission's appreciation for their years of cooperation.

6. Neel and Schull, 1956, or see Neel and Schull, 1993, where this earlier publication is reproduced in its entirety.

7. On 15 July 1948 the national government enacted a law (167) entitled "The Prevention of Venereal Disease." Section 3, Article 9 of this law stated that all pregnant women must receive a physical examination to determine whether a venereal disease was present. The law did not, however, (1) specifically require serological testing, (2) specify a time during pregnancy when this examination must be made, or (3) make punishable failure to comply with the law. Both Hiroshima and Nagasaki ostensibly refused to grant unexamined pregnant women the augmented rations that were allowed at that time if they had not complied, but it was not clear how rigorously this restriction was enforced in Hiroshima. In Nagasaki, however, ration registration was tied to the serological testing of expectant mothers.

8. Suzuki and Watanabe, 1954.
9. Neel, 1990.
10. Suzuki, 1955.
11. See Memorandum to the Research Committee from Duncan J. McDonald, dated 26 August 1952, entitled "Early Termination Program; Exposure categories," and a Memorandum to James V. Neel from Newton E. Morton, dated 30 July 1952, entitled "Preliminary analysis of early spontaneous terminations," on file in the Archives of the Radiation Effects Research Foundation.
12. Mavor, 1924.
13. Tjio and Levan, 1956.
14. Today, the study of human chromosomes rests largely on the use of short-term cultures of white blood cells derived usually from one of the blood vessels of the arm. Interestingly, in 1935, these methods had been foretold

in part in a publication by a Russian, G. K. Chrustschoff, entitled "Cytological investigations on cultures of normal human blood" that had appeared in an English scientific periodical, the *Journal of Genetics* (Chrustschoff and Berlin, 1935). Subsequent developments, notably the use of a substance known as phytohemagglutinin, extracted from the common navy bean, that stimulates cells to divide, of an organic substance, colchicine, obtained from the flower of *Colchicum autumnale,* which stops these divisions at a time propitious for the counting and study of chromosome structure, and the use of hypotonic (low) salt solutions to facilitate the spreading of the chromosomes, have revolutionized human cytogenetics. The efficacy of some of these procedural elements, such as the use of colchicine, had been known to plant cytologists for several decades, and Edwin Osgood had used phytohemagglutinin for years. But it was Peter Nowell and David Hungerford at the University of Pennsylvania who put all of the pieces together in the technique that is now used, albeit with some modifications. A detailed and entertaining account of the evolution of the techniques now employed in human cytogenetics will be found in Hsu, 1979.
15. Greulich et al., 1953.
16. Subsequently, Reynolds achieved a certain fame, or notoriety at least, through his efforts to sail his ketch, the Phoenix, into the area of the Pacific where atomic weapons were tested. Invariably, his former association with ABCC was noted by journalists, to the embarrassment of the Commission, since the tenor of the news articles often suggested that Reynolds actions had an exculpatory purpose.
17. See Reynolds, an interim report entitled "The Growth and Development Program of the Atomic Bomb Casualty Commission: Analysis of Body Measurements taken in 1951 on 4,800 Hiroshima Children." Dated 12 June 1952. RERF Library, identified as document ABCC R c.2, Appendix 3.
18. Reynolds, 1959; see also Nehemias, 1961, who analyzed these data somewhat differently. Whereas Reynolds had simply compared the exposed and not exposed, Nehemias attempted to rank order the degree of exposure, and concluded "Differences with respect to size which are statistically very significant are found among the subgroups defined on the basis of the radiation exposure index. The actual physical differences are small, however. The differences have been shown to be in the direction of decreasing size with increasing degree of radiation exposure."
19. Sutow and West, 1955.
20. Belsky and Blot, 1975.
21. Ham, 1953; see also Abelson and Kruger, 1949 for an earlier report.
22. Hirose and Fujino, 1950; Ikui.
23. Cogan et al., 1950.
24. Sinskey, 1955.
25. Otake and Schull, 1990.
26. Hall et al., 1964.
27. Cox et al., 1991.
28. Lett et al., 1991.
29. Sinskey, 1955.
30. Miller et al., 1967.

31. Memorandum dated February 1950.
32. The area within a circle is πr^2, where r is the radius of the circle in question, and π is the well-known constant, denoting the ratio of the circumference of a circle to its diameter. Thus the area within 1 km is π square km; whereas that within 2 km is $2^2\pi$, or the difference is 3π square km.
33. Folley, Borges, and Yamawaki, 1952
34. Moloney and Kastenbaum, 1954.
35. Lange et al., 1954.
36. Valentine et al., 1952.
37. Moloney and Lange, 1954a,b.
38. Lange et al., 1955.
39. Ichimaru et al., 1970.
40. Goldstein and Murphy, 1929; Murphy, 1947.
41. Plummer, 1952.
42. In the early years, it was customary to identify research studies through the use of two letters and a number. The number identified the order of approval of the research study, and the letters the department from which the study originated. Thus the designation PE-86 implies a study originating in the Pediatrics Department and the 86th research project to be approved. The surveillance of the children of exposed parents, the genetics study, was identified as the GE-3 program.
43. We set out the details on these samples for four reasons essentially. First, these details illustrate the difficulty of identifying and recruiting representative groups of individuals for study not merely of the effects of prenatal exposure but exposure more generally. Second, they indicate to some extent the ingenuity and thoroughness that have been exercised in the reconstruction of the various populations at risk. Third, they reveal the compromises forced on every study by the resources available to the investigators, and, finally, as previously written, the construction of an epidemiological sample is more than an exercise in logic or even perseverance; it involves people whose participation must be sought and whose sensitivities and limitations must be recognized. This is more true in this instance than in many others since the possibility of inadvertent stigmatization is so great. Understandably, the prenatally exposed, particularly those who have been handicapped, and their parents seek to avoid unnecessary social involvements.

    Originally, in Nagasaki the first prenatally exposed survivors were identified through the city-wide registration of pregnancies associated with the genetic studies, and the sample was assembled in connection with a study of pregnancy loss (Neel and Schull, 1956; Yamazaki, Wright and Wright, 1954). In late 1949 and early 1950, all 1,774 women then recorded in the genetics files who were of childbearing age at the time of the bombing (ATB), defined as 17 to 51, and who were stated to have been exposed within 2,000 m, were contacted to determine if pregnant at the time of exposure. Ninety-eight were. For comparative purposes, a second group of 1,774 women was randomly selected, contacted, and similarly interviewed. This group had been exposed between 4,000 and 5,000 m, a distance at which the effects of ionizing radiation were presumably negligible. One hundred and thirteen of these women had been pregnant ATB. This collection of cases and controls was designated as the PE-57 sample.

In 1953 a systematic supplementation of these groups was initiated by reviewing radiation exposure questionnaires for all females exposed within 2,000 m and a portion of those exposed beyond. About 1954, to make ascertainment more exhaustive, all birth certificates in the Nagasaki Prefectural Judicial Affairs Bureau were checked for those infants whose date of birth suggested possible exposure in utero, and the sample was redefined to include all individuals prenatally exposed within 5,000 m of the hypocenter. In 1955, to this group, which was subdivided into the heavily exposed (under 2,500 m) and the lightly exposed (2,500–5,000 m), was added a group of infants born to mothers who first entered Nagasaki after 1 January 1946. These children were age- and gender-matched to the heavily exposed group. In all, at this time, the Nagasaki sample consisted of 874 children (see the Memorandum by Clarence M. Tinsley and Isolde Loewinger, dated 1955, in the Archives of the Radiation Effects Research Foundation). In July 1959, this sample was revised, primarily to achieve economy in the use of the limited clinical facilities that were available and to better match by age, gender, and, where possible, economic status the "distally or 'lightly' exposed" (2,500–5,000 m) and "nonexposed" (10,000 m or beyond) to the group that had come to be known as the "proximally or 'heavily' exposed" (within 2,000 m). The size of the revised sample, 306 individuals, was defined by the number of pregnant women who had been proximally exposed; they numbered 102. (For a fuller description of this sample see Burrows, Hrubec and Hamilton, 1960.) It is important to note that this latter group has changed little over time. In the 1954–1955 examinations, it numbered 108, of whom 101 were examined (3 had moved; 3 refused to be examined; and 1 was unavailable); in 1955–1956, it was 116 (101 examined; 11 moved; 2 refused; 2 unavailable). As these figures indicate, the participation rates were very high, over 95%, and it would seem unlikely, therefore, that the examinations were not representative of the entire group.

In Hiroshima the sample developed differently. The first step was taken in 1950 when the PE-52 Project of Jane Borges and George Plummer was started. However, this investigation was limited to individuals who were born between 1 January and 1 May 1946, and therefore had been exposed in the first half of their prenatal development. Two hundred and five children were identified; of these, 60 were exposed at distances within 2,000 m, 65 within 2,000–2,900 m, and 80 at 3,000 m or beyond. In 1953, an additional group of children, born between 6 August and 31 December 1945, was added. Their exposure, however, had to be within 1,600 m (there were 44 such individuals), between 3,000 and 4,000 m (41 individuals), or they were not exposed (50 individuals). This new project was known as the PE-52-1 sample. In 1955, a check of the Hiroshima Master File identified 677 prenatally exposed children, including the 340 in the PE-52 and PE-52-1 samples. In August of that year, a nonexposed control group, numbering 476 individuals, was constructed from the results of the Daytime Census of Hiroshima City in 1953. It is on these two groups involving 1,153 individuals that the intelligence test scores to be described later were ob-

tained. In 1956, after the intelligence testing had ceased, the birth registrations at the Hiroshima City Office were examined to identify still other possible cases. These steps culminated in a sample of 2,016 individuals, exposed and nonexposed. Finally, in 1959, as in Nagasaki, this sample was revised, and this revision constitutes the clinical sample used in recent analyses of the data on severe mental retardation. It should be noted that the principal changes in the samples over time in both cities have involved the nonexposed comparison group, and not the prenatally exposed individuals who received doses of 0.10 Gy or greater.
44. Koga, 1937; Tanebashi, 1972.
45. Restak, 1984, 1988.
46. Omuta, 1966.

Chapter 4

1. Francis Committee Report; see Francis, 1959.
2. Epidemiologists generally employ one or the other of two basic designs in the study of causal relationships, such as that between exposure to ionizing radiation and the occurrence of cancer. One of these is known as a cohort study. It utilizes a fixed sample of individuals, a cohort, defined at some point in time, e.g., all individuals residing in Hiroshima in 1950, among whom the occurrence of a particular event, say death from stomach cancer, is then determined. This determination can rest on observations obtained through continued surveillance of the members of the cohort forward in time, or through enumeration of the events of interest that have occurred among the members in the past. The former approach is known as a prospective cohort study, and the latter as a retrospective one. Ultimately, the investigator asks whether the frequency of the event of interest, stomach cancer, increases as exposure to the potential cause, here ionizing radiation, increases.

    The other design is called a case-control study. It involves the identification of some number of individuals who have experienced a specific event, again say the occurrence of stomach cancer, and a similar or often larger number of individuals, the control, who have not experienced this event. These latter persons are so selected as to match the cases on gender, age, and such other factors, aside from the one of interest, here ionizing radiation, as may be deemed important in the occurrence of stomach cancer. One then asks whether the cases were more often exposed to ionizing radiation than were the controls, and if so a causal role to exposure may be claimed.

    Occasionally, the two designs are combined in what is termed a population-based case-control study. Such studies begin with a cohort and from this cohort are selected the cases and controls. This design has numerous advantages including the fact that it can provide information on the prevalence of the event of interest, a dose–response curve in the instance of ionizing radiation, and through the embedded case-control study the opportunity to study the interaction of radiation with other endpoints of interest.

Whatever design is chosen, epidemiological studies have their limitations and sources of uncertainty. The latter uncertainties are commonly described as confounding, selection bias, and information bias. Confounding refers to the presence of unmeasured factors that could influence the occurrence of the outcome and be distributed differently between cases and controls or in exposure groups within the cohort. Selection bias implies that either the cohort is not representative of the population one seeks to assess, or in the instance of the case-control study, the control may not be representative of the population of inference. Information bias implies the existence of systematic errors in the observations themselves. These may arise in many ways, but often they occur because the observer is aware of the exposure of an individual and examines him or her more carefully, or the person involved, the case, is aware of their own exposure and is more motivated to remember events than the unaffected control.

Thus the Francis Committee was advocating a prospective cohort study, but because of the size of the proposed cohort it could and has served as the basis for population-based case-control studies.

3. Letter to R. Keith Cannan from Robert H. Holmes, dated 22 November 1955, in the Archives of the Radiation Effects Research Foundation.
4. Memorandum entitled "Comments on the Report of the Francis Committee" addressed to Dr. Robert H. Holmes, Director, by Clarence M. Tinsley, M.D., Chief of Clinical Services, dated 18 November 1955; see the Archives of the Radiation Effects Research Foundation.
5. Memorandum entitled "Survey of ABCC from October 20th through November 9, 1955" addressed to Dr. R. Keith Cannan, Chairman, Division of Medical Sciences, National Academy of Sciences–National Research Council, by Charles H. Burnett M.D., dated 11 November 1955. See Archives of the Radiation Effects Research Foundation.
6. Thomas Parran, M.D., "Report on Visit to Japan, February 1956" submitted to the National Academy of Sciences–National Research Council; see also the accompanying appendix memorandum to Parran from the Department of Biostatistics at the School of Public Health, University of Pittsburgh—Archives of the Radiation Effects Research Foundation.
7. Beebe and Usagawa, 1968; see also Beebe, 1960, and Beebe, Fujisawa, and Yamasaki, 1960. When the Life Span Study sample was initially defined, one of the criteria for selection of an exposed individual was the presence of the individual's family register in either Hiroshima or Nagasaki or a defined, immediately surrounding area. This criterion, however, excluded a number of thousands of individuals whom it was believed could and should be studied. The first extension of the sample, therefore, involved the addition of these individuals (about 9,500 persons) to the surveillance. This increased the sample from about 100,000 to 109,000 persons, and gave rise to the expression Life Span Study—Extended Sample (or LSS-Ext). The second extension, which led to the incorporation of more individuals with doses of less than 0.10 Gy, involved only Nagasaki. Two reasons were responsible for this. First, the funds available to augment the study group, which had come from a small Japanese government grant, were not sufficient to do so in both cities, and second, since only one city sample could

be enlarged, Nagasaki seemed the better candidate. This followed from the fact that the Nagasaki sample was initially substantially smaller than the one in Hiroshima. While the Nagasaki sample presumably included most individuals with moderate to high doses, since these survivors were relatively few in number and since the survivors exposed to doses of less than 0.01 Gy were matched to this number, background rates of cancer or other possible endpoints in Nagasaki were poorly estimated. Enlarging this sample would, therefore, give more stability to the estimates of the background rates, and permit somewhat better comparison between the cities. This addition increased the sample from 109,000 to about 120,000. The latter group is designated the LSS-E85 (Life Span Study—Extended 1985).
8. Taeuber, 1958. It warrants noting too that before the first modern enumerative census of Japan in 1920, the *koseki* were the only source of demographic statistics on the country.
9. Naruge, 1956.
10. As Lord Redesdale has noted, although the Constitution promulgated in 1889 ostensibly outlawed these social distinctions, General Tamesada Kuroki told him that while other classes would serve in the same military units with the *buraku-min,* they would not eat with them (Redesdale, 1906, p. 161). This reluctance undoubtedly reflected feudal practices, when the *eta* and *hinin* were not allowed to enter the home of a person outside of their own class or to sit or to cook at the same fire.
11. These two groups are often described as the *buraku-min.* The Japanese word, *buraku,* means a hamlet, and in this case a special hamlet, for as stated in the text the pariah classes in medieval Japan were obliged to live apart from the socially accepted classes.
12. See Mabuchi et al., 1991, for a more detailed description of the operation and administration of these registries.
13. See memorandum from Robert W. Miller to Marvin A. Schneiderman dated 16 April 1974 entitled "Benefits from the NCI Contract with ABCC."
14. Retrospectively, this sum seems paltry, unlikely to have achieved its purpose, and certainly would not be a motivating factor today, when the average Japanese physician's income is at least that of his American counterpart. But as in the instance of the remuneration paid to midwives, this was not so in the middle 1950s, when the nation was still struggling with enormous economic problems.
15. Finch et al., 1965.
16. This method of sample selection ensured the largest possible representation of individuals receiving substantial doses of ionizing radiation. However, it must be borne in mind that the symptoms were identified retrospectively, by interview, and are subject to errors in recall. A definitive diagnosis of radiation sickness can only be made by a trained physician at the time of occurrence of the disease, when other causes can be ruled out. However, some of the symptoms of the disease, such as epilation, are apt to be reasonably well remembered, since they are striking events to the affected individual. But as Beebe et al. have noted, "The fact that a patient reported having suffered epilation, purpura, and oropharyngeal lesions, or any combination of them, is not necessarily good evidence of exposure to ionizing

radiation. Approximately one per cent of people at distances where they could not have received any significant amount of radiation reported one or more of these symptoms. The cause is not clear but typhoid fever and severe dysentery are suspected." (See page 46, in Beebe, Fujisawa, and Yamasaki, ABCC Technical Report 10-60.)

17. In time, similar arrangements evolved with several of the Japanese academic institutions. The Department of Radiology of the Faculty of Medicine at Kyushu University, for example, provided young radiologists on a rotating basis to the Commission and subsequently the Foundation. Similarly, in Hiroshima and Nagasaki, internists have been provided on a part-time basis by the Second Department of Internal Medicine of the Hiroshima University School of Medicine or the Nagasaki Medical School. And in both cities, the Departments of Ophthalmology have regularly provided ophthalmologists for the ocular examinations. Moreover, during the years that a Department of Dentistry existed within the Commission, dentists were recruited through the School of Dentistry at Osaka University.
18. These technical reports can be obtained, free of charge, from the Publications and Documentation Center of the Radiation Effects Research Foundation at 5-2 Hijiyama-koen, Minami-ku, Hiroshima 732, Japan.
19. Takahara, 1947.
20. Takahara et al., 1961.
21. Krooth et al., 1961.
22. Schull, 1958.
23. Schull and Neel, 1965; see Table 2.5.
24. Minutes of the Committee on Atomic Casualties.
25. See memorandum to Dr. R. Keith Cannan and the Committee on Atomic Casualties from William J. Schull entitled "Report of activities in Japan during the period 9 July 1956 to 26 September 1956" in the ABCC–RERF Archives.
26. Muller, 1950.
27. Morton, Crow, and Muller, 1956.
28. The choice of the acronym, Ni-Hon-San, was not happenstance, since the investigators were aware that it could be loosely read in Japanese as either "the Japanese three" or as "Honorable Japan."
29. Robertson et al., 1976.
30. Takeya et al., 1984.

Chapter 5

1. Kathren and Petersen, 1989.
2. There are circumstances in which this assumption is not valid, such as the dose to the skin, in or around bone, and possibly to the lens of the eye. In these instances, the actual absorbed dose may be substantially different than the energy released in or around these tissues.
3. See, e.g., ICRP, 1991.
4. Warren, 1966.
5. Ibuse, 1953.
6. Wilson, 1956.

7. Ritchie and Hurst, 1959; the doses Ritchie and Hurst used were calculated by Edwin N. York using all of the weapons effects information then available.
8. Milton and Shohoji, 1968.
9. Kerr, 1981; Loewe and Mendelsohn, 1982; Roesch, 1987.
10. Ellett, 1990.
11. Malenfant, 1984. This is but one of a series of papers, mostly highly technical, given at the 17th Midyear Topical Meeting of the Health Physics Society.
12. Kerr and Solomon, 1976.
13. Auxier, 1977.
14. This group of consultants, consisting of William T. Ham Jr., Rufus Ritchie, George S. Hurst, Payne Harris, Loren Logie, and Robert Corsbie, was sent to Japan in April, 1956 to review the progress in the shielding studies and to recommend steps that might expedite data collection. Aside from defining the priorities, they concluded that "work that has been done to date is very good and potentially significant in all aspects." See their report dated 21 June 1956 in the ABCC–RERF Archives.
15. Engel, 1964.
16. Noble, 1967.
17. Stram and Mizuno, 1989.
18. A variety of terms are used to describe the state of the fissile material in a nuclear weapon. Thus, nuclear criticality refers to that state in which one neutron from each fission causes one subsequent fission. If less than one additional fission results, the assembly is said to be subcritical, and if more than one new fission is caused, the assembly is supercritical. Almost all neutrons appear immediately in the fission process (prompt neutrons), but a few, somewhat less than 1%, do not. The latter are the delayed neutrons. The term "critical" is used to describe the situation in which the assembly uses the neutrons that do not escape by chance to maintain criticality. If the mass of fissile material is small, its surface area will be relatively large and too many neutrons will escape through the surface for the chain reaction to proceed. However as the mass (volume) increases, the relative importance of the surface area decreases. For example, the surface area per unit volume of a sphere with a diameter of one inch is three times as large as that for a sphere with a 3 inch diameter. To obtain a fission explosion, it is necessary to have assembled a highly supercritical mass—a mass of fissile material several times larger than the critical mass. Prior to its firing, the fissile material in a nuclear weapon is assembled in a subcritical state and, to produce the explosion, some of the material must be moved to achieve supercriticality.
19. Roesch, 1987. Numbers of the magnitude cited lack reality for most of us; they do not correspond to anything that can be visualized or that accords with our own experiences. Thirty million pounds of dynamite (15 kT) would fill about 150 conventional railroad coal cars, and since coal trains are generally 100 cars long, it would take one and a half such trains to haul an equivalent amount of high explosive to approach the energy released by the Hiroshima bomb. As a further point of comparison, it warrants noting that

the single most destructive air raid on Japan using conventional bombs was the March 1945 raid on Tokyo. This involved 270 planes that dropped 1,667 tons of incendiary bombs, and destroyed approximately 16 square miles of the city. Some 83,600 individuals were estimated to have lost their lives or to be missing following the bombing, and another 102,000 were injured. These are figures roughly comparable to the losses associated with the use of an atomic weapon.

20. Under normal temperature and atmospheric pressure, the blast wave would have been moving at approximately the speed of sound (1,180 feet per second), but since the pressure was increased it was actually moving more rapidly.

21. Technically, the roentgen is a measure of the ionizing effect of X- or gamma rays in air, and this effect will vary with the wavelength of the ray. For our purposes, it is sufficient to presume that one roentgen is about 0.01 Gy in soft tissue for the radiations of interest here.

22. It should be noted that the areas we described are those where the bulk of the heavier radioactive particles would have fallen. The lighter particles could be, and were carried for miles beyond these areas, but the doses accumulating when they fell to earth are well below those at which any biological effects have ever been reported.

23. Okajima, 1975.

24. Radioisotopes lose their radiant energy through the spontaneous emission of alpha particles, electrons, or gamma rays. The rapidity with which this occurs varies, and as a result it is customary to describe different isotopes in terms of the length of time it takes to lose half of the initial radioactivity. This is called the physical half-life of the isotope. This time can vary from tiny fractions of a second to minutes, hours, or many years. The physical half-lives of cesium-137 and strontium-90, two common by-products from a nuclear detonation, are about 30 and 28 years, respectively (UNSCEAR, 1977). However, the damage these isotopes can do once inhaled or ingested depends not only upon their physical half-life and the nature of the radiant energy lost in decay, but where they are deposited within the body and how long they remain there, on average. This residency time is known as the biological half-life and in the case of cesium-137 and strontium-90 these values differ markedly, from as much as 225 days for cesium-137 to many years for strontium-90. Cesium-137 is deposited largely in muscle (80%); whereas much of the ingested strontium-90 is deposited in bone.

25. National Council on Radiation Protection and Measurements, Report No. 116, see pages 54-55 and Table 19.1, 1993. The Council's recommendations are essentially the same as those of the International Commission on Radiological Protection (1991), but the National Council suggests that International Commission's values be seen as reference values rather than limits (see NCRP, 1993, p. 26).

26. A description of this device, which was constructed at Brookhaven National Laboratory, and the manner of its use to estimate the body burdens of the Marshallese on Rongelap will be found in Conard, 1993. After several years use it was replaced with a less cumbersome portable assembly of lead bricks.

27. Under the DS86 dosimetry system the computer code produces four gamma and two neutron contributions to the total dose in kerma, and six gamma and two neutron contributions to the total organ dose. These contributions are computed and stored in full precision on magnetic tapes. These tapes, which cannot be directly addressed by investigators, serve as the bases for computing a working dose tape that is available for analytic purposes. The working tape has three dose terms (gamma, neutron, and total) for kerma and 15 organs expressed in mGy. The gamma and neutron terms are the sum of the contributions mentioned above, and these sums are rounded to the nearest milligray. The total dose is the sum of the rounded gamma and neutron doses. Customarily, when these doses are used in analyses, they are further rounded to the nearest cGy (0.01 Gy). Thus an individual said to have had a total dose of less than 0.01 Gy actually had a dose less than 0.005 Gy (that is, less than 5 milligray). Individuals having doses of less than 0.005 Gy are sometimes referred to as the "zero" dose group.
28. Young, 1987.
29. Kamada et al., 1989.
30. Thiessen and Kaul, 1991. Straume et al. (1992), for example, using all of the available measurement data on thermal neutron activation including new measurements for ^{36}Cl, suggest that thermal neutron activation at about 1 km in Hiroshima was two to ten, or more times higher than that calculated based on DS86. The implications of this discrepancy with regard to risk estimates is not wholly clear, since it must be noted that low-energy neutron activation contributed little to the dose in Hiroshima. It is not presently known whether the fast neutrons—those with energies in the range ~0.1 to 1 MeV, which comprise the bulk of the neutron dose in Hiroshima, have been similarly underestimated. However, Preston et al. (1993) and Sasaki et al. (1992) have attempted to determine the possible impact of this discrepancy on risk estimates. Preston et al., based on several different assumptions regarding the neutron RBE, find that the cancer risk estimates might be in error by 2 to 20%; whereas Sasaki and his colleagues, using a series of in vitro experiments, find that the difference in chromosome aberration frequencies between the two cities is explicable if the neutron dose in Hiroshima was as large as 5% of the total dose in gray rather than the estimate of 2% or less provided by the DS86 dosimetry.

    Theoretically, the neutron fluences, and hence the neutron dose, associated with an atomic weapon can be determined in either of two ways. First, it is possible to assess the neutron fluence directly through the measurement of the neutrons released as a result of a specific nuclear detonation. Second, the neutron fluence can be estimated indirectly through activation studies, that is, through measurements of the gamma radiation stemming secondarily from the capture of neutrons in soil or any other suitable irradiated material. Insofar as the Hiroshima weapon is concerned, there is some evidence of both kinds, but neither avenue can establish the neutron fluence unequivocally at the moment. We have already cited some of the data based on neutron activation studies, past as well as recent. As has been seen, these data suggest a higher neutron yield than is currently assumed in the DS86 dosimetry; however, there are some limitations to

320 Notes

these data that need resolution. Unfortunately, the potentially relevant direct estimates are not much better. The direct evidence comes largely from a nuclear weapons test, known as Upshot Knothole Grable, conducted in the spring of 1953. This device had as its fissionable material uranium, and in this sense was like the Hiroshima bomb, but there were important differences. The casing of the Grable device was thinner; it had a pointed nose, since it was intended to be a projectile, and it was fired from a 280 mm cannon rather than dropped. Moreover, it exploded at an altitude of 160 m, which is considerably lower than the burst point of the Hiroshima weapon, and the ensuing fireball split into two parts as a result of the blast wave. Nonetheless it is generally considered to be a better approximation to the Hiroshima bomb than any other test and measurements on the neutron spectrum of this weapon do exist. It is believed that combining these measurements with the information from activation studies will ultimately lead to a better description of the neutrons released by the Thin Man weapon.

31. Kerr, 1979; Hashizume et al., 1973.
32. Fujita et al., 1987.
33. See Jablon, 1971; Gilbert, 1984; Gilbert and Ohara, 1984.
34. Antoku et al., 1991.
35. Kato et al., 1989.
36. Sawada et al., 1979.

Chapter 6

1. See "A Resumé of Measures for Atomic Bomb Survivors in Japan" published by the Ministry of Health and Welfare in April 1990. Collectively, the benefits to be paid to the survivors in 1991 were estimated to about 130,000,000,000 yen, or somewhat more than $1,000,000,000 at the exchange rate prevailing at the end of 1991.
2. Expressions such as "low," "intermediate," "high," or "very high," although useful, are arbitrary when applied to doses of ionizing radiation. To avoid ambiguity and yet give some sense of the magnitude of the exposure, we will use these terms as they have been employed by the United Nations Scientific Committee on the Effects of Atomic Radiation (1988). Low implies a dose of 0.20 Gy or less, an intermediate dose is one ranging from 0.20 to 2.0 Gy, a high dose is one of 2 to 10 Gy, and a very high dose is one above 10 Gy. To provide some perspective on the Japanese experience, based on these definitions, the average survivor of the bombing of Hiroshima or Nagasaki received an intermediate dose, about 0.25 Gy, although some individuals received high doses, possibly as much as 6 Gy.
3. Standardizing is a means to distribute an extraneous variable, such as age, equally among the groups compared. One of the groups is taken as a reference and the frequencies observed in the reference group are compared with those expected in the comparison group, if the distribution of the extraneous variable(s) were the same as in the reference group. Standardization may be direct or indirect. Of these two methods, the latter is the more common. To apply the indirect method the following must be known: (1) the crude rate, for example, for leukemia, in the group being studied, (2)

the crude rate (again for leukemia) in the standard group, (3) the proportion of individuals in the group under study in the various strata of interest (age classes, if age is the variable on which standardization is to occur); and (4) the specific rates of leukemia in the same strata for the standard group. Indirect standardization gives rise to what is commonly called the standard mortality ratio (SMR). The SMR can be calculated for deaths from all causes or from a particular cause.

The direct method differs from the indirect in that the specific rates used are those of the group being studied and not those of the standard group. Direct standardization produces what is termed the comparative mortality figure (CMF). When the age composition of a group is the object of concern, standardization occurs within the genders separately.

4. Kato and Schull, 1982; Kato et al., 1982.
5. Shimizu et al., 1991.
6. See Penn et al., 1986.
7. Scott, 1987. Recently, Stephen Epstein and his colleagues (see Speir et al., 1994) have shown that in patients who have undergone angioplasty and the vessel closed again as a result of rapid smooth muscle cell proliferation, there is an increase in p53, a tumor suppressor protein that inhibits cell cycle progression and is functionally inactivated in many human cancers. Apparently, this increase is the result of an interaction between the p53 tumor suppressor protein and the human cytomegalovirus (HCMV), which has been associated with the development of atherosclerosis. p53 is not a normal feature of the vessel wall nor is it present in primary atheromatous lesions. Similarly, HCMV was not found in any of the primary lesions but it was present in 85% of the p53-immunopositive lesions. These authors interpret their findings to imply that angioplastic injury to the vessel wall reactivates latent HCMV which, in turn, blocks p53 inhibition of cell cycle progression.
8. Robertson et al., 1976, 1979.
9. Shimizu et al., 1991.
10. Stuart Finch and John Pinkston, personal communication.
11. Sager, 1989.
12. Vogelstein, 1990; Jones et al., 1991.
13. Weinberg, 1988; this is a highly readable account of the relationship of oncogenes to tumor suppressors, or anti-oncogenes as they have also been termed.
14. Trosko et al., 1990.
15. See, e.g., Moloney and Kastenbaum, 1954.
16. Ichimaru et al., 1976.
17. Shimizu et al., 1987, 1988.
18. Tomonaga et al., 1991; see also Preston et al., 1993.
19. Moloney, 1987.
20. Harada and Ishida, 1960.
21. Wanebo et al., 1967.
22. Tokunaga et al., 1991.
23. Tokuoka et al., 1982.
24. Tokunaga et al., 1991.

25. Nakamura, 1977.
26. Matsuura et al., 1983.
27. Akiba et al., 1991.
28. Hollingsworth et al., 1963.
29. Thompson et al., 1992.
30. Anderson and Ishida, 1964.
31. Ichimaru et al., 1979.
32. Darby et al., 1984.
33. Kato et al., 1980; see also Asano et al., 1981.
34. Land et al., 1994.
35. Shimizu et al., 1988.
36. Jablon, 1971; see also Gilbert, 1984; Gilbert and Ohara, 1984.
37. Bennacerraf, 1960.
38. Balish et al., 1970.
39. Sasagawa et al., 1990.
40. Nakamura et al., 1991; see also Nakamura, Sposto, and Akiyama, 1993.
41. Gregory et al., 1967.
42. Yamamoto et al., 1978 (see Table 20).
43. See, for example, Dorn and Horn, 1941.
44. Fujiwara et al., 1990.
45. Shimizu et al., 1991.
46. Yamada, Kodama, and Wong, 1991. See also Konuma, 1967; Okunuma and Hikida, 1949; and Nishikawa and Tsuiki, 1961.
47. Matsumoto, 1969.
48. Dobson and Felton, 1983.
49. Siegel, 1964; Blot and Sawada, 1971; Blot et al., 1972; Blot et al., 1975.
50. Awa et al., 1971.
51. Awa et al., 1978.
52. Awa et al., 1984; Awa, 1989; Ohtaki, 1992.
53. Preston et al., 1988.
54. Bender et al., 1988.
55. Briefly, but more specifically, the HPRT mutation assay uses a method involving the growth of colonies of T-cells in the presence of 6-thioguanine, recombinant interleukin-2, and phytohemagglutinin. It is a simple method to identify lymphocytes with no HPRT gene activity by taking advantage of the resistance of such cells to 6-thioguanine, which kills nonresistant cells. The other two methods, the GPA and TCR assays, use a flow cytometer, which allows one to rapidly enumerate rare mutant cells. The GPA assay detects rare mutant red blood cells lacking expression of either the M or N gene products on the surface of cells from MN (heterozygous) donors using a pair of monoclonal antibodies bound to different fluorescent dyes. The TCR assay uses two monoclonal antibodies, anti-CD3 and anti-CD4. When a mutation occurs in the TCR gene of a T lymphocyte, the CD3 complex cannot be expressed on the cell surface, and such mutants are detected as CD3 negative cells among the CD4 positive T lymphocytes.
56. The molecules that are present on the red blood cell surface can be detected using specific reagents (antibodies) that will bind only to the M or the N molecule. These highly specific reagents are termed monoclonal antibod-

ies. If a fluorescent dye is bound to these reagents, a cell lacking the molecule corresponding to that antibody will not bind the antibody, and therefore will not fluoresce. These aberrant cells, although rare, can be sorted from normal ones using an automated method, known as flow cytometry, where over 1,000 cells can be examined per second and segregated into those that are NO or MO, or an MM variant. As yet it is not possible to identify NN variants in a satisfactorily reproducible manner. This is unfortunate, since one of the mechanisms that could give rise to MM variants is a rearranging of genetic material in the somatic cell; if this happens, however, an NN cell should be produced for every MM one, and a good assay for the frequency of NN variants could establish the plausibility of this mechanism. An alternative to recombination is, of course, somatic mutation but it seems improbable that this would be the entire explanation for the occurrence of MM cells for the following reason: it is known that chemically the structures of the M and N gene products differ by two amino acids. Since the monoclonal antibody used in these assays recognizes only a very small portion of the protein product, possibly no more than a stretch of several amino acids, an amino acid change in only one of the two amino acids that distinguish the N from the M protein, could be misclassified as M. Such mutants would then contribute to the frequency of mutations of the MM type. However, there are numerous other potential sites which can, by a single mutation, result in the premature termination of polypeptide synthesis, which would lead to MO type mutants. Thus, if an MM mutation was solely caused by a base pair change in the gene, the frequency would be expected to be far lower than that observed for MO, but this is not the case. The frequency of MM cells is only one half of that for MO cells, and hence some other mechanism, such as somatic recombination has to be considered.
57. Nakamura et al., 1987; Hakoda et al., 1987.
58. Kyoizumi et al., 1990.
59. Akiyama et al., 1991.
60. Yamazaki et al., 1954.
61. Note that "surviving" in this context meant "alive at the time of the examination of the children in 1950"; one child died at the age of 2.5 years and was not seen by these investigators. This child accounts for the discrepancy between the 16 mentioned as surviving and the 17 one would obtain by subtraction of the fetal, neonatal, and infant deaths (13) from 30 pregnancies.
62. Ban et al., 1990.
63. Ito et al., 1992.

Chapter 7

1. Yamazaki et al., 1954. This study, conducted in 1950, focused on the outcome of pregnancies of women who were pregnant at the time of exposure to the bombing of Nagasaki. Two groups of women were involved—98 who were exposed within 2 km of the hypocenter and 113 who were exposed at distances beyond 4000 m. The women in the former group were divided

into those who exhibited symptoms of acute radiation sickness (30) and those who did not (68). Among the 30 pregnancies to mothers with evidence of acute radiation sickness, 13 terminated in a nonviable offspring, and of those 17 infants who were born alive, 1 died of "dysentery" at 2½ years of age. Among the remaining 16 liveborn infants, 4 were found to be mentally retarded. Two other studies made shortly after the bombings, one in Nagasaki of 177 pregnant women, and one in Hiroshima of 45 pregnant women, suggested that 30% and 18%, respectively, lost their infants either through abortion or stillbirth.
2. See, for example, Walker, 1981; Watanabe, 1977.
3. More precisely, the ages at exposure were calculated as follows:

Days of pregnancy (ATB) = 280 − (Date of birth − 6 or 9 August 1945),

where the mean duration of pregnancy is taken to be 280 days, and the date of birth was obtained by interview with the individual or mother. The dates of the atomic bombings were 6 August in Hiroshima and 9 August in Nagasaki. To obtain the age after ovulation, 14 days have been subtracted from the "days of pregnancy ATB." Age in days was changed to age in weeks by dividing by 7.
4. Otake et al., 1987.
5. Miller, 1956; Miller and Blot, 1972; Miller and Mulvihill, 1976.
6. Otake and Schull, 1992.
7. Intelligence tests are usually so constructed that the distribution of test results follows an approximately symmetric curve, with some 95% of the population having scores falling within two standard deviations of the mean, commonly an IQ score of 100 (the standard deviation is customarily 12 to 15 points). Individuals whose scores lie, consistently, two standard deviations or more below the mean, that is who have IQ scores of less than 70, are described as retarded. Among the prenatally exposed in the Japanese studies, the highest score achieved by any of the severely mentally retarded children on the Koga test was 64, and most who were tested scored substantially lower.
8. Schull et al., 1988.
9. Beardsley et al., 1959.
10. See Memorandum for Record written by Jean Craig, 1957, to be found in the Archives of the Foundation.
11. See Memorandum for Record written by Jean Okumoto, 1957, to be found in the Archives of the Foundation.
12. Otake et al., 1988.
13. Dunn et al., 1988.
14. See the following references: Bergeron, 1967; Dunn et al., 1986; Layton, 1962; Mikhail and Mattar, 1978; Mueller, 1970.
15. Yoshimaru et al., 1993.
16. Strength is expressed in terms of the pressure, measured in kgs, exerted on the handle of the measuring instrument after adjustment of the grip to the size of the individual's hand.

The repetitive action test involves depressing a hand tally counter as rapidly as possible for a fixed interval of time. Tension on the counting lever

is measured, and the counter is gripped in the child's right palm. The counting lever is then pressed with the thumb while the other fingers are flexed. This is done as rapidly as possible at 5 second intervals separated by a 10 second rest period. After a trial followed by a test, two performances were scored. The procedure was repeated for the left hand. The hand of preference was, unfortunately, not routinely recorded, and since this affects the score to some degree, the average of the child's scores with the right and left hand was used to describe the performance, both for the grip and the repetitive action tests.
17. Since the two neuromuscular test scores and the four body size measurements were standardized, these standardized variables have means of 0 and standard deviations of 1. However, the estimated intercept in the multiple regression analyses is not necessarily 0, because the uterine absorbed dose was not standardized.
18. The justification for presuming that these findings reflect cerebral or cerebellar damage rests on the muscles required in the respective tests and the origin of the nerves that stimulate them. Depression of the counting lever in the repetitive action test involves primarily the stimulation of eight muscles in the thumb (Moore, 1980); whereas the squeezing action required in the measurement of grip strength embraces a larger number of muscles, including those of the digits and forearm. In both instances, the pathways of stimulation are through the brachial plexus, the spinal cord, and, ultimately, the region of the brain known as the motor cortex situated before the fissure separating the frontal from the temporal and parietal lobes. This knowledge alone does not account for what appears to be a stronger, indeed an almost two-fold greater, radiation-related effect on one of these measures of neuromuscular performance than on the other, assuming the apparent difference to be real. Possibly the most attractive explanation involves the relative number of neurons in the motor cortex required to effect fine motor control as opposed to activities requiring larger muscle masses. The ratio of the number of motor neurons, on average, supplying a muscle to the number of muscle fibers within the muscle is much smaller in the case of massive axial muscles supporting the torso than in the stimulation of the extraocular muscles (about 1 neuron to 1000 muscle fibers in the former instance, and 1 to 3 in the latter; Evarts, 1984). Thus, although the muscle mass involved in the grip test is larger than in the repetitive action of the thumb, it does not follow that the number of neurons involved in stimulation is also greater. Indeed, a disproportionate number of the neurons in the motor cortex are known to be allocated to the control of muscles involved in the most precise movements. Moreover, rapid, goal-oriented responses, such as the repetitive action test, in contrast to the grip, involve not only the motor cortex but also the cerebellum, the premotor cortex, and possibly other structures as well. Therefore, the seemingly greater sensitivity to radiation damage in the repetitive action test than in the grip strength test could reflect a larger population of neurons at risk of radiation damage. This is speculative, but it is reasonable to assume that the risk of damage is proportional to the target involved.
19. Rouma, 1919.

20. Goodenough, 1926.
21. Harris, 1963.
22. To obtain an intelligence quotient (IQ) the total score was transferred into a mental age by plotting the mean point score values made by children in successive year age groups and interpolating intermediate values. The IQ for a given child was then calculated taking the ratio of the mental age in months to the chronological age in months and multiplying by 100.
23. Bender, 1938, 1946. This test had its origins in work done by Wertheimer, who, in 1923, had constructed a series of experimental visual designs to explore "the conditions that explained how and why experiences were responded to in terms of primary configurations rather than in terms of successive steps" (Hutt, 1965). Bender selected nine of these so-called stimulus figures in constructing her test. Bender's original test involved some 30 mutually exclusive items. Scoring consisted of recording presence or absence of these items, and thus a child could have a composite score of 30.
24. The sensory cortex of the brain is located immediately posterior to the central sulcus or furrow, which divides each hemisphere into an anterior and posterior "half," and the motor cortex lies immediately in front of this sulcus.
25. Evarts, Shinoda, and Wise, 1984.
26. O'Rahilly and Muller, 1992.
27. National Center for Health Statistics, 1966, 1970, 1974, and 1977.
28. Pierce et al., 1989.
29. Nishimura, 1983.
30. Mole, 1982; see also Mole, 1990a,b.
31. Kato and Keehn, 1966.
32. See Konermann, 1989.
33. Kenneth Brizzee and his associates (1972) have examined cell recovery in the fetal brain of rats exposed to ^{60}Co radiation on gestation day 13 in single doses, ranging from 0.25 to 2 Gy in increments of 0.25 Gy, and in split doses of 1 Gy, followed 9 hours later by a second dose of 0.25 to 1.5 Gy, again in increments of 0.25 Gy. The animals were examined on day 19 of gestation. The incidence and severity of tissue alterations generally varied directly with dose and were clearly greater in single dose than in split dose groups with the same total exposure. This reduction in damage with the protraction of dose seems greater for continuous gamma-ray exposure than for serial, brief X-ray exposures, and it has been argued that this may indicate a further sparing when the protracted dose is evenly distributed over time. If true, this has important regulatory implications. Recently, Gabriela Vidal-Pergola and her coinvestigators reported results of fractionated prenatal doses on postnatal development in Sprague-Dawley derived rats (1993). Their experiment consisted of exposing pregnant females to single doses of 0.5 or 1.0 Gy, or to two doses of 0.5 Gy 6 hours apart. Offspring were subjected to four behavioral tests (negative geotaxis, reflex suspension, continuous corridor activity, and gait) on postnatal days 7–28. For all four behavioral end-points, the fractionated dose produced an effect intermediate between the 0.5 and 1.0 Gy doses and, by interpolation, could be ex-

pressed as equivalent to a single dose of about 0.7 Gy. Measurements of the various layers of the sensorimotor cortex as well as its total thickness in the dose-fractionated group revealed significantly less damage than was seen at a single dose of 1.0 Gy but more than that seen at a single dose of 0.5 Gy. Linear interpolation gave an estimated equivalent single dose of 0.7 Gy, a value similar to that seen for the behavioral end-points themselves. Finally, Reyners and his coworkers have examined the effects of protracted exposures to low doses of gamma rays on Wistar rats from day 12 to 16 postconception and found a significant reduction in brain weight at an accumulated dose as low as 160 mGy (1992). Thus all of these experimental studies suggest an effect of protracted exposures, but one that is less than is seen with acute irradiation. If these findings can be extrapolated to the human situation, they suggest that chronic exposure will have a lesser effect on the development of the human brain than instantaneous exposure.

34. Martinez Martinez, 1982; ICRP, 1986; Jacobson, 1991; O'Rahilly and Müller, 1992; Caviness, 1989; Sidman and Rakic, 1973, 1982.
35. One special variety of nerve cell, the granular or microneuron, is an exception, and does not achieve its maximum number, particularly in the cerebellum, until much later, around the time of birth.
36. Dobbing, 1981; Dobbing and Sands, 1971.
37. Rakic, 1972, 1975, 1978, 1985.
38. But see Walsh and Cepko, 1988, 1992.
39. Edelman, 1985.
40. Edelman, 1987.
41. Schwanzel-Fukuda and Pfaff, 1989; Wray et al., 1989. These neurons, known as LHRH cells, are morphologically unique and can be readily identified. They begin to migrate from their sites of origin in the placode through the olfactory bulb and into the cerebrum at about the 12th or 13th gestational day in the mouse, following a structure known as the terminal nerve. Once in the cerebrum they distribute themselves widely as single cells or as small clusters of cells. Most of the migration in the mouse is completed by the 16th gestational day. Similar events are known to occur in human beings, but at about the 12th through the 14th week following fertilization.
42. Rakic, 1988a,b.
43. Klose and Bentley, 1989; McConnell et al., 1989.
44. Galaburda and Kemper, 1979; Kemper, 1984; Geschwind and Galaburda, 1984.
45. Müller and O'Rahilly, 1983, 1984; O'Rahilly, 1983; O'Rahilly and Gardner, 1977; O'Rahilly and Müller, 1981.
46. Yokota et al., 1963; Neriishi and Matsumura, 1983. Heterotopic masses are collections of nerve cells in abnormal locations within the brain. They are due to arrested migration of immature neurons. They may be single, multiple, unilateral, bilateral, periventricular, or located deep within the white matter. The most common locations are subependymal, near the ventricles, or just below the cortex. They may be isolated, or associated with other anomalies in brain development, such as schizencephaly, a rare abnormality of the brain in which clefts extend across the cerebral hemispheres.

Individuals with isolated heterotopia can be clinically asymptomatic; however, when symptomatic, they often present with seizures in infancy or early childhood, but ectopic neuronal cells have also been associated with other neurological defects, such as the loss of vision in one half of the visual field in one or both eyes. They also have been seen accompanied by the partial or complete absence of the corpus callosum. Most heterotopias are probably microscopic, but if sufficiently large (0.5 cm or so), they can be readily visualized with either computed tomography or magnetic resonance imaging. The frequency of occurrence of isolated heterotopias, either asymptomatic or symptomatic, is not known.

47. The mamillary body can be seen in reconstructions of the embryonic brain at about 6 weeks after fertilization but it appears single at this time. Its double form can be recognized externally during the fetal period, beginning at about 12 weeks. The mamillary body is a part of the hypothalamus and hence can be assumed to be involved in autonomic regulation; however, it also has connections with the thalamus which suggests that it is a part of the limbic system, which is involved in emotions and motivation as well.
48. Schull et al., 1991.
49. Wisniewski et al., 1991.
50. Reyners et al., 1986
51. Brizzee and Ordy, 1986; Brizzee et al., 1980, 1982.
52. Driscoll et al., 1963.
53. Fushiki, Matsushita, and Schull, 1993. These authors have recently shown, using neocortical explant culture, an effect of ionizing radiation on neuronal migration at doses as low as 10 cGy together with a changing pattern of expression of the neural adhesion molecule, N-CAM.
54. Feinendegen et al., 1982, 1984.
55. Mountcastle, 1957.
56. Streissguth et al., 1978, 1980.
57. Clarren et al., 1978; Hammer et al., 1981.
58. Hicks et al., 1959; Hicks and D'Amato, 1966, 1980; D'Amato and Hicks, 1965.
59. Donoso and Norton, 1982.
60. Otake et al., 1993; for a similar analysis using the T65 dosimetry see Ishimaru et al., 1984.
61. Russell et al., 1972.
62. Miller et al., 1967.
63. Bloom et al., 1968; Ohtaki et al., 1991.
64. Duke-Elder, 1946, pages 65–74.
65. Lejeune et al., 1960; but see also Cheeseman and Walby, 1963.
66. Kawamoto et al., 1964.
67. Jablon and Kato, 1970; Kato, 1978.
68. Yoshimoto et al., 1988.
69. Blot et al., 1975.
70. Blot et al., 1972.
71. Jaenke and Angleton, 1990.
72. Freedman and Keehn, 1966.
73. See Yaar et al., 1979, 1980, 1983; Ron et al., 1982; Shore et al., 1976.

74. See, e.g., Bakwin, 1968; Rutter et al., 1978; Stott, 1968; Touwen and Prechtl, 1970.
75. Doty et al., 1984.
76. Sienkiewicz et al., 1992.

Chapter 8

1. Neel and Schull, 1956; see also Neel and Schull, 1991.
2. Ad hoc Committee Report, 10–11 July 1953.
3. Kato and Schull, 1960.
4. Schull and Neel, 1959.
5. Otake et al., 1989, 1990.
6. Neel et al., 1990. The argument can be stated more formally as follows: Let M_i be the frequency of mutation at dose i, and M_0 be the frequency at dose 0. Then the relative risk at dose i is, by definition:

   $$RR_i = M_i/M_0$$

   Now the expected frequency of mutations at dose i, assuming mutations increase linearly with dose, is

   $$M_i = M_0 + BD_i$$

   where BD_i is the increment of mutations added to the frequency at dose 0 through exposure to dose i. This equation can be recouched in terms of relative risk at dose i by dividing all terms by M_0; the result is

   $$RR_i = 1 + (B/M_0)D_i.$$

   Recall that the "doubling dose" is, by definition, that dose at which the increment of new mutations added is precisely equal to the number arising spontaneously, that is

   $$M_0 = BD_d$$

   or

   $$D_d = M_0/B.$$

   Thus the doubling dose is merely the reciprocal of the coefficient of D_i in the equation for the relative risk given above.
   To estimate the lower bound of the doubling dose we argue as follows. The large sample variance of a ratio, such as M_0/B, can be written, from general statistical theory, as

   $$V(M_0/B) = (M_0/B)^2[V(B)/B^2 + V(M_0)/M_0^2 - 2\,\mathrm{Cov}(B,M_0)/BM_0].$$

   To obtain the 95% lower bound we note that

   $$LB(D_d) = (M_0/B) - t_{0.05}V(M_0/B)$$

   where $t_{0.05}$ is the normal deviation, 1.96.
7. Yoshimoto et al., 1991.
8. Yoshimoto et al., 1988, 1989.

9. Gardner et al., 1990.
10. Little, 1991.
11. Little, 1990; Little, 1993; see also Abrahamson, 1990.
12. See for example, Jablon, Hrubec, and Boice, 1991.
13. Jones and Wheater, 1989.
14. Schull and Neel, 1962.
15. Slavin et al., 1966.
16. Omori et al., 1965.
17. Awa et al., 1988.
18. Henry G. Kunkel at the Rockefeller Institute had employed starch for similar purposes earlier, but he used starch blocks made from a slurry and this proved to be a less suitable medium for separation of the proteins than the gel Smithies pioneered.
19. Neel et al., 1988.
20. Satoh et al., 1983; see also Neel et al., 1988.
21. This number is obtained by multiplying the mutation rate per codon times the number of codons per gene times the number of genes studied (there will be two for each enzyme) times the number of individuals involved.
22. Furusho and Otake, 1978a,b, 1979, 1980, 1985.
23. Neel and Lewis, 1990; see this article for references to the experimental studies of mutations arising in the mouse following exposure to ionizing radiation.
24. Given the uncertainties and the inconclusiveness of the findings, some historians of science have questioned how reassuring to the public the genetic studies are (see, e.g., Beatty, 1991). Two observations in this regard seem warranted. First, it should be noted that the uncertainties revolve largely around the utilization of these data to estimate rates of induced mutation, and here the findings can be seen as "inconclusive." However, since the uncertainties do not involve the accuracy of the observations, the latter are certainly not irrelevant to public health concerns. Second, whereas the general public may have envisaged an epidemic of misshapen monsters in Hiroshima and Nagasaki, genetic considerations, even in 1947, made this unlikely. Given the small average gonadal dose received by the survivors, prevailing knowledge predicted the effects to be small, probably at the very limit, if not below the ability to discern them. These predictions rested largely on the genetic effects seen in the exposure of species other than man, and their pertinence to the human case required confirmation. But these observations bear on the broader public health issue of whether malformations, irrespective of their biological basis, are increased following preconception exposure to ionizing radiation. It is in this respect that the public should find the observations reassuring, since there has been no conspicuous increase in congenital defect among the children of the survivors.
25. See Abramson, 1971, for a review of the older literature, and Wilcox et al., 1988, for the results of more recent studies.
26. This process of establishing permanent cell lines involves the use of a virus, known as the Epstein–Barr virus, to transform the cell, and is sometimes referred to as "immortalizing," since the transformed cells will divide and grow indefinitely.

Chapter 9

1. Note that in terms of the excess relative risk at dose d, say ERR(d), the attributable risk, in percent, is ERR(d) divided by [1 + ERR(d)] times 100. In particular, for leukemia, this implies an attributable risk at age 16 that is in excess of the 59%, which is actually the percent seen among all exposed survivors in the Life Span Study whose average dose was 0.25 Sv.

Chapter 10

1. IAEA, 1989.
2. Chelyabinsk.
3. Ciudad Juarez.
4. IAEA, 1988.
5. IAEA, 1989.
6. Chang and Kau, 1993.
7. Kneale and Stewart, 1988, but see also Little and Charles, 1990.
8. Wing et al., 1991.
9. Gilbert et al., 1989.
10. Kendall et al., 1992; see also Little et al., 1993.
11. Muirhead and Darby, 1987a,b.
12. Committee on the Biological Effects of Ionizing Radiation, 1990.
13. Storer et al., 1988.
14. United Nations Scientific Committee on the Effects of Atomic Radiation, 1986, 1988.
15. Recently, the National Council on Radiation Protection and Measurements (1993) has extended this comparison to five populations, that is, Japan, the United States, Puerto Rico, the United Kingdom, and China. The results are similar to those obtained by UNSCEAR, namely, the transfer of the Japanese risks to these populations leads to a total probability of a fatal radiogenic cancer that varies by less than a factor of two, although, again, individual sites can and do vary.
16. See, e.g., Holm et al., 1988.
17. These two accidents clearly revealed the lack of preparedness of the responsible governments, and the failure to inform the public adequately as to what would be required of citizens in the event of an accident. Fortunately, at Three Mile Island, the design of the reactor and the nature of the accident limited its potential destructiveness, but this was not so at Chernobyl. Here the crude design of the reactor and a dissembling government led to a catastrophe whose effects are likely to be experienced for decades. The human dimension of this accident can be best illustrated through the words of one of the liquidators, the individuals involved in the removal of radioactive materials at the Chernobyl Nuclear Power Plant, who has written

> I was called up by the VOENCOMAT (Military enlistment office) from reserves on July 21, 1986. Our regiment was housed in tents in the 30 kilometer zone, and we were living 45 people in 25-bed tents. We were working on removing the radioactive waste on the south side of the ruined fourth nuclear reactor. The work span was a few minutes

a day, since the dosage of radioactive irradiation had to be not more than two roentgens. In case of a bigger dosage, the commanding officer had to write an explanation.

In the second half of August 1986, having accomplished the cleaning and blocking of the accesses to the reactor, we began to clean-up the area of the fifth substation with extremely high levels of radiation. We were working in our overcoats, which were given to us in Leningrad. The same overcoats were used as blankets in the tents. The only protective clothing that was used was a respirator, which could not always be changed, and by the way, it did not turn maroon due to the penetration into the body of radioactive particles of cesium. Our unprotected eyes were watering, and our throats were swelling; we were coughing. The doctors tried to convince us that this condition is temporary, that with time our body will get used to it. Then cases of bloody diarrhea began and three men from our regiment were sent to Kiev suspected to have dystenery. In a few days they were back with the explanation that they had a simple diarrhea.

To try to explain to incompetent people (we were under the command of reserve officers who never served in chemical war regiments) that it is a crime to let people work and sleep in the same clothing when they work with radioactive materials was senseless. We were under martial law, where military tribunals and the Special Department (KGB) was in force.

18. A description of some of these "passively safe" reactors will be found in Ahearne, 1993. This article also presents one reasoned view of the future of nuclear power in the United States. Ahearne, a former chairman of the U.S. Nuclear Regulatory Commission concludes that "America will choose nuclear power only if demand for electricity accelerates, nuclear costs are contained, and global-warming worries grow."
19. Fischhoff et al., 1978; see also Slovic, 1979.
20. In the years 1969 through 1990, some 600,000 workers were employed in one aspect or another of the nuclear industry. Of these, 120,000 would normally be expected to die of cancer based on current death rates in the United States. It has been estimated that occupational exposure to ionizing radiation will add possibly 115 cancers deaths to this number, but it might add none. The estimate of 115 rests on assumptions that have never been rigorously tested, and may, indeed, be untestable. As repeatedly noted, the most important of these assumptions is that the response to exposure is linear with dose even at very low doses and dose rates. Some biologists now question the plausibility of this assumption, largely on the basis of the emerging complexity of the events that culminate in a malignancy. It is argued that even if a somatic mutation is always the initiating event, and the probability of such a mutation is linear with dose, it does not follow that the frequency of cancer will also be linear with dose.
21. Barny et al., 1990.
22. Edelson, 1985.
23. The League of Women Voters Education Fund, 1985.
24. Swinbanks, 1991.

Glossary

ABCC (Atomic Bomb Casualty Commission)—the agency of the Japanese Ministry of Health and Welfare and the U.S. National Academy of Sciences charged with the responsibility for the study of the survivors of the atomic bombings from the inception of the investigations in 1947 until 1975.

abscopal—the effect on nonirradiated tissue of the irradiation of other tissues of an organism.

absorbed dose—see rad and gray.

achlorhydria—the absence of hydrochloric acid, a normal constituent, from the secretions of the stomach.

Adult Health Study—the program of biennial clinical examinations of the survivors of the atomic bombings of Hiroshima and Nagasaki to ascertain their health status. This program involves a fixed sample of what was initially more than 20,000 individuals.

allele—one of two or more alternate forms of a gene occupying the same locus or position on a particular chromosome.

alpha particle—a charged particle emitted from the nucleus of an atom having the same mass and charge as that of a helium nucleus stripped of its electrons.

anencephaly—a fatal developmental defect in which the cerebrum and cerebellum fail to develop properly and the bony covering of the brain does not develop.

aneuploidy—the occurrence in a cell, tissue, or individual of one, two, or a few more (or less) chromosomes than the normal complement of 46.

antibody—one of the immunoglobulins of the body produced in response to a specific antigen that is instrumental in counteracting the antigen.

antigen—a substance (often a protein), capable of stimulating an immune response.

ATB—an abbreviation of the expression "at the time of the bombings." It is most commonly used in connection with the age of the survivors at the time of exposure.

atherosclerosis—a common form of arteriosclerosis characterized by the deposition in the innermost layers of the large and medium-sized arteries of the body of yellowish plaques that contain cholesterol and other fatty materials.

autosome—a chromosome that is distinct from the sex chromosomes.

axon—a nerve fiber that is continuous with the body of a nerve cell and is the essential stimulus-conducting portion of the cell. It consists of a series of neurofibrils surrounded by a well-defined sheath, known as the axolemma; the latter is in turn encased in myelin, a mixture of lipids, and a final sheath, the neurolemma.

B-cell—one of a group of white cells that have their origin and maturation in the bone marrow.

Becquerel (Bq)—the amount of radioactivity quantified in terms of the number of disintegrations that occur per second, not by gray or sievert. The older unit for radioactivity was the Curie (Ci), equal to 3.7×10^{10} disintegrations per second; the Becquerel, introduced with the Systeme Internationale, is equal to 1 disintegration per second. Thus, 1 Ci is equal to 3.7×10^{10} Bq.

BEIR—the National Academy of Sciences' Committee on the Biological Effects of Ionizing Radiations.

brain—the mass of nervous tissue that lies within the cranium, or skull, consisting of a number of discrete parts such as the brain stem, the cerebrum, and the cerebellum.

brain stem—the portion of the brain that connects the cerebrum with the spinal column. It consists of three parts: the medulla, the pons, and the midbrain. The medulla is the expanded inch or so of tissue where the spinal cord enters the brainstem. This region controls such functions as swallowing, vomiting, breathing, talking, singing, respiration, and the control of blood pressure. The pons, which means bridge, lies between the medulla and the midbrain and contains a band of fibers that connect the cerebrum with the cerebellum. The midbrain sits between the pons and the cerebrum.

bulb, olfactory—the grayish expanded anterior extremity of the olfactory tract, lying on a sieve-like plate above the ethmoid bone and receiving the olfactory nerves.

carcinoembryonic antigen (CEA)—this antigen, which is measurable in serum, has been used as a marker for a variety of malignancies. Among these are cancers of the breast, lung, colon, and gallbladder. Initially, it was believed that an elevation of this antigen was symptomatic of cancer, but it is now known that it can be elevated due to other causes, such as smoking and liver cirrhosis. However, it is still thought that a continuing rise in its serum level can contribute to the diagnosis of some cancers.

carcinoma—a malignant growth involving those cells that cover the inner and outer surfaces of the body, the epithelium.

cataract—a loss of the translucency of the crystalline lens of the eye; a better expression in the context of radiation-related lenticular changes is opacity.

centromere—a specialized region of the chromosome where the two chromosomal arms meet; the centromere is also sometimes called the kinetochore or the spindle attachment point.

cerebellum—the posterior portion of the brain, consisting of two hemispheres connected by a narrow mass of tissue known as the vermis.

cerebrum—the largest portion of the brain; includes the cerebral hemispheres. In its earlier stages of development, the cerebrum is characterized by four major layers: the mantle (the outermost margin of the brain), the cortical plate, the intermediate or migratory zone, and the matrix or proliferative zone.

chonaikai—the system of neighborhood groups, mandatory in urban prewar Japan, used to disseminate information, stifle dissent, and mobilize the population. These groups were summarily disbanded by the postwar occupying authorities.

chromatid—one of the two daughters of a chromosome produced by replication of the DNA; the chromatids separate during cell division, one going to each of the two cells produced by the division.

chromosome—one of several small, darkly staining bodies that appear in the nucleus of a cell at the time of cell division; they contain the genes and are normally constant in number within a species. The normal number in human beings is 46—22 pairs of autosomes and two sex chromosomes (either XX or XY). One member of each pair is derived from each of the parents of an individual.

CI (confidence interval)—an estimate of the range within which the true value of a parameter is expected to occur in some predetermined proportion of trials. Commonly, this proportion is selected to be 90 or 95%.

cisterna magna—the largest of the structures (cisterns) involved in the recirculation of cerebrospinal fluid; it lies immediately behind and between the two lobes of the cerebellum.

correlation—the degree of association between two variables, such as height and weight. It is customarily measured by a unitless coefficient, known as the correlation coefficient, that varies between -1 and $+1$. A correlation coefficient of 0 implies no association, whereas a value of -1 or $+1$ implies a perfect association, one variable increasing while the other decreases in the first instance, and both increasing in the second.

Curie—see Becquerel.

deletion—a chromosomal aberration arising from the loss of a portion of a chromosome. If the lost segment is at the end of the chromosome, the deletion is described as terminal; whereas if it is in a mid-portion of the chromosome, it is said to be interstitial.

dendrite—one of the branching protoplasmic processes of the nerve cell.

differentiation—the process by which cells (or tissues) acquire characteristic attributes.

DNA—deoxyribonucleic acid, the complex carrier of the genetic information we inherit.

dose—the amount of radiation or energy absorbed by a tissue or another substance of interest. See gray.

dose equivalent—a quantity that expresses an equal biological effectiveness of a given absorbed dose on a common scale for all kinds of ionizing radiation. The dose equivalent is now most commonly expressed in sieverts (Sv), but has, in the past, been stated in rem. See rem and sievert.

dose–response curve (or relationship)—the mathematical representation of the rate of increase in a radiation-related effect with increasing dose. The most commonly used curves or models assume the increase to be either linear, linear-quadratic, or quadratic with dose.

"doubling dose"—in the context of exposure to ionizing radiation the doubling dose is that amount of irradiation that will produce precisely as many induced mutations as spontaneously occur in the population of interest in each generation.

DRREF—dose rate reduction effectiveness factor, a measure of the extent to which radiation-related damage accruing at a high dose rate is ameliorated when the dose rate is low. This value will presumably vary with the endpoint measured, but it is not known precisely for such endpoints as incidence of death due to cancer, brain damage, or any of the other effects seen among the survivors. Experimental studies suggest a value between 5 and 20, but the epidemiological data imply a lower number, possibly 1 to 2.

DS86—the currently employed system of dosimetry used to describe the exposure of the survivors of the atomic bombings of Hiroshima and Nagasaki; introduced in 1986, this system is fully described in Roesch (1987).

dyslexia—the severe, uncorrectable inability to recognize certain combinations of letters or words encountered in otherwise intelligent individuals. It appears to be due to maldevelopment of specific cortical areas of the brain.

encephalomeningocele—a developmental anomaly in which portions of the brain protrude through a defect in the skull.

ependyma—the lining of the ventricles of the brain and the central canal of the spinal cord. The expression subependymal refers to the area of the brain immediately beneath the lining.

epicenter or **"burst point"**—the position in air where the actual detonation of an atomic weapon occurred, as opposed to the hypocenter on the ground (see hypocenter).

epilation—loss of the hair of the body, but particularly the hair on the head.

erg—the amount of energy required to move 1 gram of free mass through a distance of 1 centimeter at the rate of 1 centimeter per second per second.

erythrocytes—the red blood cells of the body; responsible for the transportation of oxygen.

exposure—the amount of air ionized by radiant energy, specifically, the amount of electrical charge produced in 1 cc of air under conditions of electron equilibrium.

fallout—radioactive debris from the detonation of a nuclear weapon or other source (e.g., a nuclear reactor); the term usually implies deposited airborne particles.

ferritin—an iron–protein complex, measurable in serum, involved in the storage of iron in the body. It has been implicated in the metabolism of iron in the gastrointestinal tract.

fluence (or flux)—the number of photons (or neutrons, etc.) per unit area.

"fragile site"—one of many locations on human chromosomes that are especially liable to "spontaneous" breakage.

Francis Committee—a committee of three, under the chairmanship of Thomas Francis Jr., a viral epidemiologist, sent to Japan by the National Academy of Sciences—National Research Council in the autumn of 1955 to evaluate the ABCC scientific program and recommend changes, if these seemed warranted.

gap junction—the cells of a multicellular organism are able to influence the activities of one another through a variety of processes. Collectively, these processes are referred to as intercellular communication. Adjacent cells, for example, can transport small molecules back and forth through their cell walls by means of a specific channel, known as the gap junction. The capacity to do this, however, is influenced or modulated by a number of factors, such as the calcium levels within the cells. Regulation of this communication through gap junctions is presumed to play a role in the occurrence of cancer.

ganglion (plural: **ganglia**)—a term generally used to describe a knot or knot-like collection of nerve cell bodies outside the central nervous system.

genome—the genome of an organism is its complete sequence of genetic information, that is, the sequence of all of the bases on all of the chromosomes.

gingivitis—an inflammation of the tissue, the gingiva, surrounding the teeth. Gingivitis is said to be necrotic when the initial inflammation gives way to open sores.

glycophorin A—a gene involved in the specification of a protein whose structure determines two common red blood cell antigens, known as M and N.

gray—a unit of absorbed dose, equal to 100 rad, named after the English biophysicist, Louis Harold Gray. One gray equals one joule per kilogram.

half-life, biological—the time required for the elimination of half of a substance, such as a radioactive isotope, from the body.

half-life, radioactive—the time required for a radioactive substance to lose half of its activity through decay.

heterozygosity—having different alleles at one or more genetic loci.

HLA—see MHC

homologous—implying a correspondence in structure between different parts of the same individual, such as two chromosomes within a cell.

homozygosity—having identical alleles at a particular genetic locus.

honseki—one's legal address; more specifically the address under which one's family or household register is to be found.

HPRT—an abbreviation for the enzyme known as hypoxanthine-guanine phosphoribosyl transferase, which plays a controlling role in the synthesis of purines. An alternative abbreviation for this enzyme is HGPRT.

hypertension—a persisting abnormal elevation of the systolic or diastolic blood pressure. Conventionally, a systolic blood pressure in excess of 160 mm of mercury, or a diastolic pressure above 95 mm, or both is viewed as abnormal.

hypocenter—the point on the ground immediately beneath the point of detonation of an atomic bomb.

hypoxemia—a lessening of the oxygen in circulating blood; a tissue deprivation of oxygen.

IAEA—the International Atomic Energy Agency, one of the specialized bodies of the United Nations charged with the responsibility of overseeing and setting standards and recommendations for the operation of nuclear activities by the member states. It is headquartered in Vienna, and has played a major role in the accumulation and dissemination of the information derived from the Chernobyl accident as well as other accidents involving exposure to ionizing radiation.

ICRP—the International Commission on Radiological Protection, a nongovernmental agency headquartered in the United Kingdom, concerned with radiation protection in the workplace and of the general population. It is generally viewed as the world's leading source of authoritative statements on radiation protection.

immunoglobulin—a protein, commonly found in the serum, composed of chains of different weights; all known human antibodies are immunoglobulins.

incidence—the number of new cases of a specific disease occurring during a certain period of time. See prevalence.

initiation—the irreversible conversion of a stem cell to one that is incapable of terminal differentiation. It is now generally believed that this irreversible conversion involves the mutation of one (possibly more) of a number of specific genes.

inversion—a rearrangement of a portion of a chromosome involving a 180° rotation of the inverted part. A pericentric inversion is one in which the rotated part includes the centromere.

in utero—in the womb, that is, before birth; it is used synonymously with prenatal.

Joint Commission—actually, the "Joint Commission for the Investigation of the Effects of the Atomic Bomb in Japan." This was the collection of investigative teams from the United States Army and Navy, the Manhattan District, and the Japanese government, under the direction of Col. Ashley Oughterson, whose purpose was to provide a unified control of the various groups of investigators interested in the study of the medical and physical effects of the atomic bombings.

joule—a unit of energy equal to 10^7 ergs. The average 60 watt light bulb consumes about 60 joules per second. This unit derives its name from the 19th century English physicist, James Prescott Joule.

kerma—an acronym derived from the expression "kinetic energy released in materials"; these materials could be air, a body organ, or a structural component of a building. Kerma is expressed in gray.

kerma in free air—energy released in a small tissue sample; as used here, the amount of radiation in air. Account is taken of the distance from the hypocenter, the absorption and scattering of energy by the ground and the air intervening between the burst point and the location at which kerma is expressed, but not the effects due to shielding by buildings or other structures (see shielded kerma below).

koseki—the name of the family or household register required of Japanese citizens by law. These registers identify all of the events—marriages, births, deaths, adoptions, etc.—that affect the composition of a family. These records are maintained by the Japanese Ministry of Justice.

$LD_{50/60}$ or median lethal dose—the dose of radiation required to kill 50% of individuals within 60 days of exposure. The $LD_{95/60}$ is the dose at which 95% of individuals will die within 60 days of exposure.

LET (linear energy transfer)—the average amount of energy lost per unit of charged particle track length. Low LET, implying little loss of energy per unit of particle track length, is characteristic of high energy electrons from X-rays and gamma rays. High LET is characteristic of alpha particles and the highly energetic protons produced by neutron collisions in tissue.

leukemia—the term used to describe a group of malignant, commonly fatal blood diseases with certain common findings, notably a progressive anemia, internal bleeding, exhaustion, and a marked increase in the number of white cells (generally their immature forms) in the circulating blood.

Life Span Study—the program of mortality surveillance among the survivors of the atomic bombings of Hiroshima and Nagasaki. This study, like the one of morbidity, involves a fixed sample of survivors and comparison persons. The current study sample numbers about 120,000 individuals.

locus—the physical position of a gene in the genome.

mamillary bodies—two protuberances on the undersurface of the brain beneath the corpus callosum that are a part of the brain known as the hypothalamus.

metaplasia—the abnormal transformation of an adult, fully differentiated tissue of one kind into a differentiated tissue of another kind.

MHC (major histocompatibility complex)—a region on human chromosome 8 that determines the specificity of a group of antigens concerned with tissue compatibility.

mutation—a heritable or transmissible change in the genetic material (DNA) of an organism. Mutations arise spontaneously in all organisms, and can be induced by exposure to a variety of physical and chemical agents, such as ionizing radiation. On average, most mutations are deleterious to the organism. They can be classified in a variety of ways. One common way is to divide them into "lethals," those that are incompatible with survival; "visibles," those that produce a discernible change, usually physical, in the organism; and "detrimentals," those that produce a smaller, quantitative effect on survival or reproductivity.

mutation, somatic—a transmissible change in the DNA of a cell destined to give rise to cells of the body other than the germ cells.

NCRP—the National Council on Radiation Protection and Measurements, a nongovernmental agency based in Bethesda, Maryland, with a charter similar to that of ICRP, but focusing in particular on issues related to radiation protection in the United States.

neuroglia (or often just **glia**)—the nonnervous cellular components of the nervous system that provide support to the developing structures of the nervous system and perform important metabolic functions.

There are a variety of different glial cells, distinguishable by their morphology and function. One especially important group in the early development of the brain is the transitory set, known as radial glia cells, that guide the neurons as they migrate from the proliferative zones to their site of function.

neuron—a nerve cell, consisting of the cell body and its various processes, the dendrites, axons, and endings.

neuropil—a collective term describing the network of neuroglia, axons, and dendrites and their synapses in the brain.

oncogene—the term used to describe a mutated or deregulated proto-oncogene.

oropharyngeal lesions—a general term used to describe sores developing in the mouth or throat.

petechiae—minute hemorrhagic spots in the skin, usually no bigger than a pinhead.

phenotype—the observable properties of an organism produced by the interaction of the organism's genetic potential and the environment in which it finds itself.

placode, olfactory—a thickening of embryonic cells lying in the bottom of the olfactory pit as the pits are deepened by the growth of the surrounding nasal processes.

prevalence—the number of cases of a specific disease existing in a particular population or area at a certain time. See incidence.

progression—the process wherein a clonally expanded, initiated cell acquires those other attributes necessary for a malignancy to develop.

proliferative zone—the areas adjacent to the lumen of the primitive neural tube where the cerebral neurons arise.

promotion—in cancer biology, the process of expansion of an initiated cell, that is, one no longer capable of terminal differentiation.

proto-oncogene—one of a group of genes, some of which seem to be of viral origin, that play an important role in the origin of cancer. Normally, they are quiescent in the sense that they only reproduce at those times when the chromosome of which they are a part reproduces. They are regulated or controlled, but in some cancers they become activated.

Purkinje cell—a large nerve cell of the cerebellar cortex; these cells are the only ones in the cerebellum that carry nerve impulses out of the cerebellum itself.

purpura—the hemorrhaging into the skin of small blood vessels; it usually first appears as a red spot of varying dimension that becomes gradually darker, then purple, and eventually may fade to a brownish yellow.

PYGy—this is an abbreviation for the expression person-year-gray. It describes the number of years of follow-up of an individual multiplied by the dose obtained expressed in gray (Gy). A comparable expression used until recently is PYR, which stands for person-year-rad. When PYGy or PYR are aggregated over a number of individuals the former is usually stated in units of 10,000 and the latter in 1,000,000. Since a gray is equal to 100 rad, 1,000,000 PYR is equal to 10,000 PYGy.

PYR—an abbreviation for person-year-rad. See PYGy for further explanation.

PYSv—an abbreviation for person-year-sievert, this expression is useful when exposure involves several different qualities of ionizing radiation, as in the case of the atomic bomb survivors who were exposed to both gamma rays and neutrons.

quality factor—an administrative factor, dependent upon the linear energy transfer, by which absorbed doses are multiplied to obtain a quantity that expresses the effectiveness for radiation protection purposes of an absorbed dose on a common scale for all forms of radiation. See also RBE.

rad—a unit of absorbed dose. It is equivalent to 100 ergs of energy per gram. In the more recent radiobiological literature the rad has been replaced by the gray (1 Gy = 100 rad). See also gray.

radiation sickness, acute—the concatenation of signs or symptoms, including fever, nausea, vomiting, lack of appetite, bloody diarrhea, loss of hair, bleeding under the skin, sores in the throat and mouth, and decay and ulceration of the gums about the teeth, that can follow whole-body exposure to ionizing radiation. The onset of these symptoms can vary; usually the higher the dose, the more rapid the onset. It is generally assumed that a dose of about 1 Gy is required to produce these symptoms.

radon—a heavy radioactive gas that is formed by the disintegration of radium.

RBE (relative biological effectiveness)—a factor used to compare the biological effectiveness of absorbed radiation doses due to different types of ionizing radiation for a defined biological endpoint, e.g., the $LD_{50/60}$; this factor is experimentally determined using X- or gamma rays as the standard of comparison. Thus, if 1 Gy of fast neutrons produced the same amount of cell killing as 5 Gy of gamma rays, the RBE of neutrons for cell killing would be 5. Note the RBE will vary with the biologic endpoint used.

regression—in the present context, this word is used to imply the functional relationship between two or more variables. Regression analysis provides a means to predict values of one variable when given the values of other variables. For example, it can provide the basis for estimating the frequency (or number) of mentally retarded individuals at a given dose of ionizing radiation. The regression coefficient, in the simplest case, is equivalent to the angle of the straight line describing the rate of increase in one variable as a function of another.

rem—a unit of dose equivalent. The dose equivalent in rem is equal to the absorbed dose in rad multiplied by the quality factor, the distribution factor, and any other necessary modifying factors. See rad, sievert, and dose equivalent.

RERF (Radiation Effects Research Foundation)—the nonprofit agency sponsored by the governments of Japan and the United States that currently supervises the studies of the atomic bomb survivors; the successor in 1975 of the Atomic Bomb Casualty Commission.

risk, absolute—the excess number of deaths (or cases) above that "normally" expected in some population in the absence of exposure to ionizing radiation beyond that to which everyone is subjected because of the radiation emanating from the earth's crust or originating in outer space.

risk, attributable—the percentage of deaths or cases ostensibly assignable to a specific cause, in this instance, ionizing radiation.

risk, relative—the ratio of the risk in one population to that in another; for example, the ratio of the risk among individuals exposed to 2 Gy as contrasted with the background risk.

roentgen—the international unit of measurement of gamma or X-irradiation. It is based upon the number of ionizations produced by X-rays or gamma rays in a standard mass of air, and unlike the rad, it is not a measure of absorbed energy. It is named after Wilhelm Roentgen, the discoverer of X-rays.

sarcoma—a malignant growth involving the tissues, such as muscle, that bind together and support the various structures of the body.

SCAP—abbreviation for the Supreme Commander of the Allied Powers, the title borne by General Douglas MacArthur as the head of the allied forces occupying Japan at the conclusion of World War II. It was commonly also used to designate the various administrative offices under his command.

shielded kerma—the amount of tissue kerma within a structure, such as a Japanese home, at a given distance from the hypocenter.

sievert—a unit of dose equivalent, abbreviated Sv. It is equal to the dose in gray times a quality factor times any other factors that may modify the dose. The name is derived from the Swedish physicist, Rolf M. Sievert.

somatosensory cortex—that portion of the cortex of the brain that is involved in the processing of sensory stimuli that arise in the body outside of the brain.

standard deviation—see variance.

stem cell—those cells capable of self-renewal and which give rise to a differentiated cell type. As the fertilized egg develops, some of its undifferentiated cells are "committed" to be stem cells in the formation of specific tissues. Thus, for example, a hematopoietic stem cell is one that can give rise to other hematopoietic stem cells, or differentiated hematopoietic cells such as the white cells.

synaptogenesis—the process or processes that culminate in the formation of synapses—the places where a nerve impulse is transmitted from one neuron to another.

T-cell—one of a group of lymphocytes that early in life arise in the thymus, but later in life arise in lymphatic tissue elsewhere in the body, such as in the tonsils, spleen, and lymph nodes.

T-cell receptor gene—a gene that determines the nature of a specific site of immunological activity on the surface of the T lymphocytes.

T57—a tentative system of dosimetry introduced in 1957 based on an assessment of the yield and spectrum of radiation by York, but published by Robert Wilson in 1957. This system saw very little application in the study of the effects of exposure in Hiroshima or Nagasaki.

T65—a tentative system of dosimetry introduced in 1965, described in a technical report in 1968 by Roy C. Milton and Takao Shohoji based on an extensive experimental assessment of shielding transmission factors and the yield and nature of the radiation associated with the two atomic bombs detonated in August 1945.

thymus—a gland-like body situated in the chest that reaches its maximum development during the early years of childhood and then slowly involutes (regresses). The thymus itself is apparently essential to the establishment of a normal immune system, largely through its role in the development of lymphoid tissue.

tonari-gumi—see chonaikai.

transfection—as commonly used, this implies the incorporation of DNA from one species into that of another—most frequently, perhaps, it is used to describe the incorporation of viral or bacterial DNA into a higher organism such as the mouse or the human.

translocation—a chromosomal aberration arising from chromosome breakage and the subsequent rearrangement of the broken segments between the same or different chromosomes. A reciprocal translocation is one involving two chromosomes with the mutual exchange of the broken part of one with the broken part of the other. Robertsonian translocations arise through the fusion of the centromeres of two chromosomes; they result in a reduction in the chromosome number without loss of genetic material.

tumor suppressor gene—one of a family of genes that normally work to suppress or inhibit the multiplication of a cell; they are the brakes on the engine of cellular reproduction. Loss of these genes through mutation seemingly releases the normal constraints placed on a cell's capacity to divide, and, when accompanied by the activation of a proto-oncogene, results in increased cell multiplication and those other abnormalities characteristic of a tumor cell.

UNSCEAR—the United Nations Scientific Committee on the Effects of Atomic Radiation, one of the specialized bodies of the United Nations charged with the responsibility of evaluating the effects of exposure to atomic (ionizing) radiation on behalf of the member nations.

variance—a measure of the spread or dispersion of a variable about its mean value. It is by definition the mean of the squared deviations of the values of the variable about their mean. The standard deviation is the square root of the variance.

ventricles—as used here, the cavities within the brain derived from the lumen, or open canal, of the primitive neural tube. The term ventricular is used to describe the layer in the brain immediately adjacent to the ventricles.

X-rays—high energy photons, historically the rays produced in an electrical device, e.g., such as a diagnostic X-ray machine. Later on it was shown that these rays are identical to gamma rays, both being high-energy (shortwave-length) photons.

References

The literature is far too voluminous to make the list of references that follows exhaustive. More than 800 technical reports describing one aspect or another of the Japanese studies have been published by the Atomic Bomb Casualty Commission and its successor, the Radiation Effects Research Foundation. Emphasis has been placed, therefore, either on the most recent publications, many of which provide bibliographic references to earlier work, or on review articles summarizing various aspects of our knowledge. However, a bibliographic citation for all of the references to which attention is called in the text and end notes will be found in the following list. As a general rule, technical reports have been cited because these are the most complete publications, but where a journal article derived from these reports has been published, since the journal may be accessible to more readers, the appropriate reference will be found in brackets following the citation to the technical report.

Abelson, Philip H. and Kruger, Peter G.: Cyclotron induced cataracts. Science 110: 655–657, 1949.

Abrahamson, Seymour: Commentary: Childhood leukemia at Sellafield. Radiation Research 123: 237–238, 1990.

Abramson, Frederic D.: Spontaneous fetal death in man: A methodological and analytical evaluation. Ph. D. Dissertation. Ann Arbor, Michigan: University of Michigan, 1971.

Ahearne, John F.: The future of nuclear power. American Scientist 81: 24–35, 1993.

Akiba, Suminori; Neriishi, Kazuo; Blot, William J.; Kabuto, Michinori; Stevens, Richard G.; Kato, Hiroo; and Land, Charles E.: Serum ferritin and stomach cancer risk among a Japanese population. Cancer 67: 1707–1712, 1991.

Akiyama, Mitoshi; Kyoizumi, Seishi; Kushiro, Junichi; Kusunoki, Yoichiro; Hirai, Yuko; and Nakamura, Nori: Development of the assay systems for the detection of somatic mutation in radiation-exposed people by means of flow cytometry. In: *Flow Cytometry and Image Analysis for Clinical Applications.* Nishiya, I.; Cram, L. S.; and Gray, J. W. (eds.) New York: Elsevier, pp. 65–100, 1991.

Anderson, Robert E. and Ishida, Kenzo: Malignant lymphoma in survivors of the atomic bomb, Hiroshima. Atomic Bomb Casualty Commission Technical Report 2-64, 1964. [Annals of Internal Medicine 61: 853–862, 1964]

Antoku, Shigetoshi; Hoshi, M.; Russell, Walter J.; Fujita, Shoichiro; and Pinkston, John A.: Radiation therapy among the Life Span Study subjects, Hiroshima and Nagasaki. (In manuscript)

Arakatsu, Bunsaki: Report on radiological investigations of Hiroshima city conducted for several days after the bombing. Genshi Bakudan Saigai Chosa Hokoku Shu. Vol. 1, pp. 5–10. Tokyo, Nippon Gakujutsu Kaigai, 1953.

Arnold, Lorna: *A Very Special Relationship: British Atomic Weapon Trials in Australia.* London: Her Majesty's Stationery Office, pp. xvii and 323, 1987.

Asano, Masahide; Kato, Hiroo; Yoshimoto, Keiko; Seyama, Shinichi; Itakura, Hideyo; Hamada, Tadao; and Iijima, Soichi: Primary liver carcinoma and liver cirrhosis in atomic bomb survivors, Hiroshima and Nagasaki, 1961–75, with special reference to HBs antigen. Radiation Effects Research Foundation Technical Report 9-81, 1981. [Journal of National Cancer Institute 69: 1221–12, 1982]

Auxier, John A.: *Ichiban: Radiation Dosimetry for the Survivors of the Bombings of Hiroshima and Nagasaki.* Technical Information Center, Energy Research and Development Administration, 1977.

Awa, Akio A.: Chromosome aberrations in A-bomb survivors, Hiroshima and Nagasaki. In: *Chromosome Aberrations: Basic and Applied Aspects.* Obe, G. and Natarajan, A. T. (eds.) Berlin: Springer-Verlag, pp. 180–190, 1989.

Awa, Akio A.; Sofuni, Toshio; Honda, Takeo; Hamilton, Howard B.; and Fujita, Shoichiro: Preliminary reanalysis of radiation-induced chromosome aberrations in relation to past and newly revised dose estimates for Hiroshima and Nagasaki A-bomb survivors. In: *Biological Dosimetry.* Eisert, W. G. and Mendelsohn, Mortimer L. (eds.) Berlin: Springer-Verlag, pp. 74–82, 1984.

Awa, Akio A.; Neriishi, Shotaro; Honda, Takeo; Yoshida, M. C.; Sofuni, Toshio; and Matsui, Takashi: Chromosome-aberration frequency in cultured bloodcells in relation to radiation dose of A-bomb survivors. Atomic Bomb Casualty Commission Technical Report 27-71, 1971.

Awa, Akio A.; Sofuni, Toshio; Honda, Takeo; Itoh, Masahiro; Neriishi, Shotaro; and Otake, Masanori: Relationship between radiation dose and chromosome aberrations in atomic bomb survivors of Hiroshima and Nagasaki. Journal of Radiation Research 19: 126–140, 1978.

Awa, Akio A.; Honda, Takeo; Neriishi, Shotaro; Sofuni, Toshio; Shimba, Hachiro; Ohtaki, Kazuo; Nakano, Mimako; Kodama, Yoshiaki; Itoh, Masahiro; and Hamilton, Howard B.: Cytogenetic study of the offspring of atomic bomb survivors, Hiroshima and Nagasaki. Radiation Effects Research Foundation Technical Report 21-88, 1988.

Bakwin, H: Developmental disorders of motility and language. Pediatric Clinics of North America 15: 3–20, 1968.

Balish, Edward; Pearson, Ted A.; and Chaskes, S.: Irradiated humans: Microbial flora, immunoglobulins, complement (C'3), transferrin, agglutinins, and bacteriocidins. Radiation Research 43: 729–756, 1970.

Ban, Sadayuki; Setlow, Richard B.; Bender, Michael A.; Ezaki, Haruo; Hiraoka, Toshio; Yamane, Motoi; Nishiki, Masayuki; Dohi, Kiyohiko; Awa, Akio A.; Miller, Richard C.; Parry, Dilys M.; Mulvihill, John J.; and Beebe, Gilbert W.: Radiosensitivity of skin fibroblasts from atomic bomb survivors with and without breast cancer. Radiation Effects Research Foundation Technical Report 6-90, 1990. [Cancer Research 50: 4050–4055, 1990]

Barny, Marie Helena; ct al.: Nucléaire et opinion publique en France: données sur les déchets radioactifs. Evolutions depuis 1977. Institut de Protection et de Surete Nucléaire, DPS/SEGP, Note LSEES 90/10, 1990.

Beardsley, Richard K.; Hall, John W.; and Ward, Robert E.: Village Japan: *A Study Based on Niiike Buraku*. Chicago, Illinois: University of Chicago Press, 1959.

Beatty, John: Genetics in the Atomic Age: The Atomic Bomb Casualty Commission, 1947–1956. In: *The Expansion of American Biology*. Benson, Keith R. et al. (eds.). Newbrunswick, New Jersey: Rutgers University Press, pp. 284–324, 1991.

Beebe, Gilbert W.: Adult Health Study Reference Papers. Atomic Bomb Casualty Commission Technical Report 10-60, 1960.

Beebe, Gilbert W.: Reflections on the work of the Atomic Bomb Casualty Commission in Japan. Epidemiological Reviews 1: 184–210, 1979.

Beebe, Gilbert W. and Usagawa, Mitsugu: The Major ABCC Samples. Atomic Bomb Casualty Commission Technical Report 12-68, 1968.

Beebe, Gilbert W.; Fujisawa, Hideo; and Yamasaki, Mitsuru: ABCC-NIH Adult Health Study Reference papers. 1. Selection of the sample. 2. Characteristics of the sample. Atomic Bomb Casualty Commission Technical Report 10-60, 1960.

Belsky, John L. and Blot, William J.: Adult stature in relation to childhood exposure to the atomic bombs of Hiroshima and Nagasaki. American Journal of Public Health 65: 489–494, 1975.

Bender, Lauretta: *A Visual Motor Gestalt Test and its Clinical Use*. New York: American Orthopsychiatric Association, 1938.

Bender, Lauretta: *Bender Motor Gestalt Test: Cards and Manual of Instructions*. New York: American Orthopsychiatric Association, 1946.

Bender, Michael A.; Awa, Akio A.; Brooks, Antony L.; Evans, H. John; Groer, Peter G.; Littlefield, L. Gayle; Pereira, Carlos; Preston, R. Julian; and Wachholz, Bruce W.: Current status of cytogenetic procedures to detect and quantify previous exposures to radiation. Mutation Research 196: 103–159, 1988.

Bennacerraf, Baruj: Influence of irradiation on resistance to infection. Bacteriological Reviews 24: 35–40, 1960.

Bergeron, R. Thomas: Pneumographic demonstration of subependymal heterotopic cortical gray matter in children. American Journal of Roentgenology 101: 168–177, 1967.

Bloom, Arthur D.; Neriishi, Shotaro; and Archer, Philip G.: Cytogenetics of in utero exposed subjects Hiroshima and Nagasaki. Atomic Bomb Casualty Commission Technical Report 7-68, 1968.

Blot, William J., and Miller, Robert W.: Small head size following in utero exposure to atomic radiation, Hiroshima and Nagasaki. Atomic Bomb Casualty Commission Technical Report 35-72, 1972.

Blot, William J. and Sawada, Hisao: Fertility among female survivors of the atomic bomb, Hiroshima-Nagasaki. Atomic Bomb Casualty Commission Technical Report 26-71, 1971. [American Journal of Human Genetics 24: 613–622, 1972]

Blot, William J.; Moriyama, Iwao; and Miller, Robert W.: Reproductive potential of males exposed in utero or prepubertally to atomic radiation. Atomic Bomb Casualty Commission Technical Report 39-72, 1972.

Blot, William J.; Shimizu, Yukiko; Kato, Hiroo; and Miller, Robert W.: Frequency of marriage and live birth among the survivors prenatally exposed to the atomic bomb. Atomic Bomb Casualty Commission Technical Report 2-75, 1975.

Brizzee, Kenneth R. and Brannon, R. B.: Cell recovery in foetal brain after ionizing radiation. International Journal of Radiation Biology 21: 375–388, 1972.

Brizzee, Kenneth R. and Ordy, J. Mark: Effects of prenatal ionizing irradiation on neural function and behavior. In: *Radiation Risks to the Developing Nervous System*. Kriegel, H. et al. (eds.) Stuttgart, New York: Gustav Fischer Verlag, pp. 255–282, 1986.

Brizzee, Kenneth R.; Ordy, J. Mark; and D'Agostino, A. N.: Morphological changes of the central nervous system after radiation exposure in utero. In: *Developmental Effects of Prenatal Irradiation*. Stuttgart, New York: Gustav Fischer, pp. 145–173, 1982.

Brizzee, Kenneth R.; Ordy, J. Mark; Kaack, M. Bernice; and Beavers, Terry: Effect of prenatal ionizing radiation on the visual cortex and hippocampus of newborn squirrel monkeys. Journal of Neuropathology and Experimental Neurology 39: 523–540, 1980.

Browne, Sir Thomas: *Religio Medici*. London: Printed for Andrew Crooke, p. 183, 1643.

Brues, Austin M.: The chrysanthemum and the feather merchant: A trip to Hiroshima and Nagasaki shortly after the atomic bombs. Hiroshima Igaku Zasshi (Journal of the Hiroshima Medical Association) 21: 74–94, 1968 (in English).

Burrows, Gerald; Hrubec, Zdenek; and Hamilton, Howard B.: Study of adolescents exposed in utero, research plan. Atomic Bomb Casualty Commission Technical Report 16-60, 1960.

Caviness, Verne S. Jr.: Normal development of cerebral neocortex. Developmental Neurobiology 12: 1–9, 1989.

Chang, Wushou P. and Kau, Jih-Wen: Taiwan: Exposure to high doses of radiation. Lancet 341: 750, 1993.

Cheeseman, Edward A. and Walby, A. L.: Intra-uterine irradiation and iris heterochromia. Annals of Human Genetics 27: 23–29, 1963.

Chrustchoff, Gyorgy F. and Berlin, E. A.: Cytological investigations on cultures of normal human blood. Journal of Genetics 31: 243–261, 1935.

Clarren, Sterling K.; Alvord, Ellsworth C.; Sumi, Mark S. et al.: Brain malformations related to prenatal exposure to ethanol. Journal of Pediatrics 92: 64–67, 1978.

Cogan, David G.; Martin, S. Forrest; Kimura, Samuel J.; and Ikui, Hiroshi: Ophthalmologic survey of atomic bomb survivors in Japan, 1949. Transactions of the American Ophthalmological Society 48: 62–87, 1950.

Committee for the Compilation of Materials on Damage Caused by the Atomic Bombs in Hiroshima and Nagasaki: *Hiroshima and Nagasaki: The Physical, Medical, and Social Effects of the Atomic Bombings.* Translated by Eisei Ishikawa and David L. Swain. Tokyo: Iwanami Shoten, pp. xlv and 706, 1981. (This book was originally published in 1979 by Iwanami Shoten under the title *Hiroshima Nagasaki Genbaku Saigai*.)

Committee on the Biological Effects of Ionizing Radiation: *Health Effects of Exposure to Low Levels of Ionizing Radiation.* Washington, D.C.: National Academy Press, pp. xiii and 421, 1990.

Conard, Robert A.: *Fallout: The Experiences of a Medical Team in the Care of a Marshallese Population Accidently Exposed to Fallout Radiation.* BNL 46444 Informal Report. Upton, N.Y.: Brookhaven National Laboratory.

Cox, Ann B.; Lee, Arthur C.; Williams, George R.; and Lett, John T.: Late cataractogenesis in primates and lagomorphs after exposure to particulate radiations. Advances in Space Research 1–6, 1991.

Crease, Robert P. and Samios, Nicholas P: Managing the Unmanageable. Atlantic Monthly, pp. 80–88, January 1991.

D'Amato, Constance J. and Hicks, Samuel P.: Effects of low levels of ionizing radiation on the developing cerebral cortex of the rat. Neurology 15: 1104–1116, 1965,

Darby, Sarah C.; Nakashima, Eiji; and Kato, Hiroo: A parallel analysis of cancer mortality among atomic bomb survivors and patients with ankylosing spondylitis given X-ray therapy. Radiation Effects Research Foundation Technical Report 4-84, 1984. [Journal of the National Cancer Institute 75: 1–21, 1985]

Dobbing, John: The later development of the brain and its vulnerability. In: *Scientific Foundations of Pediatrics.* Second edition. Davis, J. A. and Dobbing, John (eds.) London: William Heineman, 1981.

Dobbing, John and Sands, Jean: Quantitative growth and development of the human brain. Archives of Diseases of Childhood 48: 757–767, 1973.

Dobson, R. Lowry and Felton, James S.: Female germ cell loss from radiation and chemical exposures. American Journal of Industrial Medicine 4: 175–190, 1983.

Donoso, J. Alejandro and Norton, Stata: The pyramidal neuron in cerebral cortex following prenatal X-irradiation. Neurology and Toxicology 3: 72–84, 1982.

Dorn, Harold F. and Horn, James I.: The reliability of certificates of death from cancer. American Journal of Hygiene 34: 12–20, 1941.

Doty, Richard L.; Shaman, Paul; and Dann, M.: Development of the University of Pennsylvania Smell Identification Test: A standardized microcapsulated test of olfactory function. Physiology and Behavior 32: 49–501, 1984.

Driscoll, Shirley G.; Hicks, Samuel P.; Copenhaver, Edward H.; and Easterday, Charles L.: Acute radiation injury in two human fetuses. Archives of Pathology 76: 125–131, 1963.

Duke-Elder, Sir W. Stewart: *Textbook of Ophthalmology.* Volume 1. The Development, Form, and Function of the Visual Apparatus. St. Louis: C. V. Mosby, pp. xxxi and 1136, 1946.

Dunn, Val; Mock, Teresa; Bell, William E.; and Smith, Wilbur: Detection of heterotopic gray matter in children by magnetic resonance imaging. Magnetic Resonance Imaging 4: 33–39, 1986.

Dunn, Kimberly; Yoshimaru, Hiroshi; Otake, Masanori; Annegers, John F.; and Schull, William J.: Prenatal exposure to ionizing radiation and subsequent development of seizures. Radiation Effects Research Foundation Technical Report 13-88, 1988. [American Journal of Epidemiology 131: 114–123, 1990]

Edelman, Gerald M.: Molecular regulation of neural morphogenesis. In: *Molecular Bases of Neural Development*. Edelman, Gerald M.; Gall, W. Einar; and Cowan, W. Maxwell (eds.) New York: Wiley, pp. 35–60, 1985.

Edelman, Gerald M.: *Neural Darwinism: The Theory of Neuronal Group Selection*. New York: Basic Books, 1987.

Edelson, Edward: *The Journalist's Guide to Nuclear Energy*. Bethesda, Md: Atomic Industrial Forum, Inc., 1985.

Eisenbud, Merril: *An Environmental Odyssey*. Seattle: University of Washington Press, pp. xi and 264, 1990.

Ellett, William H.: Neutrons at Hiroshima—How Their Disappearance Affected Risk Estimates. International Colloquium on Neutron Radiobiology, November 5–7, 1990. Radiation Research 128: S147-S152, 1991.

Engel, Heinrich: The Japanese House: A Tradition for Contemporary Architecture. Rutland, Vermont: Charles E. Tuttle, p. 493, 1964.

Evarts, Edward V.: Hierarchies and emergent features in motor control. In: *Dynamic Aspects of Neocortical Function*, Edelman, Gerald M.; Gall, W. Einar; and Cowan, W. Maxwell (eds.) New York: Wiley, pp. 557–580, 1984.

Evarts, Edward V.; Shinoda, Yoshikazu; and Wise, Steven P.: *Neurophysiological Approaches to Higher Brain Function*. New York: Wiley, pp. x and 198, 1984.

Feinendegen, Ludwig E.; Muhlensiepen, H.; Porschen, W.; and Booz, J.: Acute non-stochastic effect of very low dose whole-body exposure, a thymidine equivalent serum factor. International Journal of Radiation Biology 41(2): 139–150, 1982.

Feinendegen, Ludwig E.; Muhlensiepen, H.; Lindberg, C.; Marx, J.; Porschen, W.; and Booz, J.: Acute and temporary inhibition of thymidine kinase in mouse bone marrow cells after low-dose exposure. International Journal of Radiation Biology 45(3): 205–215, 1984.

Finch, Stuart C. and Finch, Clement A.: Summary of the studies at ABCC-RERF concerning the late hematologic effects of atomic bomb exposure in Hiroshima and Nagasaki. Radiation Effects Research Foundation Technical Report 23-88, 1988.

Finch, Stuart C.; Hoshino, Takashi; Hrubec, Zdenek; Itoga, Takashi; and Nefzger, M. Dean: Operations manual for the detection of leukemia and related disorders, Hiroshima and Nagasaki. Atomic Bomb Casualty Commission Manual 1–65, 1965.

Fischhoff, Baruch; Slovic, Paul; Liechtenstein, Sarah; Read, Stephen; and Combs, Barbara: How safe is safe enough? A psychometric study of attitudes towards technological risks and benefits. Policy Science 8: 127–152, 1978.

Folley, Jarrett H.; Borges, Wayne; and Yamawaki, Takuso: Incidence of leukemia in survivors of the atomic bombs in Hiroshima and Nagasaki, Japan. American Journal of Medicine 13: 11–21, 1952.

Francis, Thomas: Report of ad hoc committee for appraisal of ABCC program. Atomic Bomb Casualty Commission Technical Report 33-59, 1959.

Freedman, Lawrence R. and Keehn, Robert J.: Urinary findings of children exposed in utero to the atomic bombs. Atomic Bomb Casualty Commission Technical Report 14-66, 1966.

Fujita, Shoichiro; Kato, Hiroo; and Schull, William J.: The LD50 associated with exposure to the atomic bombing of Hiroshima and Nagasaki: A review and reassessment. Radiation Effects Research Foundation Technical Report 17-87, 1987. [Japanese Journal of Radiation Research Supplement, Volume 32, 154–161, 1991]

Fujiwara, Saeko; Ezaki, Haruo; Sposto, Richard; Akiba, Suminori; Neriishi, Kazuo; Kodama, Kazunori; Yoshimitsu, Kengo; Hosoda, Yutaka; and Shimaoka, Katsutaro: Hyperparathyroidism among atomic bomb survivors in Hiroshima, 1986–88. Radiation Effects Research Foundation Technical Report 8-90, 1990.

Furusho, Toshiyuki and Otake, Masanori: A search for genetic effects of atomic bomb radiation on the growth and development of the F_1 generation. 1. Stature of 15- to 17-year-old senior high school students in Hiroshima. Atomic Bomb Casualty Commission Technical Report 4-78, 1978.

Furusho, Toshiyuki and Otake, Masanori: A search for genetic effects of atomic bomb radiation on the growth and development of the F_1 generation. 2. Body weight, sitting height, and chest circumference of 15- to 17-year-old senior high school students in Hiroshima. Atomic Bomb Casualty Commission Technical Report 5-78, 1978.

Furusho, Toshiyuki and Otake, Masanori: A search for genetic effects of atomic bomb radiation on the growth and development of the F_1 generation. 3. Stature of 12- to 14-year-old junior high school students in Hiroshima. Atomic Bomb Casualty Commission Technical Report 14-79, 1979.

Furusho, Toshiyuki and Otake, Masanori: A search for genetic effects of atomic bomb radiation on the growth and development of the F_1 generation. 4. Body weight, sitting height, and chest circumference of 12- to 14-year-old senior high school students in Hiroshima. Atomic Bomb Casualty Commission Technical Report 5-78, 1980.

Furusho, Toshiyuki and Otake, Masanori: A search for genetic effects of atomic bomb radiation on the growth and development of the F_1 generation. 5. Stature of 6- to 11-year-old elementary school pupils in Hiroshima. Atomic Bomb Casualty Commission Technical Report 9-85, 1985.

Fushiki, Shinji; Matsushita, Koji; and Schull, William J.: Decelerated migration of neocortical neurones in explant culture after exposure to radiation. NeuroReport 5: 353–356, 1993.

Galaburda, Albert M. and Kemper, Thomas L.: Cytoarchitectonic abnormalities in developmental dyslexia: A case study. Annals of Neurology 6: 94–100, 1979.

Gardner, Martin J.; Snee, Michael P.; Hall, Andrew J.; Powell, Caroline A.; Downes, Susan; and Terrell, John D.: Results of case-control study of leukaemia and lymphoma among young people near Sellafield nuclear plant in West Cumbria. British Medical Journal 300: 423–429, 1990.

Geschwind, Norman and Galaburda, Albert M.: *Cerebral Dominance: The Biological Foundations*. Cambridge, Mass.: Harvard University Press, 1984.

Gilbert, Ethel S.: Some effects of random dose measurement errors on analysis of atomic bomb survivor data. Radiation Research 98: 591–605, 1984.

Gilbert, Ethel S. and Ohara, Jill L.: An analysis of various aspects of atomic bomb dose estimation at RERF using data on acute radiation symptoms. Radiation Research 100: 124–138, 1984.

Gilbert, Ethel S.; Fry, Shirley A.; Wiggs, L. D.; Voelz, George L.; Cragle, Donna L.; and Petersen, Gerald R.: Analyses of combined mortality data on workers at the Hanford site, Oak Ridge National Laboratory and Rocky Flats nuclear weapons plant. Radiation Research 120: 19–35, 1989.

Goldstein, Leopold and Murphy, Douglas P.: Etiology of the ill-health in children born after maternal pelvic irradiation. Part 2. Defective children born after post-conception pelvic irradiation. American Journal of Roentgenology 22: 322–331, 1929.

Goodenough, Florence L.: *Measurement of Intelligence by Drawings.* New York: World Book Company, pp. xi and 177, 1926.

Gregory, Peter B.; Milton, Roy C.; Johnson, Marie-Louise T.; and Taura, Tadashi: Spleen shielding in survivors of the atomic bomb. Atomic Bomb Casualty Commission Technical Report 17-66, 1966.

Greulich, William W.; Crismon, Catherine S.; and Turner, Margaret L.: The physical growth and development of children who survived the atomic bombing of Hiroshima or Nagasaki. Journal of Pediatrics 43: 121–145, 1953.

Hakoda, Masayuki; Akiyama, Mitoshi; Kyoizumi, Seishi; Awa, Akio A.; Yamakido, Michio; and Otake, Masanori: Increased somatic cell mutant frequency in atomic bomb survivors. Radiation Effects Research Foundation Technical Report 18-87, 1987. [Mutation Research 201: 39–48, 1988]

Hall, Carl W.; Miller, Robert J.; and Nefzger, M. Dean: Ophthalmologic findings in atomic bomb survivors: Hiroshima 1956-57. Atomic Bomb Casualty Commission Technical Report 12-64, 1964.

Ham, William T. Jr.: Radiation cataract. Archives of Ophthalmology 50: 618–643, 1953.

Hamilton, B. F.; Benjamin, A. Steven; Angleton, George M.; and Lee, Arthur C.: The effect of perinatal ^{60}Co gamma radiation on brain weight in beagles. Radiation Research 119: 366–379, 1989.

Hammer, Ronald P.; Scheibel, Arnold B. et al.: Morphologic evidence for a delay of neuronal migration in fetal alcohol syndrome. Experimental Neurology 74: 587–596, 1981.

Harada, Tomin and Ishida, Morihiro: Neoplasms among A-bomb survivors in Hiroshima: First report of the Research Committee on Tumor Statistics, Hiroshima Medical Association, Hiroshima, Japan. Journal of the National Cancer Institute 25: 1253–1264, 1960.

Harris, Dale B.: *Children's Drawings as Measures of Intellectual Maturity: A Revision and Extension of the Goodenough Draw-a-Man Test.* New York: Harcourt, Brace, and World, pp. xiii and 367, 1963.

Hashizume, Tadashi; Maruyama, Takashi; Nishizawa, Kanae; and Nishimura, Akihisa: Dose estimation of human fetus exposed in utero to radiations from atomic bombs in Hiroshima and Nagasaki. Journal of Radiation Research 14: 346–362, 1973.

Hayakawa, Norihiko; Munaka, Masaki; Kurihara, Minoru; and Ohkita, Takeshi: Analysis of mortality rates of survivors exposed within Japanese wooden houses in Hiroshima by exposed distance. Hiroshima Igaku Zasshi 39: 126–129, 1986 (in Japanese).

Hibakusha: Survivors of Hiroshima and Nagasaki. Translated by Gaynor Sekimori. Tokyo: Kosei Publishing Company, p. 206.

Hicks, Samuel P. and D'Amato, Constance J.: Effects of ionizing radiation on mammalian development. In: *Advances in Teratology.* Volume 1. Woollum, D. H. M. (ed.) London: Logos Press, pp. 196–250, 1966.

Hicks, Samuel P. and D'Amato, Constance J.: Effects of radiation on development, especially of the nervous system. American Journal of Forensic Medicine and Pathology 1: 309–317, 1980.

Hicks, Samuel P.; D'Amato, Constance J.; and Lowe, Mary Jane: The development of the mammalian nervous system. I. Malformations of the brain, especially of the cerebral cortex, induced in rats by radiation. II. Some mechanisms of the malformations of the cortex. Journal of Comparative Neurology 113: 435–469, 1959.

Hirose, Kinnosuke and Fujino, Takashi: Cataracts due to the atomic bomb. Acta Ophthalmologica Japonica 54: 449–454, 1950.

Hollingsworth, Dorothy R.; Hamilton, Howard B.; Tamagaki, Hideo; and Beebe, Gilbert W.: Thyroid disease: A study in Hiroshima, Japan. Medicine 42: 47–71, 1963.

Holm, Lars-Erik; Wiklund, Kerstin C.; Lundell, Goran, E.; Bergman, N. Ake; Bjelkengren, Goran; Cedarquist, Ebbe S.; Ericsson, Ulla-Brit C.; Larsson, Lars-Gunnar; Lidberg, Monika E., Lindberg, R. Sture; Wicklund, Harriet V.; and Boice, John D. Jr.: Thyroid cancer after diagnostic doses of iodine–131: A retrospective cohort study. Journal of the National Cancer Institute 80: 1132–1138, 1988.

Hsu, Tao Chiuh: *Human and Mammalian Cytogenetics: An Historical Perspective.* New York: Springer Verlag, pp. xi and 186, 1979.

Hutt, M. L.: *The Hutt Adaptation of the Bender-Gestalt Test.* Fourth Edition. Orlando, Fl.: Grune and Stratton, Inc., 1985.

Ibuse, Masuji: *Black Rain.* Translated by John Bester. Tokyo: Kodansha International, p. 300, 1953.

Ichimaru, Michito; Ishimaru, Toranosuke; Mikami, Motoko; and Matsunaga, Masako: Multiple myeloma among atomic bomb survivors in Hiroshima and Nagasaki, 1950–1976. Radiation Effects Research Foundation Technical Report 9-79, 1979. [Journal of National Cancer Institute 69: 323–328, 1982]

Ichimaru, Michito; Ishimaru, Toranosuke; Tsuchimoto, Taiso; and Kirshbaum, Jack D.: Aplastic anemia in Hiroshima and Nagasaki with special reference to increase in atomic bomb survivors. Atomic Bomb Casualty Commission Technical Report 31-70, 1970.

Ichimaru, Michito; Ishimaru, Toranosuke; Belsky, Joseph L.; Tomiyasu, Takanori; Sadamori, Naoki; Hoshino, Takashi; Tomonaga, Masao; Shimizu, Nobuhiro; and Okada, Hiromu: Incidence of leukemia in atomic bomb survivors, Hiroshima and Nagasaki 1950–1971. Radiation Effects Research Foundation Technical Report 10-76, 1976.

Inskip, Mary J. and Piotrowski, J. K.: Review of the health effects of methylmercury. Journal of Applied Toxicology 5: 113–133, 1985.

International Atomic Energy Agency: The radiological accident in Goiania. Vienna, IAEA, 1988.

International Atomic Energy Agency: Medical aspects of the Chernobyl accident. Vienna, IAEA-TECDOC-516, 1989.

International Commission on Radiological Protection: Developmental effects of irradiation on the brain of the embryo and fetus. Annals of the ICRP 16: 1–43, 1986 (See also ICRP Publication No. 49.)

International Commission on Radiological Protection: 1990 Recommendations of the International Commission on Radiological Protection. Annals of the ICRP 21: No. 1–3, 1991.

Ishida, Morihiro and Matsubayashi, Ikuzo: An analysis of early mortality rates following the atomic bomb—Hiroshima. Atomic Bomb Casualty Commission Technical Report 20-61, 1961.

Ishimaru, Toranosuke; Nakashima, Ei; and Kawamoto, Sadahisa: Relationship of height, weight, head circumference, and chest circumference at age 18, to gamma and neutron doses among in utero exposed children, Hiroshima and Nagasaki. Radiation Effects Research Foundation Technical Report 19-84, 1984.

Ito, Takashi; Seyama, Toshio; Mizuno, Terumi; Tsuyama, Naohiro; Hayashi, Tomonori; Hayashi, Yuzo; Dohi, Kiyohiko; Nakamura, Nori; and Akiyama, Mitoshi: Unique association of p53 mutations with undifferentiated but not with differentiated thyroid gland carcinomas. Radiation Effects Research Foundation Technical Report 3-92, 1992.

Jablon, Seymour: Atomic bomb radiation dose estimation at ABCC. Atomic Bomb Casualty Commission Technical Report 23-71, 1971.

Jablon, Seymour and Kato, Hiroo: Childhood cancer in relation to prenatal exposure to A-bomb radiation. Lancet ii: 1000–1003, 1970 (See also Atomic Bomb Casualty Commission Technical Report 26-70.)

Jablon, Seymour; Hrubec, Zdenek; and Boice, John D.: Cancer in populations living near nuclear facilities. Journal of the American Medical Association 265: 1403–1408, 1991.

Jacobs, L.A. and Brizzee, Kenneth R.: Effects of total-body X-irradiation in single and fractionated doses on developing cerebral cortex in rat foetus. Nature 210: 31–33, 1966.

Jacobson, Marcus: *Developmental Neurobiology*. Third Edition. New York: Plenum Press, pp. ix and 776, 1991.

Jaenke, Roger S. and Angleton, George M.: Perinatal radiation-induced renal damage in the beagle. Radiation Research 122: 58–65, 1990.

Jones, K. P. and Wheater, A. W.: Obstetric outcomes in West Cumberland Hospital: Is there a risk from Sellafield? Journal of the Royal Society of Medicine 82: 524–527, 1989.

Jones, Peter A.; Buckley, Jonathan D.; Henderson, Brian E.; Ross, Ronald K.; and Pike, Malcolm C.: From gene to carcinogen: A rapidly evolving field in molecular epidemiology. Cancer Research 51: 3617–3620, 1991.

Kamada, Nanao; Shigeta, Chiharu; Kuramoto, Atsushi; Munaka, Masaki; Yokoro, Kenjiro; Niimi, Masanobu; Aisaka, Chuichi; Ito, Chikako; and Kato, Hiroo: Acute and late effects of A-bomb radiation studied in a group of young girls with a defined condition at the time of bombing. Japanese Journal of Radiation Research 30: 218–225, 1989.

Kathren, Ronald L. and Petersen, Gerald R.: Units and terminology of radiation measurement: A primer for epidemiologists. American Journal of Epidemiology 130: 1076–1087, 1989.

Kato, Hiroo: Mortality of in-utero children exposed to the A-bomb and of offspring of A-bomb survivors. International Atomic Energy Agency-SM-224/603: 49–60. Vienna, IAEA, 1978.

Kato, Hiroo and Keehn, Robert J.: Mortality in live-born children who were in utero at the time of the atomic bombs: Hiroshima and Nagasaki. Atomic Bomb Casualty Commission Technical Report 13-66, 1966.

Kato, Hiroo and Schull, William J.: Joint JNIH-ABCC life-span study of children born to atomic bomb survivors. Research Plan. Atomic Bomb Casualty Commission Technical Report 4-60, 1960.

Kato, Hiroo and Schull, William J.: Studies of the mortality of A-bomb survivors. Report 7. Mortality, 1950–1978. Part I. Cancer Mortality. Radiation Effects Research Foundation Technical Report 12-80, 1980. [Radiation Research 90: 395–432, 1982]

Kato, Hiroo; Brown, Charles C.; Hoel, David G.; and Schull, William J.: Studies of the mortality of A-Bomb survivors. Report 7. Mortality, 1950–1978. Part II. Mortality from causes other than cancer and mortality in early entrants. Radiation Effects Research Foundation Technical Report 5-81, 1981. [Radiation Research 91: 243–264, 1982]

Kato, Hiroo; Mayumi, Makoto; Nishioka, Kusuya; and Hamilton, Howard B.: The relationship of HB surface antigen and antibody to atomic bomb radiation in the adult health study sample, 1975–1977. Radiation Effects Research Foundation Technical Report 13-80, 1980. [American Journal of Epidemiology 117: 610–620, 1983]

Kato, Kazuo; Antoku, Shigetoshi; Sawada, Shozo; and Russell, Walter J.: Organ doses to atomic bomb survivors from radiological examinations at the Radiation Effects Research Foundation. Radiation Effects Research Foundation Technical Report 19-89, 1989. [British Journal of Radiology 64: 720–727 and 728–733, 1991]

Kawamoto, Sadahisa; Fujino, Tadashi; and Fujisawa, Hideo: Ophthalmologic status in children exposed in utero. Atomic Bomb Casualty Commission Technical Report 23-64, 1964.

Kemper, Thomas L.: Asymmetrical lesions in dyslexia. In: *Cerebral Dominance: The Biological Foundations*. Geschwind, Norman and Galaburda, Albert M. (eds.) Cambridge, Mass.: Harvard University Press, pp. 75–92, 1984.

Kendall, G. M.; Muirhead, Colin R.; MacGibbon, B. H.; O'Hagan, J. A.; Conquest, A. J.; Goodill, A. A.; Butland, B. K.; Fell, T. P.; Jackson, D. A.; Webb, M. A.; Haylock, R. G. E.; Thomas, J. M.; and Silk, T. J.: Mortality and occupational exposure to radiation: First analysis of the National Registry for Radiation Workers. British Medical Journal 304: 220–225, 1992.

Kerr, George D.: Organ dose estimates for the Japanese atomic bomb survivors. Health Physics 37: 487–508, 1979.

Kerr, George D.: Review of dosimetry for the atomic bomb survivors. In: *Proceedings of the Fourth Symposium on Neutron Dosimetry, Munich-Neuherberg, Vol 1. Luxembourg: Commission of the European Communities, pp. 501–513, 1981.

Kerr, George D.: Effects of the air blast on the radiation dosimetry for atomic bomb survivors. Radiation Research (in press), 1995.

Kerr, George D. and Solomon, Daniel L.: The Epicenter of the Nagasaki Weapon: A Reanalysis of the Available Data with Recommended Values. Oak Ridge, Tennessee: Oak Ridge National Laboratory Report ORNL/TM-5139, 1976.

Kerr, George D.; Yamada, Hiroaki; and Marks, Sidney: A survey of radiation doses received by atomic-bomb survivors residing in the United States. Oak Ridge, Tennessee: Oak Ridge National Laboratory, ORNL/TM-5138, pp. 25, 1976.

Kerr, George D.; Pace, Joseph V.; Mendelsohn, Edgar; Loewe, William E.; Kaul, Dean C.; Dolatshahi, Farhad; Egbert, Stephen D.; Gritzner, Michael; Scott, William H. Jr.; Marcum, Jess; Kosako, Toshiso; and Kanda, Keiji: Transport of initial radiations in air over ground. In: *U.S.-Japan Joint Reassessment of Atomic Bomb Radiation Dosimetry in Hiroshima and Nagasaki*. Final Report. Volume 1. Hiroshima: Radiation Effects Research Foundation, pp. 66–142, 1987.

Klose, Monika and Bentley, David: Transient pioneer neurons are essential for formation of an embryonic peripheral nerve. Science 245: 982–984, 1989.

Kneale, George and Stewart, Alice: Pre-cancers and the liability to other diseases. British Journal of Cancer 37: 448–457, 1978.

Koga, Yukiyoshi: Two intelligence test methods viewed in relation to evaluated intelligence. In: *Collection of Reports in Commemoration of Dr. Matsumoto, Studies in Psychology and Arts*, pp. 923–988, 1937 (in Japanese).

Konermann, Gerhard: Postnatal brain maturation damage induced by prenatal irradiation: Modes of effect manifestation and dose-response relations. In: *Low Dose Radiation: Biological Bases of Risk Assessment*. Baverstock, Keith F. and Stather, John W., (eds.). London: Taylor and Francis, pp. 364–376, 1989.

Konuma, N.: Psychiatric atomic bomb casualties—Summary of Psychiatric Department. Hiroshima Medical Journal 20: 231–236, 1967. (In Japanese)

Krooth, Robert S.; Howell, R. Rodney; and Hamilton, Howard B.: Properties of acatalasic cells growing in vitro. Atomic Bomb Casualty Commission Technical Report 14-61, 1961.

Kumami, Takehiko: *The Atom Bomb! The Memoir of a Banker Who Barely Escaped with his Life*. Translated by Peter Tuffley. Tokyo: Seiryo Printing Co., Ltd., pp. xiii and 83, 1993.

Kyoizumi, Seishi; Akiyama, Mitoshi; Hirai, Yuko; Kusunoki, Yoichiro; Tanabe, Kazumi; and Umeki, Shigeko: Spontaneous loss and alteration of antigen receptor expression in mature $CD4^+$ T cells. Journal of Experimental Medicine 171: 1981–1999, 1990.

Land, Charles E.; Hayakawa, Norihiko; Machado, Stella G.; Yamada, Yutaka; Pike, Malcolm C.; Akiba, Suminori; and Tokunaga, Masayoshi: A case-control interview study of breast cancer among Japanese A-bomb survivors: I. Main Effects. Cancer Causes Control 5: 157–165, 1994.

Land, Charles E.; Hayakawa, Norihiko; Machado, Stella G., Yamada, Yutaka;

Pike, Malcolm C.; Akiba, Suminori; and Tokunaga, Masayoshi: A case-control interview study of breast cancer among Japanese A-bomb survivors: II. Interactions between epidemiological factors and radiation dose. Cancer Causes Control 5: 167–176, 1994.

Lange, Robert D.; Moloney, William C.; and Yamawaki, Takuso: Leukemia in atomic bomb survivors. I. General observations. Blood 9: 574–585, 1954.

Lange, Robert D.; Wright, Stanley W.; Tomonaga, Masanobu; Kurasaki, Hirotani; Matsuoka, Shigeru; and Matsunaga, Haruji: Refractory anemia occurring in survivors of the atomic bombing in Nagasaki, Japan. Blood 10: 312–324, 1955.

Langham, Wright H. (ed.): Radiobiological Factors in Manned Space Flight. Report of the Space Radiation Study Panel of the Life Sciences Committee. Washington, D.C.: National Academy of Sciences–National Research Council, 1967.

Layton, Donald D.: Heterotopic cerebral gray matter as an epileptogenic focus. Journal of Neuropathology 1(21): 244–249, 1962.

League of Women Voters Education Fund: *The Nuclear Waste Primer: A Handbook for Citizens*. New York: Nick Lyons Books, 1985.

Lejeune, Jerome; Turpin, Raymond; Rethoré, Marie-Odile; and Mayer, M.: Résultats d'une premiére enquéte sur les effets somatiques de l'irradiation foeto-embryonnaire in utero (cas particulier des hétérochromies iriennes). Revue Français Etudes Clinique et Biologique 5: 582–599, 1960.

Lett, John T.; Lee, Arthur C.; and Cox, Ann B.: Late cataractogenesis in Rhesus monkeys irradiated with protons and radiogenic cataract in other species. Radiation Research 126: 147–156, 1991.

Liebow, Averill A.: Encounter with disaster: A medical diary of Hiroshima, 1945. Yale Journal of Medicine and Biology 38: 61–239, 1965.

Lifton, Robert Jay: *Death in Life: Survivors of Hiroshima*. New York: Random House, 1967.

Little, Mark P.: A comparison between the risks of childhood leukaemia from parental exposure to radiation in the Sellafield workforce and those displayed among the Japanese bomb survivors. Journal of Radiological Protection 10: 185–198, 1990.

Little, Mark P.: Preconception exposure risks six months prior to conception in the Sellafield workforce and the Japanese bomb survivors. Berkeley, England, Nuclear Electric Technology Division TD/RPB/REP/0047, 1991.

Little, Mark P.: A comparison of the risks of leukaemia in the offspring of the Japanese bomb survivors and those of the Sellafield workforce with those in the offspring of the Ontario and Scottish workforces. Journal of Radiation Protection 13: 161–175, 1993.

Little, Mark P. and Charles, Monty W.: Selection effects in the survivors of the atomic weapons explosions in Japan. Berkeley, England, Nuclear Energy Technology Division TD/RPB/REP/0001, pp. 1–11, 1990.

Little, Mark P.; Kendall, G. M.; Muirhead, Colin R.; MacGibbon, B. H.; Haylock, R. G. E.; Thomas, J. M.; and Goodill, A. A.: Further analysis, incorporating assessment of the robustness of risks of cancer mortality in the National Registry of Radiation Workers. Journal of Radiological Protection 13: 95–108, 1993.

Loewe, William E. and Mendelsohn, Edgar: Neutron and gamma ray doses at Hiroshima and Nagasaki. Nuclear Science and Engineering 81: 325–350, 1982.

Los Alamos Laboratory Group: The Effects of Atomic Weapons. Washington, D.C.: Government Printing Office, 1950.

Mabuchi, Kiyohiko; Soda, Midori; Ron, Elaine; Tokunaga, Masayoshi; Ochikubo, Sachio; Sugimoto, Sumio; Ikeda, Takayoshi; Terasaki, Masayuki; Preston, Dale L. and Thompson, Desmond E.: Cancer Incidence in Atomic Bomb Surivors. Part I: Use of the Tumor Registries in Hiroshima and Nagasaki for Incidence Studies. Commentary and Review Series. RERF CR 3-91, 1991. [Radiation Research 137: S1–S16, 1994]

Malenfant, Richard E.: Little Boy Replication: Justification and Construction. Proceedings of the 17th Midyear Topical Meeting of the Health Physics Society, February 5–9, 1984, Pasco, Washington, pp. 8.1–8.4, 1984.

Martinez Martinez, P. F. A: *Neuroanatomy. Development and Structure of the Central Nervous System*. Philadelphia: Saunders, 1982.

Maruki, Iri and Maruki, Toshi: *The Hiroshima Murals: The Art of Iri Maruki and Toshi Maruki*. Dower, John W. and Junkerman, John (eds.). Tokyo: Kodansha International Ltd., p. 128, 1985.

Masuyama, Motosaburo: Stochastic studies on the atomic bomb casualties. Bulletin of Mathematical Statistics 5: 21–30, 1952.

Matsumae, Shigeyoshi: *My Turbulent Life in a Turbulent Century*. Tokyo: Tokai University Press, p. 241, 1982.

Matsumoto, Y. Scott: Social impact on atomic bomb survivors: Hiroshima and Nagasaki. Atomic Bomb Casualty Commission Technical Report 12-69, 1969.

Matsuura, Hiroo; Yamamoto, Tsutomu; Sekine, Ichiro; Ochi, Yoshimichi; and Otake, Masanori: Pathological and epidemiological study of gastric cancer in atomic bomb survivors, Hiroshima and Nagasaki, 1950–77. Radiation Effects Research Foundation Technical Report 12-83, 1983.

Matsuzaka, Yoshimasa: Notes on a study of former ABCC. Hiroshima Igaku Zasshi 34: 1149–1163, 1981; 35: 110–126; 35: 210–219; 35: 543–553, 1982 (in Japanese). [A printed, English translation of these Notes is available through the Foundation.]

Mavor, James W.: The production of non-disjunction by X-rays. Journal of Experimental Zoology 39: 381–432, 1924.

McConnell, Susan K.; Ghosh, Anirvan; and Shatz, Carla J.: Subplate neurons pioneer the first axon pathway from the cerebral cortex. Science 245: 978–982, 1989.

Mikhail, Mikhael. A. and Mattar, Adel G.: Malformation of the cerebral cortex with heterotopia of the gray matter. Journal of Computer Assisted Tomography 2: 291–296, 1978.

Miller, Robert J.; Fujino, Tadashi; and Nefzger, M. Dean: Lens findings in atomic bomb survivors. A review of major ophthalmic surveys at the Atomic Bomb Casualty Commission (1949–1962). Archives of Ophthalmology 78: 697–704, 1967.

Miller, Robert W.: Delayed effects occurring within the first decade after exposure of young individuals to the Hiroshima atomic bomb. Pediatrics 18: 1–18, 1956.

Miller, Robert W. and Blot, William J.: Small head size after in-utero exposure to atomic radiation. Lancet II: 784–787, 1972.

Miller, Robert W. and Mulvihill, John J.: Small head size after atomic irradiation. Teratology 14: 355–358, 1976.

Milton, Roy C. and Shohoji, Takao: Tentative 1965 radiation dose estimation for atomic bomb survivors. Atomic Bomb Casualty Commission Technical Report 1-68, 1968.

Mole, Robin H.: Consequences of prenatal radiation exposure for postnatal development: A review. International Journal of Radiation Biology 42: 1–12, 1982.

Mole, Robin H.: The effect of prenatal radiation exposure on the developing human brain. International Journal of Radiation Biology 57: 647–663, 1990.

Mole, Robin H.: Severe mental retardation after large prenatal exposures to bomb radiation. Reduction in oxygen transport to fetal brain: A possible abscopal mechanism. International Journal of Radiation Biology 58: 705–711, 1990.

Moloney, William C.: Radiogenic leukemia revisited. Blood 70: 905–908, 1987.

Moloney, William C. and Kastenbaum, Marvin A.: Leukemogenic effects of ionizing radiation on atomic bomb survivors in Hiroshima City. Science 121: 308–309, 1954.

Moloney, William C. and Lange, Robert D.: Cytologic and biochemical studies on the granulocytes in early leukemia among atomic bomb survivors. Texas Reports on Biology and Medicine 12: 887–897, 1954.

Moloney, William C. and Lange, Robert D.: Leukemia in atomic bomb survivors. II. Observations on early phases of leukemia. Blood 9: 663–685, 1954.

Moore, G. William; Hutchins, Grover M.; and O'Rahilly, Ronan: The estimated age of staged human embryos and early fetuses. American Journal of Obstetrics and Gynecology 139: 500–506, 1981.

Moore, Keith L.: *Clinically oriented anatomy*. Baltimore: Williams and Wilkins, see p. 796, 1980.

Morton, Newton E.; Crow, James F.; and Muller, Herman J.: An estimate of the mutational damage in man from data on consanguineous marriages. Proceedings of the National Academy of Sciences 42: 855–863, 1956.

Mountcastle, Vernon B.: An organizing principle for cerebral function: The unit module and the distributed system. In: *The Neurosciences: Fourth Study Program*. Schmitt, F. O. and Worden, F. G. (eds.) Cambridge, Mass.: MIT Press, pp. 21–42, 1979.

Mueller, Charles F.: Heterotopic gray matter. Radiology 94: 357–358, 1970.

Muirhead, Colin R. and Darby, Sarah C.: Modelling the relative and absolute risks of radiation-induced cancers. Journal of the Royal Statistical Society 150-A, 83–118, 1987.

Muirhead, Colin R. and Darby, Sarah C.: Distinguishing relative and absolute risk models for radiation-induced cancers. Proceedings of the British Nuclear Energy Society Conference "Health Effects of Low Dose Ionizing Radiation. Recent Advances and Their Implications." London, 1987.

Müller, Fabiola and O'Rahilly, Ronan: The first appearance of the major divisions of the human brain at stage 9. Anatomy and Embryology 168: 419–432, 1983.

Müller, Fabiola and O'Rahilly, Ronan: Cerebral dysraphia (future anencephaly) in a human twin embryo at stage 13. Teratology 30: 167–177, 1984.

Muller, Herman J.: Our Load of Mutations. American Journal of Human Genetics 2: 111–176, 1950.

Murphy, Douglas P.: Maternal pelvic irradiation. In: *Congenital Malformations* (2nd edition). Murphy, Douglas P. (ed.) Philadelphia: Lippincott, 1947.

Nagai, Takashi: *The Bells of Nagasaki*. Translated by William Johnston. Tokyo: Kodansha International, pp. xxiii and 118, 1984. Originally published by Hibiya Shuppan in 1949 under the title *Nagasaki no kane*.

Nakamura, Kuniomi: Stomach cancer in atomic bomb survivors, 1950–73. Radiation Effects Research Foundation Technical Report 8-77, 1977.

Nakamura, Nori; Sposto, Richard; and Akiyama, Mitoshi: Dose Survival of G_0 lymphocytes irradiated in vitro: A test for a possible population bias in the cohort of atomic bomb survivors exposed to high doses. Radiation Research 134: 316–322, 1993.

Nakamura, Nori; Akiyama, Mitoshi; Kyoizumi, Seishi; and Kusunoki, Yoichiro: Current summary of lymphocyte survival study. Japanese Journal of Radiation Research 32(Suppl.): 327–329, 1991.

Nakamura, Nori; Akiyama, Mitoshi; Kyoizumi, Seishi; Langlois, Richard G.; Bigbee, William L.; Jensen, Ronald H.; and Bean, Michael A.: Frequency of somatic cell mutations at the glycophorin A locus in erythrocytes of atomic bomb survivors. Radiation Effects Research Foundation Technical Report 1-87, 1987.

Naruge, Tetsuji: *Koseki no Jitsumu to sono Riron*. Tokyo: Nihon Kajo-Shuppan, p. 620, 1956.

National Center for Health Statistics: Evaluation of psychological measures used in the Health Examination Survey of children ages 6–11. Vital and Health Statistics. PHS Publication No. 1000—Series 2—No. 15. Washington, D.C.: U.S. Government Printing Office, 1966.

National Center for Health Statistics: Intellectual Maturity of Children as measured by the Goodenough-Harris Drawing Test. Vital and Health Statistics. PHS Publication No. 1000—Series 11—No. 105. Washington, D.C.: U.S. Government Printing Office, 1970.

National Center for Health Statistics: The Goodenough-Harris Drawing Test as a Measure of Intellectual Maturity of Youths 12-17 Years: United States. Vital and Health Statistics. PHS Publication No. 74-1620. Series 11, No. 138. Rockville, MD: Health Resources Administration, National Center for Health Statistics, 1974.

National Center for Health Statistics: Goodenough-Harris Test Estimates of Intellectual Maturity of Youths 12–17 years: Demographic and Socioeconomic Factors. Vital and Health Statistics. PHS Publication No. 77-1641—Series 11—No. 159. Rockville, MD: Health Resources Administration, National Center for Health Statistics, 1977.

National Council on Radiation Protection and Measurements: *Conceptual Basis for Calculations of Absorbed-dose Distributions*. Report No. 108. Bethesda, MD: NCRP, 1991.

National Council on Radiation Protection and Measurements: *Risk Estimates for Radiation Protection*. Report No. 115. Bethesda, MD: NCRP, 1993.

National Council on Radiation Protection and Measurements: *Limitation of Exposure to Ionizing Radiation*. Report No. 116. Bethesda, Md: NCRP, 1993.

Neel, James V.: *Physician to the Gene Pool: Genetic Lessons and other Stories*. New York: Wiley, pp. ix and 457, 1994.

Neel, James V. and Lewis, Susan E.: The Comparative Radiation Genetics of Humans and Mice. Annual Review of Genetics 24: 327–362, 1990.

Neel, James V. and Schull, William J.: *The Effect of Exposure to the Atomic Bombs on Pregnancy Termination in Hiroshima and Nagasaki*. Washington, D.C.: National Academy of Sciences-National Research Council Publication 461, pp. xvi and 241, 1956.

Neel, James V. and Schull, William J. (eds.): *The Children of Atomic Bomb Survivors: A Genetic Study*. Washington, D.C.: National Academy Press, pp. vi and 518, 1991.

Neel, James V.; Schull, William J.; Awa, Akio A.; Satoh, Chiyoko; Kato, Hiroo; and Yoshimoto, Yasuhiko: Implications of the Hiroshima–Nagasaki genetic studies for the estimation of the human "doubling dose." Genome 31: 853–859, 1990

Neel, James V.; Schull, William J.; Awa, Akio A.; Satoh, Chiyoko; Kato, Hiroo; Otake, Masanori; and Yoshimoto, Yasuhiko: The children of parents exposed to atomic bombs: Estimates of the genetic doubling dose of radiation for humans. American Journal of Human Genetics 46: 1053–1072, 1990.

Neel, James V.; Satoh, Chiyoko; Goriki, Kazuaki; Asakawa, Jun-ichi; Fujita, Mikio; Takahashi, Norio; Kageoka, Takeshi; and Hazama, Ryuji: Search for mutations altering protein charge and/or function in children of atomic bomb survivors: Final report. American Journal of Human Genetics 42: 663–676, 1988.

Nehemias, John V.: Multivariate analysis and the IBM704 computer applied to ABCC data on growth of surviving Hiroshima children. Atomic Bomb Casualty Commission Technical Report 22-61, 1961.

Neriishi, Shotaro and Matsumura, Hideo: Morphological observations of the central nervous system in an in-utero exposed autopsied case. Japanese Journal of Radiation Research 24: 18, 1983.

Nishikawa, T., and Tsuiki, S.: Psychiatric investigations of atomic bomb survivors. Nagasaki Medical Journal 36: 717–722, 1961. (In Japanese)

Nishimura, Hideo: Introduction. *Atlas of Human Perinatal Histology*. Tokyo: Igaku-Shoin, 1983.

Noble, Kenneth B.: Shielding survey and radiation dosimetry study plan: Hiroshima-Nagasaki. Atomic Bomb Casualty Commission Technical Report 7-67, 1967.

Ohtaki, Kazuo: G-banding analysis of radiation-induced chromosome damage in lymphocytes of Hiroshima A-bomb survivors. Japanese Journal of Human Genetics 37: 245–262, 1992.

Ohtaki, Kazuo; Sposto, Richard; Kodama, Yoshiaki; Nakano, Mimako; and Awa, Akio A.: Aneuploidy in Somatic Cells of in Utero Exposed A-Bomb Survivors in Hiroshima. Radiation Effects Research Foundation Technical Report 9-93, 1993.

Okajima, Shunzo: Fallout in Nagasaki—Nishiyama District. Japanese Journal of Radiation Research 16(Suppl): 35–41, 1975.

Okunuma, N. and Hikida, H.: Results of psychoneurological studies on atomic bomb survivors. Kyushu Neuropsychiatry 1: 50–52, 1949. (In Japanese)

Omori, Yoshiaki; Morrow, Lewis B.; Ishimaru, Toranosuke; Johnson, Kenneth G.; Maeda, Tetsuo; Shibukawa, Tetsuo; and Takaki, Kenkichi: A buccal smear survey for sex chromatin aberration in Hiroshima high school children. Atomic Bomb Casualty Commission Technical Report 11-65, 1965.

Omuta, Minoru: The microcephalic children of Hiroshima. Japan Quarterly 13: 375–384, 1966.

O'Rahilly, Ronan: The timing and sequence of events in the development of the human eye and ear during the embryonic period proper. Anatomy and Embryology 168: 87–99, 1983.

O'Rahilly, Ronan and Gardner, Ernest: The developmental anatomy and histology of the human central nervous system. In: *Handbook of Clinical Neurology*. Volume 30. Vinken, P. J. and Bruyn, G. W. (eds.) New York: North-Holland, pp. 17–40, 1977.

O'Rahilly, Ronan and Müller, Fabiola: The first appearance of the human nervous system at stage 8. Anatomy and Embryology 163: 1–13, 1981.

O'Rahilly, Ronan and Müller, Fabiola: *Human Embryology and Teratology*. New York: Wiley, pp. x and 330, 1992.

Otake, Masanori and Schull, William J.: Radiation-related posterior lenticular opacities in Hiroshima and Nagasaki atomic bomb survivors based on the DS86 dosimetry system. Radiation Research 121: 3–13, 1990.

Otake, Masanori and Schull, William J.: Radiation-related small head sizes among prenatally exposed A-bomb survivors. International Journal of Radiation Biology 63: 255–270, 1993.

Otake, Masanori; Schull, William J.; and Neel, James V.: The effects of exposure to the atomic bombing of Hiroshima and Nagasaki on congenital malformations, stillbirths and early mortality among the children of atomic bomb survivors: A reanalysis. Radiation Effects Research Foundation Technical Report 13-89, 1989. [Radiation Research 122: 1–11, 1990]

Otake, Masanori; Yoshimaru, Hiroshi; and Schull, William J.: Severe mental retardation among the prenatally exposed survivors of the atomic bombing of Hiroshima and Nagasaki: A comparison of the old and new dosimetry system. Radiation Effects Research Foundation Technical Report 16-87, 1987.

Otake, Masanori; Fujikoshi, Yasunori; Schull, William J.; and Izumi, Shizue: A longitudinal study of growth and development of stature among prenatally exposed atomic-bomb survivors. Radiation Research 134: 94–101, 1993.

Otake, Masanori; Schull, William J.; Yoshimaru, Hiroshi; and Fujikoshi, Yasunori: Effect on school performance of prenatal exposure to ionizing radiation in Hiroshima: A comparison of the T65DR and DS86 dosimetry systems. Radiation Effects Research Foundation Technical Report 2-88, 1988. [Japanese Journal of Hygiene 46: 747–754, 1991]

Oughterson, Ashley W. and Warren, Shields: *Medical Effects of the Atomic Bomb in Japan*. New York: McGraw-Hill, 1956.

Oughterson, Ashley W.; LeRoy, George V.; Liebow, Averill A.; Hammond, E. Cuyler; Barnett, Henry L.; Rosenbaum, Jack D.; and Schneider, B. Aubrey: *Medical Effects of Atomic Bombs*. USAEC, Office of Technical Information,

Technical Information Service, Oak Ridge, Tennessee, Report NP-3041, Volume 6, 1951.

Penn, Arthur; Garte, Seymour J.; Warren, Lisa; Nesta, Douglas; and Mindich, Bruce: Transforming gene in human atherosclerotic plaque DNA. Proceedings of the US National Academy of Sciences 83: 7951–7955, 1986.

Penney, Lord William George; Samuels, Dennis E. J.; and Scorgie, Guy C.: The nuclear explosive yields at Hiroshima and Nagasaki. Philosophical Transactions of the Royal Society of London: A. Mathematical and Physical Sciences 266: 357–424, 1970.

Peterson, Leif E.; Pepper, Larry J.; Hamm, Peggy B.; and Gilbert, Susan L.: Longitudinal study of astronaut health: Mortality in the years 1959–1991. Radiation Research 133: 257–264, 1993.

Pierce, Donald A. and Vaeth, Michael: Cancer risk estimation from the A-bomb survivors: Extrapolation to low doses, use of relative risk models and other uncertainties. In: *Low Dose Radiation: Biological Bases of Risk Estimation*. Baverstock, Keith F. and Stather, John W., (eds.). London: Taylor and Francis, pp. 54–69, 1989.

Pierce, Donald A.; Stram, Daniel O.; and Vaeth, Michael: Allowing for random errors in radiation exposure estimates for atomic bomb survivors data. Radiation Effects Research Foundation Technical Report 2-89, 1989. [Radiation Research 123: 275–284, 1990]

Plummer, George: Anomalies occurring in children in utero, Hiroshima. Atomic Bomb Casualty Commission Technical Report 29-C-59, 1952. [Pediatrics: 687–693, 1952]

Preston, Dale L. and Pierce, Donald A.: The impact of changes in dosimetry on cancer mortality risks in the atomic bomb survivors. Radiation Effects Research Foundation Technical Report 9-87, 1987. [Radiation Research 114: 437–466, 1988]

Preston, Dale L.; Pierce, Donald A.; and Vaeth, Michael. Neutrons and Radiation Risk: A Commentary. RERF Update, Winter 1992–1993, p. 5, , 1993.

Preston, Dale L.; McConney, Mary E.; Awa, Akio A.; Ohtaki, Kazuo; Itoh, Masahiro; and Honda, Takeo: Comparison of the dose-response relationships for chromosome aberration frequencies between the T65D and DS86 dosimetries. Radiation Effects Research Foundation Technical Report 7-88, 1988.

Preston, Dale L.; Kusumi, Shizuyo; Tomonaga, Masao; Izumi, Shizue; Ron, Elaine; Kuramoto, Atsushi; Kamada, Nanao; Dohy, Hiroo; Matsuo, Tatsuki; Nonaka, Hiroaki; Thompson, Desmond E.; Soda, Midori; and Mabuchi, Kiyohiko: Cancer Incidence in Atomic Bomb Survivors. Part III: Leukemia, Lymphoma, and Multiple Myeloma, 1950–87. Radiation Effects Research Foundation Technical Report 24-92, 1992. [Radiation Research 137: S68–S97, 1994.]

Putnam, Frank W.: Hiroshima and Nagasaki Revisited: The Atomic Bomb Casualty Commission and the Radiation Effects Research Foundation. Perspectives in Biology and Medicine 37(4): 515–545, 1994.

Rakic, Pasko: Mode of cell migration to the superficial layers of fetal monkey neocortex. Journal of Comparative Neurology 145: 61–83, 1972.

Rakic, Pasko: Neurons in Rhesus monkey visual cortex: Systematic relation

between time of origin and eventual disposition. Science 183: 425–427, 1974.
Rakic, Pasko: Cell migration and neuronal ectopias in the brain. In: *Morphogenesis and Malformation of the Face and Brain*. Bergsma, Daniel (ed.) New York: Alan Liss, pp 95–129, 1975.
Rakic, Pasko: Neuronal migration and contact guidance in the primate telencephalon. Postgraduate Medical Journal 54(Suppl) 1: 25–40, 1978.
Rakic, Pasko: Neuronal-glial interaction during brain development. Trends in Neuroscience 4: 184–187, 1981.
Rakic, Pasko: Mechanisms of neuronal migration in developing cerebellar cortex. In: *Molecular Bases of Neural Development*. Edelman, Gerald M.; Gall, W. Einar; and Cowan, W. Maxwell (eds.) New York: Wiley, pp. 139–160, 1985.
Rakic, Pasko: Limits of neurogenesis in Primates. Science 227: 1054–1055, 1985.
Rakic, Pasko: Specification of cerebral cortical areas. Science 241: 170–176, 1988.
Rakic, Pasko: Defects of neuronal migration and the pathogenesis of cortical malformations. Progress in Brain Research 73: 15–37, 1988.
Rakic, Pasko and Sidman, Richard L.: Histogenesis of cortical layer in human cerebellum, particularly the lamina dissecans. Journal of Comparative Neurology 139: 473–500, 1970.
Redesdale, Lord (Algernon Mitford): *The Garter Mission to Japan*. London: Macmillan and Company, Ltd., pp. ix and 280, 1906.
Restak, Richard: *The Brain*. New York: Bantam Books, pp. x and 371, 1984.
Restak, Richard: *The Mind*. New York: Bantam Books, pp. xvi and 328, 1988.
Reyners, H.; Gianfelici de Reyners, E.; and Maisin, Jean R.: The role of the GLIA in late damage after prenatal irradiation. In: *Radiation Risks to the Developing Nervous System*. Kriegel, H., Schmahl, W.; Gerber, George B. and Stieve, F.-E. (eds.) Stuttgart: Gustav Fischer, pp. 118–131, 1986.
Reyners, H.; Gianfelici de Reyners, E.; Poortman, F.; Cramtz, A.; Coffigny, H.; and Maisin, Jean-R.: Brain atrophy after foetal exposure to very low doses of ionizing radiation. International Journal of Radiation Biology 62: 619–626, 1992.
Reynolds, Earle L.: Growth and Development of Hiroshima Children Exposed to the Atomic Bomb: Three-Year Study (1951–1953). Atomic Bomb Casualty Commission Technical Report 10-59, 1959.
Rhodes, Richard: *The Making of the Atomic Bomb*. New York: Simon and Schuster, Inc., p. 886, 1986.
Ritchie, Rufus H. and Hurst, G. Samuel: Penetration of weapons radiation: Application to the Hiroshima-Nagasaki studies. Health Physics 1: 390–404, 1959.
Robertson, Thomas L.; Kato, Hiroo; Rhoads, George G.; Kagan, Abraham; Marmot, Michael; Syme, S. Leonard; Gordon, Tavia; Worth, Robert; Belsky, Joseph L.; Dock, Donald S.; Miyanishi, Michihiro; and Kawamoto, Sadahisa: Epidemiologic studies of coronary heart disease and stroke in Japanese men living in Japan, Hawaii, and California: Incidence of myocardial infarction and coronary heart disease death. Radiation Effects Research Foundation Technical Report 2-76, 1976.

Robertson, Thomas L.; Shimizu, Yukiko; Kato, Hiroo; Kodama, Kazunori; Furonaka, Hiroshi; Fukunaga, Yasuo; Lin, Chow L.; Danzig, Michael D.; Pastore, John O.; and Kawamoto, Sadahisa: Incidence of stroke and coronary heart disease in atomic bomb survivors living in Hiroshima and Nagasaki, 1958–1974. Radiation Effects Research Foundation Technical Report 12-79, 1979.

Roesch, William C. (ed.): *US-Japan Joint Reassessment of Atomic Bomb Radiation Dosimetry in Hiroshima and Nagasaki. Final Report. Volume 1.* Hiroshima: The Radiation Effects Research Foundation, p. 434, 1987.

Ron, Elaine; Modan, Baruch; Floro, S; Harkedar, I.; and Gurkewitz, R: Mental function following scalp irradiation during childhood. American Journal of Epidemiology 116: 149–160, 1982.

Ron, Elaine; Preston, Dale L.; Mabuchi, Kiyohiko; Thompson, Desmond E. and Soda, Midori: Cancer Incidence in Atomic Bomb Survivors. Part IV: Comparison of Cancer Incidence and Mortality. Radiation Effects Research Foundation Technical Report 11-93, 1993.

Rotblat, Josef: Acute radiation mortality in a nuclear war. In: *The Medical Implications of Nuclear War.* Washington, D.C.: pp. 233–250, 1986.

Rouma, Gustavo: *El Lenguaje Grafico del Nino.* (translated by De Someillan, A.) Havana: Gutierrez and Company, 1919.

Russell, Walter J.; Keehn, Robert J.; Ihno, Yu; Hattori, Fumio; Kogure, Takashi; and Imamura, Kunihiro: Bone maturation in children exposed in utero to the atomic bomb. Atomic Bomb Casualty Commission Technical Report 1-72; 1972.

Rutter, Michael; Graham, Philip; and Yule, William: *A Neuropsychiatric Study of Childhood.* Philadelphia: J. B. Lippincott, 1978.

Sager, Ruth: Tumor suppressor genes: The puzzle and the promise. Science 246: 1406–1412, 1989.

Sasagawa, Sumiko; Yoshimoto, Yasuhiko; Toyota, Emiko; Neriishi, Shotaro; Yamakido, Michio; Matsuo, Miyo; Hosoda, Yutaka; and Finch, Stuart C.: Phagocytic and bactericidal activities of leukocytes in whole blood from atomic bomb survivors. Radiation Research 124: 103–106, 1990.

Sasaki, M.S.; Saigusa, S.; Kimura, I.; Kobayashi, T.; Ikushima, T.; Kobayashi, K.; Saito, I.; Sasuga, N.; Oka, Y.; Ito, T.; and Kondo, Sohei: Biological effectiveness of fission neutrons: Energy dependency and its implication for risk assessment. Proceedings of an International Conference on Radiation Effects and Protection, Mito, Japan, March 18–20, 1992, pp. 31–35 (1992).

Satoh, Chiyoko; Neel, James V.; Yamashita, Akiko; Goriki, Kazuaki; Fujita, Mikio; and Hamilton, Howard B.: Frequency among Japanese of heterozygotes for deficiency variants of 11 enzymes. Radiation Effects Research Foundation Technical Report 2-83, 1983. [American Journal of Human Genetics 35: 656– 674, 1983]

Sawada, Shozo; Land, Charles E.; Otake, Masanori; Russell, Walter J.; Takeshita, Kenji; Yoshinaga, Haruma; and Hombo, Zenichiro: Hospital and clinic survey estimates of medical x-ray exposures in Hiroshima and Nagasaki: Part I. RERF population and the general population. Radiation Effects Research Foundation Technical Report 16-79, 1979.

Schull, William J.: Empirical risks in consanguineous marriages: Sex ratio, malformations and viability. American Journal of Human Genetics 10: 294–343, 1958.

Schull, William J.: *Song Among the Ruins*. Cambridge, Mass.: Harvard University Press, pp. viii and 305, 1990.

Schull, William J.: Biology and Society: The Legacy of Trinity. The Grant Taylor Memorial Lecture: 1991. RERF Update, Supplement/Autumn, S1-S4, 1991.

Schull, William J. and Neel, James V.: Atomic bomb exposure and the pregnancies of biologically related parents: A prospective study of the genetic effects of ionizing radiation in man. American Journal of Public Health 49: 1621–1629, 1959.

Schull, William J. and Neel, James V.: Maternal radiation and mongolism. Lancet i: 537–538, 1962.

Schull, William J. and Neel, James V.: *The Effects of Inbreeding on Japanese Children*. New York: Harper and Row, pp. xii and 419, 1965.

Schull, William J.; Otake, Masanori; and Neel, James V.: Genetic effects of the atomic bombs: A reappraisal. Science 213: 1220–1227, 1981.

Schull, William J.; Otake, Masanori; and Yoshimaru, Hiroshi: Effect on intelligence test score of prenatal exposure to ionizing radiation in Hiroshima and Nagasaki: A comparison of the T65DR and DS86 dosimetry systems. Radiation Effects Research Foundation Technical Report 3-88, 1988.

Schull, William J.; Otake, Masanori; and Yoshimaru, Hiroshi: Radiation-related damage to the developing human brain. In: *Low Dose Radiation: Biological Bases of Risk Assessment*. Baverstock, Keith F. and Stather, John W. (eds.) London: Taylor and Francis, pp. 28–41, 1989.

Schull, William J.; Neel, James V.; Otake, Masanori; Awa, Akio A.; Satoh, Chiyoko; and Hamilton, Howard B.: Hiroshima and Nagasaki: Three and a Half Decades of Genetic Screening. In: *Environmental Mutagens and Carcinogens*. Sugimura, Takashi; Kondo, Sohei and Takebe, Hiraku (eds.) Tokyo: University of Tokyo Press, pp. 687–700, 1982.

Schull, William J.; Nishitani, Hiromu; Hasuo, Kanehiro; Kobayashi, Takuro; Goto, Ikuo; and Otake, Masanori: Brain abnormalities among the mentally retarded prenatally exposed survivors of the atomic bombing of Hiroshima and Nagasaki. Radiation Effects Research Foundation Technical Report 13-91, 1991.

Schwanzel-Fukuda, Mariene and Pfaff, Donald W.: Origin of luteinizing hormone-releasing hormone neurons. Nature 338: 161–164, 1989.

Scott, James: Oncogenes in atherosclerosis. Nature 325: 574–575, 1987.

Shimizu, Sakae: Field survey immediately after the Hiroshima atomic bombing and study of the fallout from the hydrogen bomb test in the Central Pacific. ATOMKI Kozlemenyek 24: 77–94, 1982.

Shimizu, Yukiko; Kato, Hiroo; and Schull, William J.: Life Span Study Report 11. Part 2. Cancer mortality in the years 1950–1985 based on the recently revised doses (DS86). Radiation Effects Research Foundation Technical Report 5-88, 1988. [Radiation Research 121: 120–141, 1990]

Shimizu, Yukiko; Kato, Hiroo; Schull, William J.; and Hoel, David G.: Life Span Study Report 11. Part 3. Non-cancer mortality in the years 1950–1985 based on the recently revised doses (DS86). Radiation Effects Research Foundation Technical Report 2-91, 1991. [Radiation Research 130: 249–266, 1992]

Shimizu, Yukiko; Kato, Hiroo; Schull, William J.; Preston, Dale L.; Fujita, Shoichiro; and Pierce, Donald A.: Life Span Study Report 11. Part 1. Comparison of risk coefficients for site-specific cancer mortality based on the DS86 and T65D shielded kerma and organ doses. Radiation Effects Research Foundation Technical Report 12-87, 1987. [Radiation Research 118: 502–524, 1989]

Shirabe, Raisuke: *Nagasaki Genbaku Taiken* (Nagasaki A-Bomb Experiences). Tokyo: University of Tokyo Press, 1982 (in Japanese).

Shirabe, Raisuke; Fujii, Hiroshi; Ishimaru, Nobumasa; and Sato, Takemasa: Statistical observation of casualties caused by atomic bomb in Nagasaki. Part 1. Rate of death caused by atomic bomb. (Original typewritten manuscript. ABCC Library Files, Hiroshima, 1953.)

Shirabe, Raisuke; Kido, Riichi; Sato, Junichiro; Ichinose, Kengo; and Takahashi, Shoshiro: Statistical observation of casualties caused by atomic bomb in Nagasaki. Part 3. Surgical injuries caused by atomic bomb. (Original typewritten manuscript. ABCC Library Files, Hiroshima, 1953.) See also: Shirabe, Raisuke: Medical Survey of Atomic Bomb Casualties. Military Surgeon 113: 251–263, 1953.

Shirabe, Raisuke; Kido, Riichi; Sato, Junichiro; Ichinose, Kengo; and Takahashi, Shoshiro: Statistical observation of casualties caused by atomic bomb in Nagasaki. Part 4. On the radiation sickness caused by atomic bomb. (Original typewritten manuscript. ABCC Library Files, Hiroshima, 1953.) See also: Shirabe, Raisuke: Medical Survey of Atomic Bomb Casualties. Military Surgeon 113: 251–263, 1953.

Shirabe, Raisuke; Suyama, Hirofumi; Kamei, Terumi; Akabane, Kaku; and Kubota, Tadashi: Statistical observation of casualties caused by atomic bomb in Nagasaki. Part 2. Time of death of persons receiving injuries from atomic bomb explosion. (Original typewritten manuscript. ABCC Library Files, Hiroshima, 1953.) See also: Shirabe, Raisuke: Medical Survey of Atomic Bomb Casualties. Military Surgeon 113: 251–263, 1953.

Shohno, Naomi: *The Legacy of Hiroshima: Its Past, our Future*. Tokyo: Kosei Publishing Company, p. 150, 1986.

Shore, Roy E.; Albert, Roy E.; and Pasternack, Bernard S.: Follow-up study of patients treated by x-ray for tinea capitis: Resurvey of post-treatment illness and mortality experience. Archives of Environmental Health 31: 21–28, 1976.

Sidman, Richard L. and Rakic, Pasko: Neuronal migration, with special reference to developing human brain: A review. Brain Research 62: 1–35, 1973.

Sidman, Richard L. and Rakic, Pasko: Development of the human central nervous system. In: *Histology and Histopathology of the Nervous System*. Haymaker, W. and Adams, R. D. (eds.) Springfield, Ill.: C. C. Thomas, pp. 3–145, 1982.

Siegel, David G.: Frequency of live births among atomic bomb survivors, Hiroshima-Nagasaki. Atomic Bomb Casualty Commission Technical Report 25-64, 1964. [Radiation Research 28: 278–288, 1966]

Sienkiewicz, Zenon J.; Saunders, Richard D.; and Rutland, Barbara K.: Prenatal irradiation and spatial memory in mice: Investigation of critical period. International Journal of Radiation Biology 62: 211–219, 1992.

Sinskey, Robert M.: The status of lenticular opacities caused by atomic radiation. American Journal of Ophthalmology 39: 285–293, 1955.

Slavin, Richard E.; Kamada, Nanao; and Hamilton, Howard B.: A cytogenetic study of 92 cases of Down's syndrome Hiroshima and Nagasaki. Atomic Bomb Casualty Commission Technical Report 2-66, 1966.

Slovic, Paul: Images of Disaster: Perception and Acceptance of Risks from Nuclear Power. In: *Proceedings of the Fifteenth Annual Meeting of the National Council on Radiation Protection and Measurements.* Washington, D.C.: National Council on Radiation Protection and Measurements, pp. 34–56, 1979.

Snell, Fred M.; Neel, James V.; and Ishibashi, Koichi: Hematologic studies in Hiroshima and a control city two years after the atomic bombing. Archives of Internal Medicine 84: 569–604, 1949.

Soka Gakkai: *Cries for Peace: Experiences of Japanese Victims of World War II.* Tokyo: The Japan Times Limited, pp. 234, 1978.

Speir, Edith; Molai, Rama; Huang, Eng-Shang; Leon, Martin B.; Shawl, Fayaz; Finkel, Toren and Epstein, Stephen E.: Potential role of human cytomegalovirus and p53 interaction in coronary restonosis. Science 265: 391–394, 1994.

Sposto, Richard and Preston, Dale L.: Correcting for catchment area non-residency in studies based on tumor-registry data. Radiation Effects Research Foundation Commentary and Review 1–92, 1992.

Storer, John B.; Mitchell, Toby J.; and Fry, R. J. Michael: Extrapolation of the relative risk of radiogenic neoplasms across mouse strains and to man. Radiation Research 114: 331–353, 1988.

Stott, Denis H.: A general test of motor impairment for children. Developmental Medicine and Child Neurology 5: 323–330, 1968.

Stram, Daniel O. and Mizuno, Shoichi: Analysis of the DS86 atomic bomb radiation dosimetry methods using data on severe epilation. Radiation Research 117: 93–113, 1989.

Straume, Tore; Egbert, S. D.; Woolson, William A.; Finkel, R. C.; Kubik, P. W.; Gove, H. E.; Sharma, P.; and Hoshi, M.: Neutron discrepancies in the DS86 Hiroshima dosimetry system. Health Physics 63: 421–426, 1992.

Streissguth, Ann P.; Herman, Cynthia S.; and Smith, David W.: Intelligence, behavior and dysmorphogenesis in the fetal alcohol syndrome. Journal of Pediatrics 92: 363–367, 1978.

Streissguth, Ann P.; Landesman-Dwyer, Sharon; Martin, Joan C. et al.: Teratogenic effects of alcohol in humans and laboratory animals. Science 209: 353–361, 1980.

Sutow, Wataru W. and West, Emory: Studies on Nagasaki (Japan) children exposed in utero to the atomic bomb. A roentgenographic survey of the skeletal system. American Journal of Roentgenology, Radium Therapy, and Nuclear Medicine 74: 493–499, 1955.

Suzuki, Masamichi: Ovarian function in latent genital tuberculosis. Fertility and Sterility 6: 259–270, 1955.

Suzuki, Masamichi and Watanabe, Teruo: Ovarian function following induced abortion. American Journal of Obstetrics and Gynecology 67: 596–604, 1954.

Swinbanks, David: Japan promotes nuclear power. Nature 353: 782, 1991.

Tabuchi, Akira; Hirai, Tsuyoshi; Nakagawa, Shigeru; Shimada, Katsunobu; and Fujito, Junro: Clinical findings on in utero exposed microcephalic children. Atomic Bomb Casualty Commission Technical Report 28-67, 1967.

Taeuber, Irene: *The Population of Japan*. Princeton, New Jersey: Princeton University Press, 1958.

Takahara, Shigeo: Three cases of progressive oral gangrene due to lack of catalase in blood. Read at the 56th Chugoku District Assembly of the Japanese Otorhinolaryngeal Society, July 11, 1947. See also Nihon Jibi Inkoka Gakkai Kaiho 51: 163–164, 1948.

Takahara, Shigeo; Hamilton, Howard B.; Neel, James V.; Kobara, Thomas Y.; Ogura, Yoshio; and Nishimura, Edwin T.: Hypocatalasemia: A new carrier state. Journal of Clinical Investigation 39: 610–619, 1961.

Takeya, Yo; Popper, Jordan S.; Shimizu, Yukiko; Kato, Hiroo; Rhoads, George G.; and Kagan, Abraham: Epidemiologic studies of coronary heart disease and stroke in Japanese men living in Japan, Hawaii, and California: Incidence of stroke in Japan and Hawaii. Radiation Effects Research Foundation Technical Report 6-84, 1984.

Tanebashi, Masanori: Intelligence and intelligence tests. In: *Outline of Educational Psychology*. 20th edition. Koga, Yukiyoshi (ed.) Tokyo: Kyodo Shuppansha, pp. 128–158, 1972. (in Japanese).

Taylor, H. Grant: *Remembrances and Reflections*. Houston: University of Texas Health Science Center, 1991.

Thiessen, Joop W. and Kaul, Dean C.: The dosimetry system 1986 (DS86) and the tentative dosimetry system 1965 (T65D): How do they compare, what is left to do? Japanese Journal of Radiation Research, Supplement to Volume 32 entitled "A Review of Forty-five Years of Study of Hiroshima and Nagasaki Atomic Bomb Survivors," pp. 1–10, 1991.

Thompson, Desmond E.; Mabuchi, Kiyohiko; Ron, Elaine; Soda, Midori; Tokunaga, Masayoshi; Ochikubo, Sachio; Sugimoto, Sumio; Ikeda, Takayoshi; Terasaki, Masayuki; Izumi, Shizue; and Preston, Dale L.: Cancer Incidence in Atomic Bomb Survivors. Part II: Solid tumor incidence in A-bomb survivors, 1958–87. Radiation Effects Research Foundation Technical Report 5-92, 1992. [Radiation Research 137: S17–S64, 1994]

Tjio, Joe-Hin and Levan, Albert: The chromosome number of man. Hereditas 42: 1–6, 1956.

Tokunaga, Masayoshi; Land, Charles E.; and Tokuoka, Shoji: Follow-up studies of breast cancer incidence among atomic bomb survivors. Japanese Journal of Radiation Research Supplement 32: 201–211, 1991.

Tokuoka, Shoji; Asano, Masahide; Yamamoto, Tsutomu; Tokunaga, Masayoshi; Sakamoto, Goi; Hartmann, William H.; Hutter, Robert V. P.; and Henson, Donald E.: Histological review of breast cancer in atomic bomb survivors, Hiroshima and Nagasaki. Radiation Effects Research Foundation Technical Report 11-82, 1982.

Tomonaga, Masao; Matsuo, Tatsuki; Carter, Randy; Bennett, John M.; Kuriyama, Kazutaka; Imanaka, Fumio; Kusumi, Shizuyo; Mabuchi, Kiyohiko; Kuramoto, Atsushi; Kamada, Nanao; Ichimaru, Michito; Pisciotta, Anthony V.; and Finch, Stuart C.: Differential effects of atomic bomb irradiation in inducing major leukemia types (AML, ALL and CML): Analyses of open-city cases including Life Span Study cohort based upon up-dated diagnostic systems and the new dosimetry system (DS86). Radiation Effects Research Foundation Technical Report 9-91, 1991.

Touwen, Bert C. L. and Prechtl, Heinz F. R.: *The Neurologic Examination of the Child with Minor Dysfunction*. Philadelphia: J. B. Lippincott, 1970.

Trosko, James E.; Cheng, Chia C.; and Madhukar, Burra V.: Modulation of intercellular communication during radiation and chemical carcinogenesis. Radiation Research 123: 241–251, 1990.

Trumbull, Robert: *Nine who Survived Hiroshima and Nagasaki*. New York: E. P. Dutton and Company, Inc., pp. 148, 1957.

United Nations: *Sources and Effects of Ionizing Radiation*. United Nations Scientific Committee on the Effects of Atomic Radiation, 1977 Report to the General Assembly, with annexes. New York: United Nations, 1977.

United Nations: *Ionizing Radiation: Sources and Biological Effects*. United Nations Scientific Committee on the Effects of Atomic Radiation, 1982 Report to the General Assembly, with annexes. New York: United Nations, 1982.

United Nations: *Genetic and Somatic Effects of Ionizing Radiation*. United Nations Scientific Committee on the Effects of Atomic Radiation, 1986. Report to the General Assembly, with annexes. New York: United Nations. 1986.

United Nations: *Sources, Effects and Risks of Ionizing Radiation*. United Nations Scientific Committee on the Effects of Atomic Radiation, 1988 Report to the General Assembly, with annexes. New York: United Nations, 1988.

United States Environmental Protection Agency: Estimating Radiogenic Cancer Risks. Document Number EPA 402-R-93-076, 1994.

Vaeth, Michael and Pierce, Donald A.: Calculating excess lifetime risk in relative risk models. Radiation Effects Research Foundation, Commentary and Reviews 3–89, 1989.

Valentine, William N.; Beck, W. S.; Follette, J. H.; Mills, H.; and Lawrence, John S.: Biochemical studies in chronic myelogenous leukemia, polycythemia vera and other idiopathic myeloproliferative disorders. Blood 7: 959–977, 1952.

Vidal-Pergola, Gabriela M.; Kimler, Bruce F.; and Norton, Stata: Effect of in utero radiation dose fractionation on rat postnatal development, behavior, and brain structure: 6 hour interval. Radiation Research 134: 369–374, 1993.

Vogel, Friedrich: What can we learn from Hiroshima and Nagasaki? Berzelius Symposium XV, Umea, pp. 119–130, 1988.

Vogelstein, Bert: A deadly inheritance. Nature 348: 681–682, 1990.

Wakabayashi, Toshiro: Kato, Hiroo; Ikeda, Takayoshi; and Schull, William J.: Life Span Study Report 9. Part 3. Tumor registry data, Nagasaki 1959–78. Radiation Effects Research Foundation Technical Report 6-81, 1981. [Radiation Research 93: 112–146, 1983]

Walker, Joan M.: Histological study of the fetal development of the human acetabulum and labrum: Significance in congenital hip disease. Yale Journal of Biology and Medicine 54: 225–263, 1981.

Walsh, Christopher and Cepko, Constance L.: Clonally related cortical cells show several migration patterns. Science 241: 1342–1345, 1988.

Walsh, Christopher and Cepko, Constance L.: Widespread dispersion of neuronal clones across functional regions of the cerebral cortex. Science 255:434–440, 1992.

Walsh, Christopher and Cepko, Constance L.: Clonal dispersion in proliferative layers of developing cerebral cortex. Nature 362: 632–635, 1993.

Wanebo, Clifford K.; Johnson, Kenneth G.; Sato, Kazuyoshi; and Thorslund, Todd W.: Breast cancer in the ABCC-JNIH Adult Health Study, Hiroshima-Nagasaki, 1950–66. Atomic Bomb Casualty Commission Technical Report 13-67, 1967.

Warren, Stafford L.: The Role of Radiology in the Development of the Atomic Bomb. See Chapter xxxvii, pp. 831–921. In: *Radiology in World War II*. Ahnfeldt, Arnold L. (ed.) Washington, D.C.: Office of the Surgeon General, Department of the Army, 1966.

Watanabe, Ronald S.: Embryology of the human hip. Clinical Orthopedics 98: 8–26, 1977.

Watson, James D. and Crick, Francis H. C.: Genetical implications of the structure of deoxyribonucleic acid. Nature: 171: 964–967, 1953.

Weinberg, Robert A.: Finding the Anti-Oncogene. Scientific American 259: 44–51, 1988.

Wertheimer, M.: Studies in the theory of Gestalt psychology. Psychologische Forschung 4: 301–350, 1923.

Wilcox, Allen J.; Weinberg, Clarice R.; O'Connor, John F.; Baird, Donna D.; Schlatterer, John P.; Canfield, Robert E.; Armstrong, E. Glenn; and Nisula, Bruce C.: Incidence of early loss of pregnancy. New England Journal of Medicine 319: 189–194, 1988.

Williams, Roger S.: Cerebral malformations arising in the first half of gestation. Developmental Neurobiology 12: 11–20, 1989.

Wilson, Robert R.: Nuclear radiation in Hiroshima and Nagasaki. Radiation Research 4: 349–359, 1956.

Wing, Steve; Shy, Carl, M.; Wood, Joy L.; Wolf, Susanne; Cragle, Donna L.; and Frome, Edward L.: Mortality among workers at Oak Ridge National Laboratory. Journal of the American Medical Association 265: 1397–1402, 1991.

Wisniewski, K. E.; Segan, S. M.; Miezejeski, C. M.; Sersen, E. A.; and Rudelli, R. D.: The FRA(X) Syndrome: Neurological, electrophysiological, and neuropathological abnormalities. American Journal of Medical Genetics 38: 476–480, 1991.

Wood, James W.; Johnson, Kenneth G.; and Omori, Yoshiaki: In utero exposure to the Hiroshima atomic bomb: Follow-up at twenty years. Atomic Bomb Casualty Commission Technical Report 9-65, 1965.

Wood, James W.; Johnson, Kenneth G.; Omori, Yoshiaki; Kawamoto, Sadahisa; and Keehn, Robert J.: Mental retardation in children exposed in utero, Hiroshima and Nagasaki. Atomic Bomb Casualty Commission Technical Report 10-66, 1966. [American Journal of Public Health 57: 1381–1390, 1967]

Wray, Susan; Nieburgs, Andra; and Elkabes, Stela: Spatiotemporal cell expression of luteinizing hormone-releasing hormone in the prenatal mouse: evidence for an embryonic origin in the olfactory placode. Developmental Brain Research 46: 309–318, 1989.

Yaar, Israel; Ron, Elaine; Modan, Baruch; Modan, Michaela; and Perets, H.: Long-term cerebral effects of small doses of x-irradiation in childhood as manifested in adult visually evoked responses. Transactions of the American Neurological Association 104: 264–268, 1979.

Yaar, Israel; Ron, Elaine; Modan, Michaela; Perets, H.; and Modan, Baruch: Long-term cerebral effects of small doses of x-irradiation in childhood as manifested in adult visually evoked responses. Annals of Neurology 8: 261–268, 1980.

Yaar, Israel; Ron, Elaine; Modan, Baruch; Rinott, Yoseff; Yaar, Mina; and Modan, Michaela: Long-lasting cerebral functional changes following moderate dose x-radiation treatment to the scalp in childhood: An EEG power spectral study. Journal of Neurology, Neurosurgery, and Psychiatry 45: 166–169, 1983.

Yamada, Michiko; Kodama, Kazunori; and Wong, F. Lenny: The long-term psychological sequelae of atomic-bomb survivors in Hiroshima and Nagasaki. In: *The Medical Basis for Radiation-Accident Preparedness*. New York: Elsevier; pp. 155–163, 1991.

Yamamoto, Tsutomu; Moriyama, Iwao M.; Asano, Masahide; and Guralnick, Lillian: The Autopsy Program and the Life Span Study, January 1961–December 1975. Radiation Effects Research Foundation Technical Report 18-78, 1978.

Yamasawa, Yoshimichi: Hematologic studies of irradiated survivors in Hiroshima, Japan. Archives of Internal Medicine 91: 310–314, 1953.

Yamazaki, James N.; Wright, Stanley W.; and Wright, Phyllis M.: Outcome of pregnancy in women exposed to the atomic bomb in Nagasaki. American Journal of Diseases of Children 87: 448–463, 1954. [See also Atomic Bomb Casualty Commission Technical Report 24-A-59, 1959.]

Yokota, Seichiro; Tagawa, Daisaburo; Otsuru, Shin; Nakagawa, K.; Neriishi, Shotaro; Tamaki, Hideo; and Hirose, Kozumi: Tainai hibakusha ni mirareta shotosho no ichi bokenrei. Nagasaki Medical Journal 38: 92–95, 1963.

York, Edwin N. in communication from M. Morgan, AFSWC, to G. Samuel Hurst, ORNL, ORNL-CF-57-11-144 (1957) [cited by Ritchie and Hurst, 1959].

Yoshimaru, Hiroshi; Otake, Masanori; and Schull, William J.: Further observations on damage to the developing brain following prenatal exposure to ionizing radiation. Radiation Effects Research Foundation Technical Report (submitted), 1989.

Yoshimoto, Yasuhiko; Kato, Hiroo; and Schull, William J.: Risk of cancer among in utero exposed to A-bomb radiation: 1950–84. Radiation Effects Research Foundation Technical Report 4-88, 1988. [Lancet ii: 665–669, September 17, 1988]

Yoshimoto, Yasuhiko; Schull, William J.; Kato, Hiroo; and Neel, James V.: Mortality among the offspring (F_1) of atomic bomb survivors, 1946–1985. Radiation Effects Research Foundation Technical Report 1-91, 1991. [Japanese Journal of Radiation Research 32: 327–351, 1991]

Yoshimoto, Yasuhiko; Neel, James V.; Schull, William J.; Kato, Hiroo; Mabuchi, Kiyohiko; Soda, Midori; and Eto, Ryozo: The frequency of malignant tumors during the first two decades of life in the offspring of atomic bomb survivors. Radiation Effects Research Foundation Technical Report 4–89, 1989. [American Journal of Human Genetics 46: 1041–1052, 1990]

Young, Robert W.: Human mortality from uniform low-LET radiation. A report presented at the NATO RSG V meeting on LD_{50} at Gosport, United Kingdom on 11 May 1987.

Author Index

Abelson, Philip H., 310, 347
Abrahamson, Seymour, xi, 330, 347
Abramson, Frederic D., 330, 347
Ahearne, John F., 332, 347
Aisaka, Chuichi, 356
Akabane, Kaku, 369
Akiba, Suminori, 155, 322, 347, 353, 358, 359
Akiyama, Mitoshi, xi, 182, 322, 323, 347, 354, 356, 358, 362
Albert, Roy E., 369
Allen, Leroy R., 300
Alvord, Ellsworth C., 350
Anderson, Ray L., 40
Anderson, Robert E., 322, 348
Angleton, George M., 328, 354, 356
Annegers, Fred F., 352
Antoku, Shigetoshi, 320, 348, 357
Arakatsu, Bunsaki, 304, 348
Archer, Philip G., 349
Armstrong, E. Glenn, 373
Arnold, Lorna, 348
Asakawa, Jun-ichi, 363
Asano, Masahide, 322, 348, 371, 374
Auxier, John A., 109, 317, 348
Awa, Akio A., xi, 254, 322, 330, 348, 349, 354, 363, 365, 368

Baird, Donna D., 373
Bakwin, H., 329, 348
Balish, Edward, 322, 349
Ban, Sadayuki, 109, 186, 323, 349
Barnett, Henry L., 364
Barny, Marie Helena, 332, 349
Barton, Sara E., xi
Beadle, George, 48
Bean, Michael A., 362

Beardsley, Richard K., 324, 349
Beatty, John, 330, 349
Beavers, Terry, 350
Beck, W. S., 372
Beebe, Gilbert W., xi, 77, 87, 306, 314–316, 349, 355
Bell, William E., 352
Belsky, Joseph L., 355, 366
Bender, Lauretta, 66, 206, 209, 326, 349
Bender, Michael A., 322, 349
Benjamin, A. Steven, 354
Bennacerraf, Baruj, 349
Bennett, John M., 371
Bentley, David, 327, 358
Bergeron, R. Thomas, 324, 349
Bergman, N. Ake, 355
Berlin, E. A., 310, 350
Bigbee, William L., 362
Bjelkengren, Goran, 355
Bloch, Melvin, 18, 297
Bloom, Arthur D., 328, 349
Blot, William J., 238, 310, 322, 324, 328, 347, 349, 350, 361
Bock, Frederick, 1
Boerwinkle, Eric, xi
Bohr, Niels, 10
Boice, John D. Jr., 330, 355
Booz, J., 352
Borges, Jane, 66, 312
Borges, Wayne, 59, 61, 62, 311, 352
Bourgin, Cherie, xi
Boyle, Robert, 107
Brannon, R. B., 350
Brewer, Richard, 40, 111
Brizzee, Kenneth R., 326, 328, 350, 356
Bronk, Detlev, 307
Brooks, Antony L., 349

Brown, Charles C., 357
Browne, Thomas, 1
Brues, Austin M., 18, 297, 306, 307, 350
Buckley, Jonathan D., 356
Bugher, John, 76
Burnett, Charles, 75, 314
Burrows, Gerald, 97, 312, 350
Butland, B. K., 357

Canfield, Robert E., 373
Cannan, R. Keith, 25, 73, 76, 89, 299, 307, 314, 316
Carter, Randolph L., xi, 371
Cavagnaro, Louise, 22
Caviness, Verne S. Jr., 327, 350
Cedarquist, Ebbe S., 355
Cepko, Constance L., 327, 372, 373
Chang, Washou P., 331, 350
Charles, Monty W., 331, 359
Chaskes, S., 349
Cheeseman, Edward A., 328, 350
Cheng, Chia C., 372
Chrustchoff, Gyorgy, 350
Clarren, Sterling K., 328, 350
Coffigny, H., 366
Cogan, David G., 56, 310, 351
Cologne, John B., xi
Combs, Barbara, 352
Conard, Robert A., 318, 351
Conquest, A. J., 357
Copenhaver, Edward H., 351
Cox, Ann B., 310, 351, 359
Cragle, Donna L., 354, 373
Craig, Cecil, 48
Craig, Jean, 324
Cramtz, A., 366
Crease, Robert P., x, 303, 351
Crick, Francis H. C., 268, 373
Crismon, Catherine S., 354
Crow, James F., 33, 34, 98, 300, 316, 361

D'Agostino, A. N., 350
D'Amato, Constance J., 328, 351, 355
Daiger, Stephen, xi
Dann, M., 351
Danzig, Michael D., 366
Darby, Sarah C., 322, 331, 351, 361
Darling, George, 76, 77, 89, 299, 300
Day, Geoffrey, 57
Decoursey, Elbert, 76
Denit, Guy, 12
Dobbing, John, 327, 351
Dobson, R. Lowry, 322, 351
Dock, Donald D., 366
Dohi, Kiyohiko, 349, 356
Dohy, Hiroo, 365
Dolatshahi, Farhad, 358
Donaldson, David, 57

Donoso, J. Alejandro, 232, 328, 351
Dorn, Harold F., 322, 351
Doty, Richard L., 329, 351
Down, Langdon, 253
Downes, Susan, 353
Driscoll, Shirley G., 229, 328, 351
Duke-Elder, W. Stewart, 328, 351
Dunn, Kevin, xi
Dunn, Kimberley, xi, 352
Dunn, Val, 324

Easterday, Charles L., 351
Edelman, Gerald M., 220, 327, 352, 366
Edelson, Edward, 332, 352
Edington, Charles W., xi
Egbert, Stephen D., 358, 370
Eisenbud, Merril, 25, 307, 352
Elkabes, Stella, 373
Ellett, William H, xi, 317, 352
Engel, Heinrich, 317, 352
Epstein, Stephen E., 321, 330
Ericsson, Ulla-Brit C., 355
Eto, Ryuzo, 374
Evans, H. John, 349
Evarts, Edward V., 325, 326, 352
Ezaki, Haruo, 349, 353

Feinendegen, Ludwig E., 328, 352
Fell, T. P., 357
Felton, James S., 322, 357
Finch, Clement A., 352
Finch, Stuart C., xi, 86, 89, 315, 321, 352
Finkel, R. C., 370
Fischoff, Baruch, 332, 352
Floro, S., 367
Follette, J. H., 372
Folley, Jarrett H., 62, 311, 352
Forrestal, James, 18, 19, 297, 306
Francis, Thomas, 73, 75, 76, 313, 337, 353
Freedman, Lawrence R., 89, 328, 353
Frome, Edward L., 373
Fry, R. J. Michael, 284, 370
Fry, Shirley A., 354
Fujii, Hiroshi, 369
Fujiki, Norio, 96
Fujikoshi, Yasunori, 364
Fujino, Tadashi, 56, 298, 310, 355, 357, 360
Fujisawa, Hideo, 314, 316, 349, 357
Fujita, Mikio, 363, 367
Fujita, Shoichiro, xi, 305, 320, 348, 353, 368
Fujito, Junro, 370
Fujiwara, Saeko, 301, 322, 353
Fukunaga, Yasuo, 366
Furonaka, Hiroshi, 366
Furusho, Toshiyuki, 260, 330, 353
Fushiki, Shinji, 328

Author Index

Galaburda, Albert M., 327, 353, 357
Gardner, Ernest, 327, 364
Gardner, Martin J., 330, 353
Garte, Seymour J., 364
Geschwind, Norman, 327, 353, 357
Ghosh, Anirvan, 360
Gilbert, Ethel S., 320, 322, 331, 354
Gilbert, Susan L., 365
Goldstein, Leopold, 65, 311, 354
Goodenough, Florence L., 66, 206, 209, 326, 354, 362
Goodill, A. A., 357, 359
Goodpasture, Earnest, 25
Gordon, Tavia, 366
Goriki, Kazuaki, 363, 367
Goto, Ikuo, 368
Gove, H. E., 370
Graham, Philip, 367
Grayson, Kevin, xi
Gregory, Peter B., 165, 322, 354
Greulich, William W., 52, 54, 310, 354
Gritzner, Michael, 358
Groer, Peter G., 349
Guralnick, Lorraine, 374
Gurkewitz, R., 367

Hachiya, Michihiko, 7
Hakoda, Masayuki, 323, 354
Hall, Andrew J., 353
Hall, Carl W., 310, 354
Hall, John W., 349
Ham, William T. Jr., 310, 317, 354
Hamada, Tadao, 348
Hamilton, B. F., 354
Hamilton, Howard B., xi, 312, 348, 350, 355, 357, 358, 367, 368, 369, 371
Hamm, Peggy B., 365
Hammer, Ronald P., 328, 354
Hammond, E. Cuyler, 364
Hanis, Craig, xi
Harada, Tomin, 153, 299, 321, 354
Harkedar, I., 367
Harris, Dale B., 206, 209, 317, 326, 354, 362
Hartmann, William H., 371
Hashizume, Tadashi, 320, 354
Hasuo, Kanehiro, 368
Hattori, Fumio, 367
Hayakawa, Norihiko, 305, 354, 358
Hayashi, Ichiro, 84
Hayashi, Tomonori, 356
Hayashi, Yuzo, 356
Haylock, R. G. E., 357, 359
Hazama, Ryuji, 363
Henderson, Brian E., 356
Henshaw, Paul, 18, 297, 306
Henson, Donald E., 371
Herman, Cynthia S., 370

Hicks, Samuel P., 232, 328, 351, 355
Hikida, H., 322
Hill, A. Bradford, 277
Hirai, Tsuyoshi, 373
Hirai, Yuko, 358
Hiraoka, Toshio, 349
Hirose, Kinnosuke, 56, 298, 310, 355
Hirose, Kozumi, 374
Hodgkin, Thomas, 157
Hoel, David G., 357, 368
Hollerith, Herman, 68
Hollingsworth, Dorothy R., 156, 299, 322, 355
Hollingsworth, James W, 88, 89
Holm, Lars-Erik, 331, 355
Holmes, Robert, 298, 307, 314
Hombo, Zenichiro, 367
Honda, Takeo, 348, 365
Horn, James I., 322, 351
Hoshi, M., 348, 370
Hoshino, Takashi, 352, 355
Hosoda, Yutaka, 353, 367
Howell, R. Rodney, 358
Hrubec, Zdenek, 312, 330, 350, 352
Hsu, Tao Chiuh, 310, 355
Hungerford, David, 310
Hurst, G. Samuel, 317, 366
Hutchins, Grover M., 361
Hutt, M. L., 326, 355
Hutter, Robert V.P., 371

Ibuse, Masuji, 303, 316, 355
Ichimaru, Michito, 311, 321, 322, 355, 371
Ichinose, Kengo, 369
Ihno, Yu, 367
Iijima, Soichi, 348
Ikeda, Takayoshi, 84, 360, 371, 372
Ikui, Hiroshi, 56, 298, 351
Ikushima, T., 367
Imamura, Kunihiro, 367
Imanaka, Fumio, 371
Inskip, Mary J., 355
Ishibashi, Koichi, 370
Ishida, Kenzo, 348
Ishida, Morihiro, 153, 299, 305, 321, 322, 354, 356
Ishimaru, Nobumasa, 369
Ishimaru, Toranosuke, 328, 355, 356, 363
Itakura, Hideyo, 348
Ito, Chikako, 356
Ito, Koichi P., 101
Ito, Takashi, 323, 356, 367
Itoga, Takashi, 352
Itoh, Masahiro, 348, 365
Izumi, Shizue, 364, 365, 371

Jablon, Seymour, xi, 73, 87, 320, 322, 328, 330, 356

Jackson, D. A., 357
Jacobs, L. A., 356
Jacobson, Marcus, 327, 356
Jaenke, Roger S., 328, 356
Jefferson, Thomas, 295
Jensen, Ronald H., 362
Johnson, Kenneth G., 89, 363, 373
Johnson, Marie-Louise T., 354
Joji, Kenji, xi
Jones, K. P., 321, 330, 356
Jones, Peter A., 356

Kaack, M. Bernice, 350
Kabuto, Michinori, 347
Kagan, Abraham, 366, 371
Kageoka, Takeshi, 363
Kamada, Nanao, 319, 356, 365, 369, 371
Kamei, Terumi, 369
Kanda, Keiji, 358
Kastenbaum, Marvin, 62, 311, 321, 361
Kathren, Ronald L., 316, 357
Kato, Hiroo, xi, 305, 320, 321, 322, 326, 328, 329, 347, 348, 350, 351, 353, 356, 357, 363, 366, 368, 371, 372, 374
Kato, Kazuo, 357
Kau, Jih-Wen, 331, 350
Kaul, Dean C., 319, 358, 371
Kawamoto, Sadahisa, 328, 356, 357, 366, 367, 373
Keehn, Robert J., 326, 328, 353, 357, 367, 373
Kemper, Thomas L., 327, 353, 357
Kendall, G. M., 331, 357, 359
Kerr, George D., xi, 317, 320, 357, 358
Kido, Riichi, 369
Kimler, Bruce E., 372
Kimura, T., 367
Kimura, Samuel J., 56, 351
Kirk, Norman, 18
Kirshbaum, Jack D., 355
Kitamura, Saburo, 49, 50
Klose, Monika, 327, 358
Kneale, George, 331, 358
Kobara, Thomas Y., 371
Kobayashi, K., 367
Kobayashi, T., 367
Kobayashi, Takuro, 368
Kodama, Kazunori, 322, 353, 366
Kodama, Yoshiaki, 348, 363
Kodani, Masuo, 40, 51, 171
Koga, Yukiyoshi, 66, 195, 208, 233, 313, 324, 358, 371
Kogure, Takashi, 367
Komai, Taku, 37
Kondo, Sohei, 367, 368
Konermann, Gerhard, 326, 358
Konuma, N., 322
Kosako, Toshio, 358

Krooth, Robert, 94, 316, 358
Kruger, Peter G., 310, 347
Kubik, P. W., 370
Kubota, Tadashi, 369
Kudo, Akio, 101
Kumami, Takehiko, 303
Kuramoto, Atsushi, 356, 365, 371
Kurasaki, Hirotani, 359
Kurata, Robert S., 41, 44
Kurihara, Minoru, 354
Kuriyama, Kazutake, 371
Kuroki, Tamesada, 315
Kushiro, Junichi, 347
Kusumi, Shizuyo, 365, 371
Kusunoki, Yoichiro, 347, 358, 362
Kyoizumi, Seishi, 323, 347, 354, 358, 362

Land, Charles E., xi, 162, 322, 347, 358, 367, 371
Lange, Robert D., 63, 64, 311, 359, 361
Langham, Wright H., 304, 359
Langlois, Richard G., 362
Larsson, Lars-Gunnar, 355
Lawrence, John S., 59, 371
Layton, Donald D., 324, 359
Lee, Arthur C., 351, 354, 359
Lejeune, Jerome, 328, 359
LeRoy, George V., 364
Lett, John T., 310, 351, 359
Levan, Albert, 51, 309, 371
Lewis, Susan E., 265, 330, 363
Lidberg, Monika E., 355
Liebow, Averill A., 307, 359, 364
Liechtenstein, Sarah, 352
Lifton, Robert Jay, 169, 303, 359
Lin, Chow L., 366
Lincoln, Abraham, 305
Lindberg, C., 352
Lindberg, R. Sture, 355
Little, Mark P., 330, 331, 353
Littlefield, L. Gayle, 349
Loewe, William E., 317, 358, 360
Lowe, Mary Jane, 355
Lundell, Goran E., 355
Lyons, Mary, 262, 359

Mabuchi, Kiyohiko, 315, 360, 365, 371, 374
MacArthur, Douglas, 10, 11, 19, 25, 343
MacGibbon, B. H., 357, 359
Machado, Stella G., 358
Machle, Willard, 25, 26, 307
Madhukar, Burra V., 372
Maeda, Tetsuo, 364
Magura, Beth, xi
Maisin, Jean-R., 366
Maki, Hiroshi, 27, 298
Malenfant, Richard E., 317, 360

Marcum, Jess, 358
Marks, Sidney, 358
Marmot, Michael, 366
Mitchell, Thomas, 284, 370
Martin, Joan C., 370
Martin, S. Forrest, 56, 351
Martinez Martinez, P. F. A., 327, 360
Maruki, Iri, 303, 360
Maruki, Toshi, 303, 360
Maruyama, Takashi, 354
Marx, J., 352
Masuyama, Motosaburo, 305, 360
Matsubayashi, Ikuzo, 40, 78, 305, 356
Matsui, Takashi, 348
Matsumae, Shigeyoshi, 304, 360
Matsumoto, Y. Scott, 170, 322, 358, 360
Matsumura, Hideo, 327, 363
Matsunaga, Haruji, 359
Matsunaga, Masako, 355
Matsuo, Miyo, 367
Matsuo, Tatsuki, 365, 371
Matsuoka, Shigeru, 359
Matsushita, Koji, 328
Matsuura, Hiroo, 322, 360
Matsuzaka, Yoshimasa, 37, 308, 360
Mattar, Adel G., 324, 360
Mavor, James W., 51, 309, 360
Mayer, M., 359
Mayumi, Makoto, 357
McConnell, Susan K., 327, 360
McConney, Mary E., 365
McDonald, Duncan J., 309
McDowell, Arthur, 87
Meiji, Emperor, 21, 199
Mendelsohn, Edgar, 317, 348, 358, 360
Mendelsohn, Mortimer L., xi
Menzies, John, 28
Miezejeski, C. M., 373
Mikami, Motoko, 355
Mikhail, Mikhael A., 324, 360
Miller, Richard C., 349
Miller, Robert J., 310, 328, 354, 360
Miller, Robert W., xi, 192, 315, 324, 350, 360, 361
Mills, H., 372
Milton, Roy C., 317, 344, 354, 361
Mindich, Bruce, 364
Mitchell, Toby J., 370
Miyanishi, Michihiro, 366
Mizuno, Shoichi, 317, 370
Mizuno, Terumi, 356
Mock, Teresa, 352
Modan, Baruch, 367, 373, 374
Modan, Michaela, 373, 374
Mole, Robin H., 326, 361
Moloney, William C., xi, 62, 63, 64, 311, 321, 359, 361
Monzen, Tetsuo, 84

Moore, Felix, 73
Moore, Keith L., 325, 361
Moore, G. William, 361
Moriyama, Iwao, 350, 374
Morrow, Lewis B., 363
Morton, John, 298
Morton, Newton E., 98, 309, 316, 361
Mountcastle, Vernon B., 230, 328, 361
Mueller, Charles F., 324, 361
Muhlensiepen, H., 352
Muirhead, Colin R., 331, 357, 359, 361
Muller, Fabiola, 326, 364
Muller, Herman J., 98, 316, 361, 362
Mulvihill, John J., 324, 349, 361
Munaka, Masaki, 354, 356
Murphy, Douglas P., 65, 311, 354, 362

Nagai, Isamu, 27, 299
Nagai, Takashi, 303
Nakagawa, K., 374
Nakagawa, Shigeru, 370
Nakaidzumi, Masanori, 27, 298
Nakamura, Kumiomi, 155, 362
Nakamura, Nori, xi, 322, 323, 348, 356, 362
Nakano, Mimako, 348, 365
Nakashima, Eiji, 351, 356
Naruge, Tetsuji, 315, 362
Neel, James V., xi, 18, 26, 40, 48, 258, 264, 265, 297, 306, 307, 309, 311, 316, 329, 330, 363, 364, 368, 370, 371, 374
Nefzger, M. Dean, 352, 354, 360
Nehemias, John V., 310, 363
Neriishi, Kazuo, 347, 353
Neriishi, Shotaro, 327, 348, 349, 363, 367, 374
Nesta, Douglas, 364
Nieburgs, Andra, 373
Niimi, Masanobu, 356
Nirenberg, Marshall, 268
Nishikawa, T., 322
Nishiki, Masayuki, 349
Nishimori, Issei, 84
Nishimura, Akihisa, 354
Nishimura, Edwin T., 371
Nishimura, Hideo, 326, 363
Nishina, Yoshio, 10
Nishioka, Kusuya, 357
Nishitani, Hiromu, 368
Nishizawa, Kanae, 354
Nisula, Bruce C., 373
Noble, Kenneth B., 111, 317, 363
Nonaka, Hiroaki, 365
Norton, Stata, 232, 328, 351, 372
Nowell, Peter, 310

O'Connor, John F., 373
O'Hagan, J. A., 357

O'Rahilly, Ronan, 326, 327, 361, 362, 364
Ochi, Yoshimichi, 360
Ochikubo, Sachio, 360, 371
Ogura, Yoshio, 371
Ohara, Jill L., 320, 322, 354
Ohkita, Takeshi, 354
Ohkura, Koji, 96
Ohtaki, Kazuo, 322, 328, 348, 363, 365
Oka, Y., 367
Okada, Hiromu, 355
Okajima, Shunzo, 318, 363
Okamoto, Naomasa, 44
Okamoto, Yoshio, 111
Okunuma, N., 322
Omori, Yoshiaki, 330, 363, 373
Omuta, Minoru, 67, 303, 313, 364
Ordy, J. Mark, 328, 350
Osgood, Edwin, 310
Otake, Masanori, xi, 260, 310, 324, 328, 330, 348, 352, 353, 354, 360, 363, 364, 367, 368, 374
Otsuru, Shin, 374
Oughterson, Ashley, 11, 12, 297, 303, 304, 339, 364

Pace, Joseph V., 358
Parran, Thomas, 76, 314
Parry, Dilys M., 349
Pasternack, Bernard S., 369
Pastore, John O., 367
Pearson, Ted A., 349
Penn, Arthur, 321, 364
Penney, William, 10, 11, 304, 365
Pepper, Larry J., 365
Pereira, Carlos, 349
Perets, H., 373, 374
Petersen, Gerald R., 316, 354, 357
Peterson, Lief, xi, 365
Pfaff, Donald W., 327, 368
Pierce, Donald A., 326, 365, 368
Pike, Malcolm C., 356, 358, 359
Pinkston, John A., 321, 348
Piotrowski, J. K., 355
Pisciotta, Anthony V., 371
Plummer, George, 311, 312, 365
Poortman, F., 366
Popper, Jordan S., 371
Porschen, W., 352
Powell, Caroline A., 353
Prechtl, Heinz F. R., 329, 372
Preston, Dale, xi, 152, 319, 321, 322, 360, 365, 368, 370, 371
Preston, R. Julian, 349
Putnam, Frank W., xi, 306, 365
Rakic, Pasko, 230, 327, 365, 366, 369
Rappaport, Michael, xi
Read, Stephen, 352

Redesdale, Lord (Algernon Mitford), 315, 366
Renwick, James, 28
Restak, Richard, 313, 366
Rethore, Marie-Odile, 359
Reyners, H., 327, 328, 366
Reynolds, Earle L., 53, 310, 366
Reynolds, George T., 10
Rhoads, George G., 366, 371
Rhodes, Richard, 366
Rinott, Yoseff, 374
Ritchie, Rufus H., 317, 366
Rivers, Thomas, 19
Robertson, Horace C. H., 24
Robertson, Thomas L., 316, 321, 366
Roesch, William C., 317, 336, 367
Ron, Elaine, 328, 360, 365, 367, 371, 373, 374
Rosenbaum, Jack D., 364
Ross, Ronald K., 356
Rotblat, Josef, 305, 367
Rouma, Gustavo, 325, 367
Rudelli, R. D., 373
Russell, Walter J., 127, 328, 348, 357, 367
Russell, William, 265
Rutland, Barbara K., 352, 369
Rutter, Michael, 329, 367

Sadamori, Naoki, 355
Sager, Ruth, 31, 367
Saigusa, S., 367
Saito, I., 367
Sakamoto, Goi, 367
Samios, Nicholas P., x, 304, 365
Samuels, Dennis E. J., 11, 304, 365
Sanders, Fred, 109
Sands, Jean, 327, 351
Sasagawa, Sumihiko, 322, 367
Sasaki, M. S., 319, 367
Sasuga, N., 367
Sato, Junichiro, 369
Sato, Kazuyoshi, 373
Sato, Takemasa, 369
Satoh, Chiyoko, xi, 259, 330, 363, 367, 368
Saunders, Richard D., 360, 369
Sawada, Hisao, 49, 50, 320, 322, 350
Sawada, Shozo, 357, 367
Scheibel, Arnold B., 354
Schneider, B. Aubrey, 364
Schlatterer, John P., 373
Schull, William J., 1, 305, 306, 309, 310, 311, 316, 321, 324, 328, 329, 330, 352, 353, 357, 363, 364, 367, 368, 372, 374
Schwanzel-Fukuda, Marienne, 327, 368
Scorgie, Guy C., 11, 304, 365
Scott, James, 321, 368
Scott, William H. Jr., 358

Author Index

Segan, S. M., 373
Sekine, Ichiro, 360
Serber, Robert, 10
Sersen, E. A., 373
Setlow, Richard B., 349
Seyama, Shinichi, 348
Seyama, Toshio, 356
Shaman, Paul, 351
Sharma, P., 370
Shatz, Carla J., 360
Sheldon, William, 53
Shibukawa, Tetsuo, 364
Shigematsu, Itsuzo, 300
Shigeta, Chiharu, 356
Shimada, Katsunobu, 370
Shimaoka, Katsutaro, 353
Shimba, Hachiro, 348
Shimizu, Nobuhiro, 355
Shimizu, Sakae, 304, 368
Shimizu, Yukiko, xi, 31, 322, 350, 366, 368, 371
Shinoda, Yoshikazu, 326, 352
Shirabe, Raisuke, 9, 303, 304, 305, 369
Shohno, Naomi, 303, 369
Shohoji, Takao, 317, 344, 361
Shore, Roy E., 328, 369
Shy, Carl M., 373
Sidman, Richard L., 327, 366, 369
Siegel, David G., 322, 369
Sienkiewicz, Zenon J., 329, 369
Silk, T. J., 357
Sinskey, Robert M., 310, 369
Slavin, Richard E., 330, 369
Slovic, Paul, 332, 352, 370
Smith, David W., 370
Smith, Wilbur, 352
Smithies, Oliver, 256, 330
Snee, Michael P., 353
Snell, Fred M., 308, 370
Snyder, Laurence H., 48
Soda, Midori, 360, 365, 371, 374
Sofuni, Toshio, 348
Solomon, Daniel L., 317, 358
Speir, Edith, 321
Sposto, Richard, 322, 353, 363, 370
Stern, Curt, 48
Stevens, Richard G., 347
Stewart, Alice, 331, 351, 358
Storer, John, 284, 331, 370
Stott, Dennis H., 329, 370
Stram, Daniel O., 317, 365, 370
Straume, Tore, 319, 370
Streissguth, Ann P., 328, 370
Sugimoto, Sumio, 360, 361
Sumi, Mark S., 350
Sutow, Wataru W., 53, 54, 310, 370
Suyama, Hirofumi, 369
Suzuki, Masamichi, 47, 171, 309, 370

Swinbanks, David, 332, 370
Syme, S. Leonard, 366
Szathmary, Emoke J. E., xi

Tabuchi, Akira, 370
Taeuber, Irene, 315, 371
Tagawa, Daisaburo, 374
Takahara, Shigeo, 93, 316, 371
Takahashi, Norio, 363
Takahashi, Shoshiro, 369
Takaki, Kenkichi, 364
Takeshima, Koji, 41, 44
Takeshita, Kenji, 367
Tamagaki, Hideo, 355
Tamaki, Hideo, 374
Tamaki, Masao, 300
Tanabe, Kazumi, 358
Tanebashi, Masanori, 313, 371
Taura, Tadashi, 354
Taylor, H. Grant, 25, 28, 29, 298, 307, 371
Terasaki, Masayuki, 360, 371
Terrell, John D., 353
Tessmer, Carl, 25, 28, 297, 298
Thiessen, Joop W., xi, 319, 371
Thomas, J. M., 357, 359
Thomas, Robert E., xi
Thompson, Desmond E., 322, 360, 365, 371
Thorslund, Todd W., 373
Tiselius, Arne, 256
Tjio, Joe-Hin, 51, 309, 371
Tokunaga, Masayoshi, 321, 358, 359, 360, 371
Tokuoka, Shoji, 84, 321, 371
Tomiyasu, Takanori, 355
Tomonaga, Masanobu, 359
Tomonaga, Masao, 321, 355, 365, 371
Touwen, Bert C. L., 329, 372
Toyota, Emiko, 367
Trosko, James E., xi, 321, 372
Truman, Harry S, 19, 297, 306
Trumbull, Robert, 304, 372
Tsuchimoto, Taiso, 355
Tsuiki, S., 322
Tsunoo, Susumu, 9
Tsuyama, Naohiro, 356
Tsuzuki, Masao, 11, 304
Turner, Margaret L., 252, 262, 354
Turpin, Raymond, 359

Ullrich, Frederick M., 18, 297
Umeki, Shigeko, 358
Usagawa, Mitsugu, 314, 349

Vaeth, Michael, 365
Valentine, William N., 311, 372
Vidal-Pergola, Gabriela M., 326, 372

Voelz, George L., 354
Vogel, Friedrich, 308, 372
Vogelstein, Bert, 321

Wachholz, Bruce W., 349
Wagner, Louis, xi
Wakabayashi, Toshiro, 372
Walby, A. L., 328, 350
Walker, Joan M., 324, 372
Walsh, Christopher, 327, 372, 373
Wanebo, Clifford K., 154, 300, 321, 373
Ward, Robert E., 349
Warren, Lisa, 364
Warren, Shields, 76, 303, 304, 364
Warren, Stafford L., 304, 316, 373
Watanabe, Ronald S., 324, 373
Watanabe, Teruo, 47, 309, 370
Watson, James, 268, 373
Webb, M. A., 357
Wedemeyer, William, 44
Weed, Lewis, 18, 24
Weinberg, Clarice C., 373
Weinberg, Robert A., 321, 373
Wells, Warner, 28
Wertheimer, M., 326, 373
West, Emory, 54, 310, 370
Wheater, A. W., 330, 356
Wicklund, Harriet V., 355
Wiggs, L. D., 354
Wiklund, Kerstin C., 355
Wilcox, Allan J., 330, 373
Williams, George R., 351
Williams, Roger S., 273
Wilson, Robert R., 105, 316, 344, 373
Wing, Steve, 331, 373
Winternitz, Milton, 24
Wise, Stephen P., 326, 352
Wisniewski, K. E., 328, 373
Wolf, Susanne, 373
Wong, F. Lenny, 322

Wood, James W., 373
Wood, Joy L., 373
Woodbury, Lowell, 111
Woolson, William A., 370
Worth, Robert, 366
Wray, Susan, 327, 373
Wright, Phyllis M., 184, 311, 374
Wright, Stanley W., 184, 311, 359, 374

Yaar, Israel, 328, 373, 374
Yaar, Mina, 374
Yamada, Hiroaki, 111, 358
Yamada, Michiko, 322
Yamada, Yutaka, 358
Yamakido, Michio, 354, 367
Yamamoto, Tsutomu, 322, 360, 371, 374
Yamane, Motoi, 349
Yamasaki, Mitsuru, 314, 316, 349
Yamasawa, Yoshimichi, 308, 374
Yamashita, Akiko, 300, 367
Yamawaki, Takuso, 59, 61, 62, 298, 311, 352, 359
Yamazaki, James N., 65, 184, 191, 304, 311, 323, 374
Yanase, Toshiyuki, 96
Yokoro, Kenjiro, 356
Yokota, Seichiro, 327, 374
Yorichika, Edward, xi
York, Edwin N., 317, 344
York, Herbert, 105
Yoshida, M. L., 348
Yoshimaru, Hiroshi, 324, 352, 364, 368, 374
Yoshimitsu, Kengo, 353
Yoshimoto, Keiko, 348
Yoshimoto, Yasuhiko, 328, 329, 363, 367, 374
Yoshinaga, Haruma, 367
Young, Robert W., 319, 374
Yule, William, 367

Subject Index

Abortion, 28, 47–51, 97, 267, 324, 370
  induced, 47, 370
  spontaneous, 28, 49–51, 97, 267
Academy (see U.S. National Academy of Sciences)
Acatalasemia, 93, 94
Acatalasia, 94
Accidents, 35, 135, 146, 156, 240, 249, 274, 275, 292–295, 305, 331, 338, 356
  nuclear, 292–295
Achlorhydria, 155, 156, 333
Acid rain, 294
Acute meningitis, 227
Acute radiation sickness, 7, 10, 12, 14, 55, 64, 87, 149, 260, 304, 324
Adenine, 180, 268
Adenoma, 168
Adenomyomatoses, 158
Adult Health Study, 79, 87–90, 93, 99, 100, 141, 142, 154, 155, 165, 167, 168, 187, 234, 299, 301, 333, 349, 357, 373
Aga, 27
Age, 54, 62, 70, 134, 154, 156, 185, 189, 190, 192, 198, 199, 204, 205, 210–212, 245, 281
  developmental, 54, 190, 212
  effect of age at exposure on radiation-related risk, 62, 70, 134, 154, 156, 185, 189, 190, 210, 211, 245, 281
  gestational, 190, 192, 198, 199, 204, 205
Aging, 137, 187, 239, 240
  premature, 137
  process, 137, 187
  signs of, 137

Alcohol, 160, 188, 231, 354, 370
Alkaline phosphatase, 64
Alpha particles, 174, 318, 333, 340
Amino acid(s), 180, 256, 257, 268, 323
Anemia, 60, 64, 65, 340, 355, 359
  aplastic, 64, 65, 355
Anencephaly, 225, 246, 333, 362
Aniridia, 252
Ankylosing spondylitis, 158, 185, 288, 351
Anomaly, developmental, 41, 44, 189, 227, 229, 235, 255, 327, 336, 365
Anomaly, minor, 41
Anthropometrics, 52, 54, 96, 204, 232, 353, 356
Antibiotics, 14, 276
Antibody, 143, 144, 322, 323, 333, 339, 357
  monoclonal, 322, 323
  secreting cell(s), 143
Antigen, 143, 158, 333, 334, 338, 340, 348, 357, 358
  carcinoembryonic, 158, 334
Argininosuccinate synthetase, 270, 271
Assembly of Life Sciences, 306
Association for the Aid to Crippled children, 94
Atheroma(s), 140, 141
  formation of plaques, 140
Atherosclerosis, 100, 139, 140, 321, 334, 368
  circle of Willis, 100
Atomic Bomb Casualty Commission, 18–38, 40, 44–49, 52, 53, 59, 61–64, 66, 68, 72–78, 82, 83, 85–87, 89, 92–95, 97, 99, 125, 127, 195, 198, 211, 236, 243, 245, 278, 297–300, 303, 304, 306–308, 310, 314–317, 332,

383

Atomic Bomb Casualty Commission
  (*cont.*) 333, 337, 338, 339, 343,
  347–350, 352–358, 360, 361, 363–
  367, 369, 370, 373, 374
  administrative and scientific planning,
    24
  budget, annual operating, 32, 34, 308
  creation, 19
  data and sharing, 92
  dissolution, 32
  early administration, 19–25
  educational program, 35, 75
  enabling agreements, 77
  founding of Hiroshima Laboratory, 21
  founding of Nagasaki Laboratory, 21
  labor union, 29
  labor-management relations, 29
  legal status, 28, 36
  naming, 18
  possible early termination .25
  program evaluation, 34, 73–77
  research strategy, 34, 36, 38, 72, 73
  security clearance, 22
  staff, reduction in, 32
  vestige of Occupation, 21
Atomic Bomb Sufferers Medical Treatment Law, 35, 127, 130, 131
  definition of a survivor, 130, 131
  health handbook, 131
Atomic Energy Commission, 11, 23, 94, 97
Australia, 24, 348
  exclusionary policy toward Asians, 24
Autopsies, 30, 44, 73, 86, 139, 141, 166, 226, 228, 230, 309, 374
  confirmation of cause of death, 166
  detection of cause of death, 166
Axon(s), 215, 334, 341, 360

Becquerel, 334, 336
  definition, 334
Biochemical studies, 256–259
Biostatistics, 37, 73, 101, 134
Birth certificates, 48, 312
Birth defects, 188
Birth injuries, 45, 202
Birth rate, drop in, 46–48
Black rain, 105, 355
Blast, 9, 11, 12, 115, 118, 304, 318, 320
  damage, 11, 12, 304
  wave, 9, 11, 115, 118, 304, 318, 320
Bleeding, 12, 60, 87, 100, 340, 342
Board of Directors, 33, 35
Bock's Car, 1, 2, 10
Bomb, time of detonation, 117
Bombing, biologic and physical effects, 10–17
  early assessment, 10–17

Bone, 14, 39, 46, 53, 54, 59, 60, 64, 126,
  142, 143, 151, 153, 157, 168, 170,
  181, 189, 212, 229, 232, 305, 316,
  318, 334, 352, 367
  formation, centers of, 54, 232
  marrow, 14, 39, 59, 60, 64, 126, 142,
    143, 151, 153, 157, 170, 181, 212,
    229, 305, 334, 352
Brain, 55, 67, 100, 121, 137, 158, 185,
  188–190–192, 195, 196, 198, 200–
  204, 208, 209, 212–217, 219, 220,
  222, 223, 225–232, 235, 239–242,
  273, 325–328, 333–337, 340, 341,
  343, 345, 350–356, 358–361, 365,
  366, 368, 369, 372–374
  abnormalities, 202, 368
  atrophy, 230, 239, 366
  caudate nucleus, 227, 232
  cerebellum, 191, 214–216, 225, 227–
    230, 241, 325, 327, 333–335, 341,
    354, 366
  cerebrum, 190, 191, 214–216, 218, 222,
    225, 227, 228, 232, 241, 327, 333–
    335, 337, 345
    cortex, 191, 208, 214, 219, 225, 226,
      351, 355, 356, 360, 372, 373
    hemispheres, 216, 231, 242, 327, 335
    ventricles, 218, 227, 228, 232, 327,
      337, 345
  cingulate gyrus, 228
  cisterna magna, 228, 229, 335
  corpus callosum, 227, 228, 232, 328,
    340
  cortical function, 196, 225, 231, 273
  damage to developing, 55, 189, 201,
    214, 215, 225–232, 374
  definitive architecture, 190
  development, 201, 203, 215–225, 232,
    240, 327, 366
  ectopic gray matter, 202, 203, 226, 228,
    231, 232
  ependyma, 337
  hippocampus, 214, 242, 350
  intermediate zone, 216–220
  LHRH cells, 327
  mantle, 225
  marginal zone, 216
  migration layer, 216
  motor cortex, 209, 325, 326
  premotor cortex, 325
  proliferative zone, 191, 219, 220, 226,
    228, 341
  somatosensory cortex, 208, 230, 343
  stem, 215, 225, 241, 334
  subventricular zone, 218
  thalamus, 208
  tumors, 158, 185
  uncertainties in risk estimates, 209–214

Subject Index

ventricular zone, 217, 222
vulnerable periods in development, 191, 201, 202, 213, 235
weight, 227, 230, 327, 354
British Commonwealth Forces, 24, 28, 31
Budget, annual operating, 32, 34, 308
Buildings, conventional Japanese, 124
Bulb, olfactory, 241, 327
Burns, flash, 165
Burst point, 3, 11, 105, 108, 124, 320, 337, 339

Calcium, loss of, 168
Cancer, 19, 38, 46, 55, 59, 61, 62, 65, 85, 86, 89, 99, 100, 125, 128, 131, 134, 135, 137–140, 142–168, 170–172, 178, 186, 189, 211, 229, 235–240, 242, 243, 249, 250, 252, 263, 265, 266, 272–274, 277–281, 283–289, 293, 299–301, 304, 313, 315, 319, 331–337, 341, 343, 347–349, 351, 352, 354–360, 362, 365, 368, 369, 371, 373, 374
  age of onset, 159
  attributable to exposure, 160
  baseline rates, 157, 165
  biological bases, 142–148
  breast, 46, 125, 128, 135, 142, 154, 155, 159, 161–163, 166, 186, 272, 285, 286, 300, 334, 349, 358, 359, 371, 373
  carcinoma, 153, 156, 186, 229, 299, 335, 348, 356
  cervix of the uterus, 65, 166, 229
  childhood, 235–237, 252
  colon, 157, 158, 272, 285, 287, 334
  detection, 89
  effect of age at onset, 159
  epidemiology, 85
  except leukemia, 151, 153–160, 164, 250, 281
  factors affecting risk, 159
  gallbladder, 155, 157, 158, 166, 334
  host factors, 162
  incidence, 59, 62, 85, 128, 149, 153, 155, 157, 158, 160–168, 237, 273, 284, 288, 300, 301, 352, 355, 360, 365, 371
  kidney, 236, 239, 240, 252
  liver, 159, 161
  lung, 99, 128, 142, 153, 154, 159, 163, 272, 285–287, 334
  multiple myeloma, 157, 272, 278, 355, 365
  neuroblastoma, 252
  osteosarcoma, 273
  ovary, 157, 170, 272
  pancreas, 158
  prenatally exposed, 235–238, 241, 242
  rates, 157, 165
    background, 157
  retinoblastoma, 252
  sarcoma, 153, 343
  secondary, 128
  small bowel or duodenum, 158
  stomach, 155, 156, 158, 161, 272, 285–287, 313, 333, 347, 362
  thyroid, 156, 157, 161, 166, 168, 186, 272, 274, 285, 299, 355, 356
    background incidence, 157
  treatment, 142, 166
  uncertainties in risk estimates, 162–167
  urinary bladder, 157, 163, 272
Carcinogenesis, 146, 147, 162, 186, 239, 242, 284, 291, 372
  chemical, 162, 372
  models of, 146, 147, 162
    conversion, 147
    initiation, 147
    initiator, 147
    progression, 147
    promoter, 147
    promotion, 147
  radiation, 242
Catalase, 93, 94, 371
Cell adhesion molecule, 220, 328
  L1, 220
  N-CAM, 328
Cell(s), division, 51, 146, 154, 172, 173, 175, 177, 190, 216, 225, 253, 255, 335, 339
  anaphase, 172
  centromere, 173, 175, 255, 335, 339
  metaphase, 172
  endocrine, 158
  plasma, 157
  premalignant, 147
  red blood, 39, 59, 93, 180, 182, 269, 272, 322, 337, 338, 362
  stem, 60, 143, 144, 154, 170, 180, 182, 216, 230, 237, 339, 343, 344
  terminal differentiation, 147, 339, 341
  white blood, 10, 13, 14, 39, 59, 60, 64, 143, 153, 165, 173, 233, 258, 269, 270, 309, 334, 344, 367
Censorship of scientific findings, 23
Census(es), 61, 62, 66, 68, 77–80, 87, 93, 131, 236, 298, 312, 315
  Hiroshima, daytime, 78, 312
  interim, 78
  local, early, 78
  national, 66, 68, 77, 78, 93, 236, 298
  Rice Ration, 78
  special, 78
Centrifugal Fast Analyzer, 257

Cesium, 119, 120, 275, 318, 332
Chelyabinsk, 35, 331
Chernobyl, 17, 35, 240, 274, 292, 293, 305, 331, 338, 356
Child Health Survey, 94–97, 101
Child Welfare Law, 40
Chinzei school, 125
Cholesterol, 141, 158, 334
Chonaikai, 8, 74, 335, 344
Chondrodystrophy, 252
Chromosome(s), aberration, 163, 172, 178, 187
 abnormality(ies), 51, 52, 60, 144, 147, 172, 175, 178, 181, 233, 234, 252–255, 272, 304
  among F1, 252–256
  among prenatally exposed, 233, 234
 aneuploidy, 253, 262, 263, 333, 363
 autosome(s), 172, 253, 334, 335
 deletion, 51, 173, 174, 175, 235, 336
  interstitial, 175
  terminal segments, 174
 dicentric, 175
 fragile site(s), 337
 fragments, 51
 G-banding, 363
 inversion, 175, 255, 339
  pericentric, 255, 339
 nondisjunction, 51, 253
 normal number, 52, 344, 371
 rearrangements, 253–256
  balanced, 253–256
  unbalanced, 253, 254
 translocation(s), 173, 344
  Robertsonian, 344
 trisomy(ies), 192, 246, 253, 254
  chromosome 21, 192, 246, 253
Chugoku Shinbun, 8, 67
Ciudad Juarez, 275, 331
Civil planning, 7
Clinton Laboratory, 18
Club foot, 246
Cobalt, 109, 232, 265, 275
Codon, 257, 259, 330
Colchicine, 310
Colon, 157, 158, 272, 285, 287, 334
Commission on Life Sciences, 306
Committee on Atomic Casualties, 25, 26, 52, 59, 76, 77, 95, 244, 298, 316
Comprehensive Health Center, 36
Confidence interval, 237, 335
Congenital syphilis, 44
Contaminated soil, 119
 food raised in, 119
Contaminated water, 7
Conversion, 54, 70, 147, 151, 339
Crossroads, 114
Crow Committee, 34

Crystal lattice, 107
Curie, 334, 336
 definition, 334
Cyclotron, 55, 347
Cytosine, 268

Data, analysis, 70
 collection, 83, 88, 317
 noninstitutional requests, 92
 privacy and confidentiality, 49, 74, 82, 83, 92
 processing equipment, 23
 supervision, 88
 verification, 44–46, 69, 77, 114, 116, 244, 258
Death(s), attributable to causes other than cancer, 137
 cancer, 134, 139, 142, 153–157, 160, 161, 164–167, 211, 235–237, 249–250, 263, 280–283, 286, 288, 289
 competition, 135
 certificate(s), 62, 72, 73, 75, 82, 83, 86, 138, 153, 159, 161, 166, 236
  reliability of diagnosis, 166, 351
 confirmation of cause, 166
 crude rate, 320, 321
 embryonic, 267
 infectious disease, 144
 neonatal period, 244, 247, 309
 premature, 136, 137, 165, 243, 267
 rates, 72, 285, 286, 305, 355, 356
 reporting of causes, 86
 schedules, 81, 82
 unknown cause, 251
Defense Nuclear Agency, 304
Development, mental, 42, 66, 206–209, 233, 313, 324, 349, 354, 355, 358, 371
 skeletal evaluation standards, 54
 tests of, 66, 195, 206–209, 233, 313, 324, 349, 354, 355, 358, 371
  Bender-Gestalt, 66, 206, 209, 349, 355
  Draw-a-man, 66, 206, 207, 209, 354
  Koga, 66, 195, 208, 233, 313, 324, 358, 371
  Tanaka-Binet, 208
Diarrhea, 7, 12, 249, 332, 342
 bloody, 7, 12, 332, 342
 infantile, 249
Diet, 141, 244
 changes, 141
 staples, 244
Disease, 1, 7, 14, 28, 39, 42, 43, 46, 50, 59, 60, 63, 64, 70, 75, 76, 85, 94, 95, 97–101, 137–144, 146, 148, 149, 157, 158, 167–169, 187, 192, 211, 246, 249, 272, 277, 286, 288 299,

Subject Index

309, 315, 339, 341, 351, 355, 366, 367, 371, 372
allergic, 97
ankylosing spondylitis, 158, 185, 288, 351
cardiac insufficiency, 139
cardiovascular, 98, 100, 101, 141, 168, 187, 272
cerebrovascular, 94, 98–100, 141
  frequency, Japan, 99, 141
  infarction, 100
coronary artery, 99, 100, 141
coronary heart, 99, 139, 141, 366, 367, 371
defenses against, 142
hypertensive heart, 139
infectious, 14, 144, 286
intractable, 142
Division of Medical Sciences, 18, 24, 73, 306, 314
DNA (deoxyribonucleic acid), 103, 140, 143, 146, 148, 152, 180, 181, 257, 267–271, 335, 336, 340, 344, 365, 373
  damage, 146, 148
  double helix, 268
  from arterial plaques, 140
  from human bladder cancer, 140
  lesions, 146
  restriction fragments, 269, 271
  sequencing, 269, 271
Doctor draft, 90
Dose, 14, 17, 20, 39, 55, 57, 58, 61, 62, 64, 79, 80, 92, 102–130, 132–135, 137, 140–142, 146, 147, 149–152, 154–166, 168, 169, 171, 175, 177–179, 181–184, 189, 191, 192, 195, 196, 198–200, 202–205, 209–214, 227, 229, 232–235, 237–240, 242, 246–251, 254, 255, 258, 260–265, 267, 273, 274, 276–280, 288, 289, 300, 301, 304, 305, 313, 316–320, 325–327, 329–333, 336, 338, 339, 342, 343, 348, 352, 354, 356–359, 361–363, 365, 368, 372, 374
  absorbed, 102, 103, 160, 171, 192, 196, 200, 203, 205, 210, 211, 237, 274, 288, 316, 325, 333, 336, 338, 342, 343
  annual limit, 120
  bone marrow, 181
  doubling, 250, 262–265, 267, 273, 329, 336, 363
  equivalent, 104, 214, 248, 336, 343
  gonadal, 105, 254, 258, 333
  gray, definition, 338
  improbably high, 123
  individual DS86, 123

infinity, 119
kerma, 103–106, 108, 116, 117, 121, 125, 126, 177, 181, 233, 242
medial lethal, 339
organ, 103, 104, 122, 123, 125, 319, 357
random errors in estimates, 179
rate, low, 264, 279, 288
reference man, 122
reference woman, 123
regulatory concern, 124
response function, 152
response relationship, 126
shielded kerma, 103, 125, 126, 233, 242, 339, 343, 369
sievert, definition, 343
specific individual, 121
threshold, 20, 57, 58, 133, 140, 147, 149, 191, 192, 213, 214
uterus, 195
Dose rate, 336
  reduction factor, 336
Dosimetry, DS86, 105, 122, 123, 125, 126, 139, 153, 156, 160, 165, 177, 178, 195, 196, 204, 213, 232, 233, 247, 249, 300, 319, 336, 364, 365, 368–371
  T-57, 87, 105, 299, 344
  T-65, 105, 106, 125, 126, 141, 155, 232, 234, 260, 298, 299, 328, 344
Drosophila, 51, 265
Dyslexia, 226, 336, 353, 357

El Salvador, 275
Electric power, 293, 295
Electrocardiograph, 99
Electrons, 104, 107, 173, 318, 333, 340
  within a crystal, 107
Electrophoresis, 256, 257, 269–271
  denaturing gradient gel, 270, 271
  free, moving boundary, 256
  starch gel, 256
Endonucleases, 269
Enola Gay, 1, 108
enzymes, 243, 257–259, 268, 269, 367
Epicenter, 105, 108, 121, 337, 358
  locating, 108
Epidemiological detection network(s), .73, 76
Epilation, 12, 19, 39, 62, 64, 87, 163, 184, 304, 315, 337, 370
Epilepsy, 202
Epiloia, 252
Epithelium, 55, 241, 335
Epstein-Barr virus, 330
Erythrocyte(s), 39, 59, 64, 93, 94, 153, 180, 182, 258, 269, 272, 322, 337, 338, 362
Esophagus, 157, 272, 286

Eta, 81, 315
Eugenics Committee, 49
Executive Committee, 33, 35, 36
Executive Research Committee, 24
Exons, 186, 268
Exposure, 102–129, 149, 245, 250, 258, 259, 312, 316–320, 330, 332, 357, 359
  delayed radiation, 118
  distal, 245, 258, 259
  fluoroscopy, 127
  in factories, 124
  leukemogenic chemicals, 149
  medical, 129
  occupational, 332, 357
  preconception, 250, 330, 359
  prenatal, 188, 189, 201, 203, 205, 232, 235, 240, 311, 352, 356, 361, 364, 368, 374
  proximal, 245, 312
  therapeutic, 128, 129
Eye (see lens), 182, 235
  color blindness, 182
  segmental iris heterochromia.235

F1 mortality study, 248–252, 254
Fallout, 10, 21, 31, 105, 118–120, 130, 300, 337, 351, 363, 368
  Koi-Takasu, 118, 119
  Nishiyama Reservoir, 118, 119, 300
Family or household register, 79–82, 95, 249, 314, 338, 339
Fat, muscle and bone, absolute thickness, 53
Fat, muscle and bone, relative thickness of, 53
Fels Research Institute, 53
Ferritin, 156, 337, 347
Fertility, exposed women, 50, 170, 171
Fertility, prenatally exposed, 238, 239, 242
Fetal alcohol syndrome, 231, 354, 370
Fibroblasts, 94, 186, 349
Fine motor coordination, 201
Fire breaks, 7
Fire-bombing, 3
Fireball, 104, 109, 115, 118, 320
  radiation from, 118
  rate of growth, 115
Fires, 3, 5, 7, 9, 12, 14, 21, 120, 121
  in the critical areas, 121
Firestorm, 7
Fission, 104, 109, 317, 367
Flash burns, 165
Flow cytometry, 323, 348
Fluence(s), 125, 319, 337
Francis Committee, 34, 73–77, 86, 92, 298, 299, 313, 314, 337
  recommendations, 73–74

Fruit fly, 51, 145
Fukuoka, 101
Funds, 23, 34, 35, 85, 94, 300, 314
Fundus, 55
Fusuma, 111

G-banding, 363
Gallbladder, 155, 157, 158, 166, 334
  cancer, 166
  polyps, 158
Gametogenesis, differences in, 170
Ganglion, 337
Gap junction, 148, 337
General Council of Trade Unions of Japan, 29
Genetic damage, 181, 244, 262, 263, 266, 270
Genetic effects, 24, 73, 243, 244, 256, 265, 270, 330, 353, 368
Genetic variability, 98
Genetics Program, 41, 42, 48, 51, 93, 96, 98, 298
Gingivitis, 13, 338
Glutamate pyruvate transaminase, 271
Greulich-Pyle standards, 54
Grip strength, 204, 325
Ground zero, 108, 153, 304
Growth and development, 21, 24, 34, 46, 52–54, 60, 90, 95, 96, 98, 115, 140, 142, 145, 147, 153, 168, 191, 195, 204, 215, 220, 225, 232, 244, 259, 260, 263, 272, 279, 310, 322, 335, 341, 343, 351, 353, 354, 356, 363, 364, 366
  among F1, 259, 260
  among prenatally exposed, 232, 233
  anthropometrics, 52, 54, 96, 204, 232, 353, 356
  Greulich-Pyle standards, 54
  of exposed children, 52–54
  retardation, 52–54, 195, 232, 233, 247, 260
  standards for evaluating, 54
Glycophorin, 180–182, 338, 362
Guanine, 180, 268, 338

Half-life, 318, 338
  biologic, 318
  physical (radioactive), 318, 338
Haptoglobin, 259
Head size, 66, 189, 192–195, 231, 272, 350, 361, 364
  normal variability in, 194
  small, 66, 189, 192–195, 231, 272, 350, 361, 364
Health and Safety Laboratory, 25
Health Examination Survey, 87, 362
Health handbook, 131

Subject Index

Healthy worker effect, 165
Hematocrit, 39
Hematologic studies, early, 38, 39
Hemoglobin, 39, 59, 93
  concentration, 39
Heterotopia, 231, 328, 360
Hinin, 81, 315
Hippocampus, 214, 242, 350
Hirado Health Study, 97
Hiro, 2, 185
Hiroshima Atomic Bomb Casualty Council, 36
Hiroshima Bunri Daigaku, 27
Hiroshima Central Telephone Office Building, 125
Hiroshima Eugenics Committee, 49
Hiroshima Prefectural Medical College, 27
Hiroshima University, 27, 74, 94, 316
Hittosha, 80
HLA system, 258
Home visits, 41, 42
Honolulu Heart Study, 99, 100
Honseki, 81, 87, 88, 338
Hookworms, 97
Hospital, 7, 8, 21, 40, 41, 46
  Daiichi, 8
  Daini, 8
  Hiroshima Army Welfare, 8
  Hiroshima Communications, 7
  Hiroshima Red Cross, 7, 21, 40, 41
  Mitsubishi, 46
  Nagasaki Medical School, 46
  Tokyo Red Cross Maternity, 46
Host factors, 162
Household censuses, 80
Household head, 80
Housing, 23, 24, 108–113, 116, 306, 307
  fusuma, 111
  materials, 108–109
  roof tiles, 116
  scale drawings, 110–113
  shielding factors, 108
  sizes, 111
  tatami, 111
  types and distribution, 109
Hydrochloric acid, 155, 333
Hydrogen bomb, 31, 368
Hydrogen peroxide, 93
Hyperparathyroidism, 168, 169, 301, 353
Hypertension, 139, 141, 168, 239, 338
Hypocenter, 3, 8, 9, 21, 40, 56, 61, 62, 66, 79, 82, 105, 108, 109, 111, 117, 118, 120–122, 124, 125, 130, 131, 149, 170, 184, 210, 245, 260, 299, 312, 323, 337–339, 343
Hypothalamus, 328, 340

Hypoxanthine-guanine phosphoribosyl transferase, 180, 182, 322, 338
Hypoxemia, 212, 338

ICHIBAN, 109, 298, 348
Immune system, 126, 143, 144, 164, 344
  competence, 39, 143, 144, 164, 165, 236
  surveillance, 144
Immunoglobulins, 144, 145, 333, 339, 349
Inbreeding, 97, 101, 368
Intelligence, 66, 195–198, 201, 206, 208, 215, 227, 241, 312, 313, 324, 326, 354, 358, 368, 370, 371
  definitions, 195, 196
  tests of, 66, 195, 196, 198, 206, 208, 324, 371
    Koga, 66, 195, 208, 233, 313, 324, 358, 371
    Tanaka-Binet, 208
International Atomic Energy Agency (IAEA), 338, 356, 357
International Commission on Radiological Protection (ICRP), 278, 318, 338, 356
Intestine, 142, 155, 158
  brush border, 158
  large, 142
  small, 155, 158
Introns, 268
Iodine, 156, 274, 355
  radioactive, 274
Ionizations, 174, 343
  density, 174
Ionizing radiation (see radiation, ionizing)
IQ, 97, 193, 194, 201, 208, 231, 233, 272, 324, 326
Iron, storage in the body, 156
Irradiation, 50, 51, 54, 55, 57, 65, 108, 127–129, 144, 151, 152, 154, 162, 163, 165, 170, 171, 173, 185, 189, 210, 214, 230, 232, 236, 237, 240, 244, 248, 251, 253, 265, 327, 332, 333, 336, 343, 349–351, 354, 356, 358, 359, 361, 362, 367, 369, 371, 373, 374
  diagnostic and therapeutic, 127–129
  prenatal, 65–68, 350, 358, 369
Iwakuni, 3, 5

Japan Science Council, 11
Japanese American employees, 24
Japanese Bureau of Judicial Affairs, 82, 83
Japanese Bureau of the Census, 68
Japanese encephalitis, 28, 192
Japanese Institute of Nutrition, 15
Japanese Ministry of Education, 27

Japanese Ministry of Foreign Affairs, 28
Japanese Ministry of Health and Welfare, 26, 27, 33, 34, 36, 81, 131, 299, 320
Japanese Ministry of Justice, 80, 82, 339
Japanese National Census, 66, 68, 77, 78, 93, 236, 298
Japanese National Institute of Health, 26, 37, 52, 77, 298, 299
Japanese National Institute of Radiological Sciences, 33, 107
Japanese National Personnel Authority, 35
Japanese Science and Technology Agency, 295
Joint Commission, 11, 12, 14, 18, 61, 76, 125, 211, 303, 304, 307, 339
Joint Symposium on the Late Effects of Atomic Bomb Injuries, 37
Joule, 102, 103, 338, 339

Keloids, 14
Kerma, 103–104, 107, 108, 116, 117, 121, 125, 126, 177, 181, 233, 242, 319, 339, 343, 369
  definition, 339
  free-in-air, 107, 108, 121, 125, 126, 339
  shielded, 103, 125, 126, 233, 242, 339, 343, 369
Kidney, 236, 239, 240, 252
Klinefelter syndrome, 252, 254, 255, 262
Koi-Takasu, 118, 119
Kokura, 2
Korea, 27, 28, 46, 294, 303, 307
  conscripted labor, 303
Korean A-bomb survivors, 81
Korean War, 28, 307
Kure, 20, 21, 26, 28, 39, 52, 276
Kyoto University, 94
Kyshtym, 275
Kyushu University, 94, 97, 127, 184, 316

LD50/60, 17, 305, 339, 342
LD95/60, 126, 339
Large intestine, 142
Larynx, 158
Lens, 54–58, 163, 188, 189, 227, 233, 272, 298, 316, 335, 347, 354, 355, 360, 364, 369
  capsule, 55
  cataract(s), 55, 56, 335, 347, 354, 355, 359
  embedded glass, 57
  germinative epithelium, 55
  opacities, 54–58, 163, 188, 189, 227, 233, 272, 298, 364, 369
  prenatally exposed, 233
  zone of specular reflection, 58

Leukemia, 19, 55, 58–64, 75, 85–87, 93, 144, 145, 147–154, 157, 159, 160, 164, 165, 185, 188, 236–238, 249–252, 272, 273, 278, 281, 282, 285, 298, 320, 321, 331, 340, 347, 352, 355, 359, 361, 365, 371, 372
  acute, 60, 149, 152
  acute lymphatic, 60, 152
  acute lymphocytic, 149, 152
  acute myelogenous, 149, 152
  adult T-cell, 61, 152
  ascertainment of cases, 85, 86, 236
  chronic lymphocytic, 62, 151, 152, 273, 285
  chronic myelogenous, 63, 64, 152, 372
  classification of, 151
    French-American-British system, 60, 151
  preclinical signs, 63
  Registry, 86, 87, 149
Leukocyte, 10, 13, 14, 39, 59, 60, 64, 143, 153, 165, 173, 233, 258, 269, 270, 309, 334, 344, 367
  B-cell, 143, 144, 334
  function, 143
  mobility, impairment, 143
  survival time, 165
  T-cell, 61, 143–145, 147, 180, 182, 183, 322, 344, 358
    differentiation, 182
    function, 144
    mutants, 183, 322
    receptor, 144, 145, 180, 182, 183, 344
Leukopenia, 59
Liaison committees, 35
Life Span Study, 64, 79–83, 85, 90, 91, 93, 99, 126, 128, 136, 137, 151, 153, 154, 160, 161, 164–167, 284, 299, 300, 314, 315, 331, 340, 348, 368, 371, 372, 374
  strengths, 136
Limbic system, 328
Limited Test Ban Treaty, 109
Linear energy transfer (LET), 340
Lipids, 99, 141, 158, 334
  blood levels of, 99
Liver, 158, 159, 161, 166, 236, 272, 285, 335, 348
  cancer, primary, 159, 161
  cirrhosis, 159, 272, 335, 348
Los Alamos National Laboratory, 106, 304
Lucky Dragon (Fukuryu maru), 31
Lung, 99, 128, 142, 153, 154, 159, 163, 272, 285–287, 334
Lymph, 60, 143, 157, 229, 344
  nodes, 60, 143, 157, 229, 344

Lymphatic system, 143, 157
Lymphocyte(s), 59–61, 143, 144, 165, 173, 177, 180, 182, 187, 234, 256, 270, 272, 322, 344, 362, 363
Lymphoma, 145, 157, 252, 272, 348, 353, 365
Lyonization, 262

Magnetic resonance imaging, 57, 226, 227, 239, 241, 328, 352
Major histocompatibility complex (MHC), 180, 183, 338, 340
Malformations, 38, 41–46, 65, 97, 188, 189, 202, 225, 226, 243, 244, 246, 247, 249, 252, 263, 271, 330, 333, 350, 355, 362, 364, 366, 367, 373
 anencephaly, 225, 246, 333, 362
 aniridia, 252
 auricular appendages, 41
 chondrodystrophy, 252
 cleft lip, 188, 246
 cleft palate, 188, 246
 club foot, 246
 congenital, 41, 43, 44, 46, 95, 97, 202, 243, 246, 247, 263, 330, 362, 364
 congenital dislocation of the hip, 46
 congenital heart disease, 42, 46, 246
 congenital teeth, 46
 encephalomeningocele, 225, 336
 funnel chest, 46
 hydrocele, 45, 46
 hydrocephalus, 46
 major, 42, 46
 minor, 41
 polydactyly, 249
 schizencephaly, 327
 syndactyly, 246
Mamillary bodies, 227, 228, 340
Mammography, 155
Manhattan District, 10, 304
Manhattan Project, 10, 18, 20, 55
Marriage, 80, 94, 95, 97, 98, 101, 169, 238, 239, 339, 350, 361, 367
 consanguineous, 94, 95, 97, 98, 101, 361, 367
  effects of on children of first cousins, 97
Marshall Islands, 120, 318, 351
Master File, 66, 68–71, 83, 312
Medical education, Japanese, 75
Medical treatment of survivors, 29, 35, 127, 130, 131
Meltdown, 292, 295
Meningitis, acute, 227
Metaplasia, 155, 158, 340
Metastasis, 159
Methemoglobin, 93
Methylmercury, 355

Microcephaly, 193
Midwives, 41, 44, 45, 245, 246, 308, 309, 315
 participation in Genetics Program, 41
Mihara, 5
Military service, compulsory, 80
Miscarriage, 49, 184
Mitsubishi Heavy Industries, 2, 8
Models(s), additive, 280, 281, 286
 constant relative risk, 282, 284
 linear, 132
 linear relative risk, 164
 linear-quadratic, 133, 134
 multiplicative, 280, 283, 286
 risk projection, 279–282, 287, 291
 threshold, 133
Mongolian spots, 42
Monte Carlo method, 116, 123
Morbidity, surveillance, 79
Mortality (see death), 72, 136, 155, 285, 286, 305, 321, 355, 356
 comparative, 321
 rates, 72, 285, 286, 305, 355, 356
 standardized rate, 136, 155, 321
Mortality surveillance, 66, 73, 78, 79, 81, 90, 137, 163, 299, 340
 purposes of, 79
 samples, 73, 79, 81
Multiple myeloma, 157, 272, 278, 355, 365
Mutation(s), 38, 40, 98, 133, 140, 141, 144–146, 148, 172, 179–184, 186, 234, 235, 243, 246, 249, 250, 252, 255–267, 269–272, 300, 322, 323, 329, 330, 332, 339, 340, 344, 348, 349, 356, 362, 363, 367
 deficiency variants, 257–259, 367
 germinal, 250
 null, 257–259, 367
 point, 267
 spontaneous, 264
 transmissible, 243, 246
 X-linked lethal, 260–262, 339, 340
Mutation(s) somatic, 140, 141, 148, 179–184, 234, 235, 272, 322, 323, 332, 348
 assays for, 322
  glycophorin (GPA).322
  HPRT, 322
  T-cell receptor, 322
 frequency of, 179–184
 prenatally exposed, 234, 235
Myocardial infarction, 100, 141, 366

Nagasaki, Arsenal, 2
 Chinzei school, 125
 Shiroyama school, 125
Nasopharyngeal ulcers, 13

National Academy of Sciences, 18, 19, 24, 32, 33, 37, 73, 305, 306, 308, 314, 333, 359, 361, 363, 365
  Assembly of Life Sciences, 306
  Commission on Life Sciences, 306
  Division of Medical Sciences.18, 24, 73, 306, 314
National Personnel Authority, 35
National Research Council, 18, 19, 25, 26, 52, 59, 73, 76, 77, 89, 95, 244, 298, 265, 282, 306, 307, 314, 316, 331, 337, 351, 359, 363
  Board of Radiation Effects Research, 306
  Committee on Atomic Casualties, 25, 26, 52, 59, 76, 77, 95, 244, 298, 316
  Committee on the Biological Effects of Ionizing Radiation, 265, 282, 331, 351
Necrotic gingivitis, 13
Neuroblastoma, 252
Neurofibromatosis, 252
Neuroglia, 190, 218, 219, 230, 340, 341
  radial, 218, 219, 230, 341
Neuromuscular development, 201, 203–205, 272
  examinations, 201, 203, 204
    repetitive action test, 205, 324, 325
Neuron(s), 121, 190–192, 196, 204, 213, 214–223, 225–232, 232, 241, 325, 327, 328, 334, 336, 341, 344, 351, 354, 358, 360, 365, 366, 368, 369
  axons, 215, 334, 341, 360
  dendrites, 336
  differentiation, 190
  granular or micro-neurons, 327
  migration, 217, 220, 228, 230, 328, 354, 366, 369
  production of, 190, 191, 213, 225
Neurophysiologic tests, 96, 97, 242
Neuropil, 341
Neutron(s), 56, 103–108, 116, 117, 122, 125, 126, 160, 164, 178, 292, 304, 317, 319, 320, 337, 342, 352, 365, 367
  activation studies, 125, 319, 320
  biologic effects of, 178
  delayed, 317
  effectiveness of, 126
  fast, 104, 319, 342
  prompt, 104, 317
  scatter, 106
  thermal, 104, 125, 164, 304
  transport, 122, 164
Nevada Test Site, 108, 109, 114
Ni-Hon-San Study, 98–101
Ninoshima, 8
Nishiyama Reservoir, 118, 119, 300

Nose, 93, 106, 157, 206, 241, 250–253, 320, 347, 353, 356, 359
Notes Verbale, 28, 298, 300, 306
Nuchal hemangioma, 41
Nuclear accident(s), 17, 35, 156, 240, 274, 275, 292, 293, 305, 331, 338, 356, 359
  Chelyabinsk, 35, 331
  Chernobyl, 17, 35, 240, 274, 292, 293, 305, 331, 338, 356
  Ciudad Juarez, 275, 331
  Kyshtym, 275
  Sellafield, 250–252, 347, 353, 356, 359
  Three Mile Island, 292
Nuclear power, 274, 278, 292–295, 331, 332, 347, 370
  advocates, 294
  generation, 274
  reactors, 109, 274, 278, 292–295, 331, 332, 337
    accidents, 292, 295
    employees, 278
    moderator, 292
Nuclear reaction, 114, 269, 270, 271, 317
  chain, 114, 269, 270, 271
  criticality, 317
  subcritical, 114, 317
  supercritical, 114, 317
Nuclear waste, 294, 359
Nuclear weapons, 23, 91, 103, 109, 114, 119, 274, 275, 320, 354
  atmospheric testing, 109, 119, 274
  tests, 31, 105, 109, 114, 120, 297, 304, 368
    Bravo, 31, 120
    Crossroads, 114
    Minor Scale, 304
    Plumb bob, 109
    Trinity, 105, 114, 297, 368
Nucleoside phosphorylase, 259
Nucleotide(s), 257, 268
  triplets, 268
Nutrition, postwar, 15, 52, 54, 233, 278
Nutritional status, 15

Oak Ridge National Laboratory, 109, 111, 257, 265, 278, 354, 358, 373
Okayama University, 93
Olfactory bulb, 241, 327
Olfactory placode, 222, 327, 341, 373
Oncogene, 145, 341, 345, 373
Oregon League of Women Voters, 293
Oropharyngeal lesions, 62, 87, 184, 315, 341

p53, 145, 146, 186, 321, 356
Pancreas, 158
Pancytopenia, 13, 14, 64

Paranoia of the McCarthy years, 23
Parasites, intestinal, 97
Parathyroid glands, 168, 272
Parathyroidism, 168
Partial albinism, 46
Participation, 10, 30–32, 36, 41, 43, 53, 86–89, 96, 311, 312
 rates, 96, 312
Particle histories, 116
Pearl Harbor, 28
Pelvis, socket of, 189
Perceptuo-motor maturation, 69, 206–209, 349, 354, 355
 Tests, 66, 206, 207, 209, 349, 354, 355
  Bender-Gestalt, 66, 206, 209, 349, 355
  Draw-a-man, 66, 206, 207, 209, 354
Petechiae, 13, 184, 341
Pharynx, 158
phosphoglucomutase, 258
Phosphogluconate dehydrogenase, 259
Plutonium, 3, 104, 114, 295
Polycythemia, 59, 372
Polyps, 158
Positron emission tomography, 242
Pregnancy, mean duration of, 324
 outcome study, 39–49, 244–248
 radiation therapy during, 65
 registration, 95
 termination, artificial, 47, 370
 wastage, 184
Presidential Directive, 19
Prevalence, 28, 76, 99, 158, 159, 168, 188, 228, 235, 313, 339, 341
Print media, 2, 31, 91
Prostate, 158, 166
Proto-oncogene, 140, 145, 341, 345
 activation, 140, 145, 345
Protons, 104, 173, 340, 359
Pseudogenes, 269–271
Psychological concerns, 131, 169, 170
Psychometric studies, 96
Public Health and Welfare Section (SCAP), 18, 52
Public relations, 21, 74, 92, 365, 368
 RERF Update, 92, 365, 368
Purkinje cell, 341
Purpura, 13, 62, 184, 315, 341

Racial prejudice, 24
Radiation, ionizing, 12, 20, 21, 36, 38, 51–55, 65, 70, 73, 79, 92, 102–105, 107–132, 134, 135, 137, 143, 146–148, 151, 153–155, 160, 162, 163–165, 170, 173–175, 178, 182–186, 188–190, 192, 194, 195, 201–205, 209, 210, 215, 222, 225, 228–232, 234, 235, 239–241, 243, 244, 246–248, 253, 260, 262, 265, 267, 270, 272, 274–278, 280, 282, 284, 286, 291, 293, 300, 305, 311, 313, 315, 317, 318, 320, 327, 328, 330–333, 336, 338, 340–342, 343, 345, 348, 350–352, 355, 356, 358, 361, 363, 364, 366, 368, 370, 372, 374
 alpha particles, 174, 318, 333, 340
 delayed, 105, 118
 diagnostic and therapeutic, 127–129
 dosimetry, 73, 102, 300, 348, 358, 363, 367, 370
 estimation, 121–126
 fallout, 118–120
 fluoroscopy, 127
 gamma rays, 103, 104, 107, 108, 122, 126, 160, 174, 178, 265, 318, 327, 340, 342, 343, 345
 low LET, definition, 340
 prenatal exposure to, 188, 189, 201, 203, 205, 232, 235, 240, 311, 352, 356, 361, 364, 368, 374
 prompt, 105, 107–118
 prompt neutrons, 104, 317
 quality factor, 342, 343
 relative biological effectiveness, 126, 342
 residual, 21, 105, 120, 121, 131
 transmission, 126
  gamma rays, 126
  neutrons, 126
 transport of neutrons, 122, 164
 units, definition, 338, 341–343
  gray, 338
  person-year-gray, 341
  person-year-rad, 342
  person-year-sievert, 342
  rad, 342, 343
  rem, 343
  sievert, 343
 X rays, 38, 42, 54, 65, 103, 104, 127, 185, 293, 340, 343, 345
Radiation, protection, 120, 124, 132, 135, 213, 275, 278, 279, 280, 284, 289, 291, 318, 331, 339, 340, 342, 359, 362, 363, 370
 occupational, 278, 289
 public, 278, 289
 regulatory concern, 124, 132, 135, 280, 291
Radiation, sickness, acute, 7, 10, 12, 14, 55, 64, 87, 149, 260, 304, 315, 324, 342, 369
 symptoms, 10, 64, 260, 324
Radiation Effects Research Foundation (RERF), 22, 23, 30, 32–37, 68, 73, 82, 83, 85 92, 94, 100, 127–129, 186, 187, 241, 243, 270, 298, 300,

Radiation Effects Research Foundation
(RERF) (cont.) 306–310, 312, 314,
316, 317, 324, 343, 347–349, 351–
358, 360, 362–372, 374
  Act of Endowment, 32, 33, 308
  Board of Directors, 30, 35
  creation, 32
  labor-management relations, 35
  programmatic stability, 73
  public corporation enacted by law, 32
  Science council, 33
Radioactive contamination, 20
Radioactive dust, 118
Radioactive particles, 119, 163, 318, 332
Radioisotopes, 156, 318
Radionuclides, 3, 65, 104, 109, 114, 118–
120, 163, 229, 232, 265, 275, 292,
293, 295, 318, 320, 332, 342, 370
  cesium, 119, 120, 275, 318, 332
  cobalt, 109, 232, 265, 275
  decay, 104
  inhalation or ingestion, 118
  plutonium, 3, 104, 114, 295
  radium, 65, 229, 342, 370
  strontium, 318
  uranium, 3, 104, 114, 163, 292, 293,
  295, 320
Radiosensitivity, 157, 178, 181, 268
  of the young, 157
  variation among individuals, 178, 181,
  268
Radium, 65, 229, 342, 370
Radon, 342
Rationing, 40, 48, 49, 97, 244, 308, 309
  clothing, 308
  food, 308
    supplemental, 40
  registration, 40, 97, 309
  regulations, 244
Records, medical, centralized library, 84
Recruitment, 22, 24, 27, 28, 41, 77, 86, 89,
311
  participants, 86
  physicians, 41
  representative groups, 311
  research and administrative personnel,
  77
Rectum, 158
Reports, technical, 91, 92, 316, 347
Reports, weekly health, 74
Residual radiation, 21, 105, 120, 121, 131
Retardation, developmental, 54
Retardation, mental, 66, 184, 188, 189–
196, 198, 201–205, 210, 212–215,
228–231, 239, 252, 272, 273, 298,
313, 361, 364, 373
  frequency, 191
  radiation risk, 191, 192

Retardation, physical (see Growth and
development)
Retinoblastoma, 252
Ribonucleic acid (RNA), 268, 270
  messenger (mRNA), 268
Risk, 124, 126, 161, 166, 167, 242, 278,
285, 286, 319, 352, 362
  absolute, 135, 152, 157, 159, 282, 287,
  361
  amelioration of, 187
  attributable, 135, 151, 156, 161, 237,
  273, 280, 281, 331
  constant relative, 282, 284
  early knowledge, 15, 38, 51, 56, 59, 64,
  65, 148, 173, 188, 233, 236, 311
  estimation of, 132–136, 213
  excess deaths, 135, 140, 151, 166, 281,
  282, 286
  excess relative risk, 134, 151, 152, 156,
  157, 159, 161, 248–250, 263, 280,
  281, 286, 331
  excess risk, 162, 211, 237, 248
  factors, 94, 141, 162
  genetic, 262–267
  individual specific, 148
  lifetime, 289–291
  perception of, 292–294
  projection, 167, 278–284, 287, 288, 291
    low dose and dose rate.278–284, 288,
    289
  racial variation, 285–288
  relative, 134, 135, 142, 151, 152, 156,
  157, 159, 161, 164, 166, 168, 203,
  236, 237, 248–250, 263, 277, 280–
  282, 284, 286
  temporal variation in, 151
  uncertainties in estimates, 162–167,
  209–214
Roundworms, 97
Rules on fraternization between races, 24
Russia, 35

Salivary gland, 157, 272
  tumors, 157, 272
Samples, 34, 66, 73, 79, 87, 88, 90, 93,
128, 194, 210, 211, 241, 313, 315
  clinical, 66, 194, 210, 211, 241, 313
  eligibility for inclusion, 79
  fixed, 34, 73, 93
  health surveillance, 128
  master, 79
    proper, 79
    reserve, 79
  matched schedule groups, 88
  selection, 79, 87, 315
  special mortality, 90
San Francisco, 28, 56, 99, 100
Sarcoma(s), 153, 343

Sasebo, 20, 21, 26, 52
Scabies, 97
Schizencephaly, 327
School, 125, 195, 198–201, 364
  academic performance in, 195, 199–201, 364
  attendance, 198
  Chinzei, 125
  Shiroyama, 125
Sclerosis, small blood vessel, 100
Security clearance, 22
Segmental iris heterochromia, 235
Seizures, 201–203, 210, 213, 214, 241, 272, 328, 352
  febrile, 202
  unprovoked, 201, 203, 214
Sellafield, 250–252, 347, 353, 356, 359
Senile dementia, 100, 187
Sentinel phenotype, 252–254
Sex chromosome(s), 172, 234, 253, 254, 260, 263, 335
  abnormalities, 254
  aneuploidy, 253, 263
Sex ratio, 48, 260–262, 367
Sexual maturation, evaluation of, 53, 238
Shielding, 76, 103, 108, 109, 111, 114, 118, 122, 124–126, 317, 339
  by buildings, 103, 108, 109, 111, 118, 122, 124–126, 339
  by terrain, 124
  histories, 114, 124
  studies, 76, 317
Shiroyama school, 125
Shizuoka, 101
Shock wave, 2
Skin, 10, 13, 14, 43, 97, 104, 105, 118, 137, 146, 158, 161, 165, 186, 231, 252, 269, 272, 316, 341, 342, 349
Smoking, 141, 160, 163, 278, 335
Social class, 81, 82, 315
  buraku-min, 315
  discrimination, 315
  eta, 81, 315
  hierarchy, 81
  hinin, 81, 315
  kawatta seikatsu, 82
  suiheisha, 82
Social customs, Japanese, 8, 74, 80–82, 91, 249, 315, 335, 339, 344, 362
  chonaikai, 8, 74, 335, 344
  hittosha, 80
  ie (household system), 80
  koseki, 80–82, 91, 249, 315, 339, 362
  koseki-ka, 81
  koseki-yakuba, 81
  tokushu hojin, 27
  tonari-gumi, 74
Spinal cord, 215, 325, 334, 337

Spleen, 60, 165, 344, 354
Staff, reduction in, 32
Stem cell(s), 60, 143, 144, 154, 170, 180, 182, 216, 230, 237, 339, 343, 344
Sterility, 19, 171, 189, 239, 273, 370
Stillbirth(s), 246, 247, 309, 324, 364
Stomach, 155, 156, 158, 161, 272, 285–287, 313, 333, 347, 362
Stroke, 99, 100, 139, 141, 366, 367, 371
  hemorrhagic, 100
  thromboembolic, 100
Strontium, 318
Suicide, 169, 249
Sulfonamides, 15, 276
Supreme Commander of the Allied Powers, 11, 306, 343
Survivors, distally exposed, 245, 258, 259
  prenatally exposed, 66, 67, 188, 209, 210, 214, 229, 230, 234, 235, 237–240, 311, 364, 368
  proximally exposed, 245, 312
Synaptogenesis, 190, 191, 208, 225, 241, 344
Syndactyly, 246
Syndrome, 192, 246, 252–255, 262
  Down, 192, 246, 253
  Klinefelter, 252, 254, 255, 262
  Turner, 252, 254, 262

T-cell, 61, 143–145, 147, 180, 182, 183, 322, 344, 358
  mutants, 183, 322
  receptor, 144, 145, 180, 182, 183, 344
Tatami, 111
Tests, nuclear weapons, 31, 105, 109, 114, 120, 297, 304, 368
  Bravo, 31, 120
  Crossroads, 114
  Minor Scale, 304
  Plumb bob, 109
  Trinity, 105, 114, 297, 368
Thalamus, 208
Thalidomide, 263
Thermal pulse, 2
Thermal radiation, 14, 108
Thermoluminescence, 106, 107, 125
  dosimetry, 106, 107
  measurements, 125
Thioguanine, 322
Threshold, 20, 57, 58, 133, 140, 147, 149, 191, 192, 213, 214
  dose, 20, 57, 58, 133, 140, 147, 149, 191, 192, 213, 214
  model, 133
Thumb, muscles, 325
Thymidine, 268, 352
Thymus, 60, 61, 121, 143, 182, 344

Thyroid, 156, 157, 161, 166, 168, 186, 272, 274, 285, 299, 355, 356
Tissue and tumor registries, 83–87, 153, 155, 156, 157, 166–168, 236, 252, 301, 360
  ascertainment of cases, 85, 86, 236
  establishment of uniform principles, 83
  sponsorship, 83
Tokushu hojin, 27
Tokyo Imperial University, 11
Tokyo Dental and Medical University, 94, 97
Tonari-gumi, 74
Transcription, 268
Transfection, 344
Trinity Test, 105, 114, 297, 368
Triosephosphate isomerase, 259
Tuberculosis, 50, 165, 171, 370
  urogenital, 50
Tumor(s), 84, 128, 140, 142, 145, 146, 148, 153, 157–168, 172, 178, 179, 184, 187, 236, 249, 272, 273, 299, 321, 344, 367, 374
  benign, 140, 148, 158, 167, 168, 249
  central nervous system, 158
  incidence, 160–167
  lymph nodes, 157
  malignant, 84, 128, 140, 142, 148, 153, 157, 167, 172, 178, 179, 184, 187, 236, 272, 273, 299, 374
  suppressor gene(s), 145, 146, 321, 344, 367
  unspecified, 159, 249
Typhoid fever, 7, 316
Typhoon, 5, 8, 130, 305

U.K. Atomic Energy Agency, 11, 338, 356, 358
U.K. Ministry of Home Security, 304
U.K. National Registry for Radiation Workers, 278
U.S. Agency for International Development, 35
U.S. Atomic Energy Commission, 11, 23, 94, 97
U.S. Defense Nuclear Agency, 304
U.S. Department of Energy, 34
U.S. Energy Research and Development Agency, 32
U.S. National Academy of Sciences, 19, 23, 32–34, 37, 73, 76, 282, 297, 298, 305, 306, 308, 314, 333, 334, 337, 351, 359, 361, 363, 365
U.S. National Cancer Institute, 85, 289, 300, 348, 351, 354, 355
U.S. National Center for Health Statistics, 87, 326, 362
U.S. National Heart Institute, 73, 99, 299

U.S. National Institute for Neurological Diseases and Stroke, 99
U.S. National Research Council, 18, 19, 25, 73, 89, 306, 307, 314, 337, 359, 363
U.S. Naval Technical Mission to Japan, 304
U.S. Public Health Service, 76, 89, 308
Ulcerations, mouth, 93
Umbilical hernia, 42, 45
Unified Study Program, 32, 76, 149, 298
University of California at Berkeley, 56, 89, 99
University of Chicago, 18, 149
University of Kyoto medical team, 5
University of Michigan, 73, 96, 97, 347
University of Nagasaki Medical School, 44
University of Rochester, 59
Unregistered births, 45, 246
UNSCEAR, 286, 289, 318, 331, 345
Untoward pregnancy outcome, 246–248, 250, 252, 263
Urakami Valley, 9
Uranium, 3, 104, 114, 163, 292, 293, 295, 320
  fuel, 295
  mining, 163, 293
Urinary bladder, 157, 163, 272
Uterus, 47, 65, 123, 158, 161, 166, 171, 203, 227, 229, 237

Venereal disease, 43, 309
Verification of data, 44, 69, 77, 114, 116, 244, 258
Veterans' Follow-up Agency, 73, 89
Viral disease, 28, 59, 192
Virus, 61, 152, 159, 330
  Epstein-Barr, 330
  hepatitis B, 159
  HTLV-I, 152
Visiting scientist program, 85
Vital events, 80, 215

Water, contaminated, 7
Water, drinking, 118, 119
Watt, 102, 103, 339
Weapons, nuclear, 23, 91, 103, 109, 114, 119, 274, 275, 320, 354
  atmospheric testing, 109, 119, 274
  tests, 31, 105, 109, 114, 120, 297, 304, 368
    Bravo, 31, 120
    Crossroads, 114
    Minor Scale, 304
    Plumb bob, 109
    Trinity, 105, 114, 297, 368
Whipworms, 97

White cell(s), 10, 13, 14, 39, 59, 60, 64,
    143, 153, 165, 173, 233, 258, 269,
    270, 309, 334, 344, 367
Wilms' tumor, 236, 252

X rays, 38, 42, 54, 65, 103, 104, 127, 185,
    293, 340, 343, 345
X-ray machines, 23, 38, 104, 127, 345
X-ray therapy, 351
Xeroderma pigmentosum, 146, 148

Y chromosome, 253, 260
Yale University, 89
Yen, reevaluation of, 32
Yield, atomic bombs, 114, 115, 125, 126
    controversy, 125, 126
    Hiroshima, 114, 115
    Nagasaki, 114, 115

Zaidan ho-jin, 32